Feature Papers for Land Systems and Global Change Section

Editors

Le Yu
Pengyu Hao

Basel • Beijing • Wuhan • Barcelona • Belgrade • Novi Sad • Cluj • Manchester

Editors
Le Yu
Tsinghua University
Beijing
China

Pengyu Hao
Food and Agriculture
Organization of the United
Nations
Rome
Italy

Editorial Office
MDPI AG
Grosspeteranlage 5
4052 Basel, Switzerland

This is a reprint of articles from the Special Issue published online in the open access journal *Land* (ISSN 2073-445X) (available at: https://www.mdpi.com/journal/land/special_issues/FP_Land_system).

For citation purposes, cite each article independently as indicated on the article page online and as indicated below:

Lastname, A.A.; Lastname, B.B. Article Title. *Journal Name* **Year**, *Volume Number*, Page Range.

ISBN 978-3-7258-1877-8 (Hbk)
ISBN 978-3-7258-1878-5 (PDF)
doi.org/10.3390/books978-3-7258-1878-5

© 2024 by the authors. Articles in this book are Open Access and distributed under the Creative Commons Attribution (CC BY) license. The book as a whole is distributed by MDPI under the terms and conditions of the Creative Commons Attribution-NonCommercial-NoDerivs (CC BY-NC-ND) license.

Contents

Alem Oyarmoi, Stephen Birkinshaw, Caspar J. M. Hewett and Hayley J. Fowler
The Effect of Papyrus Wetlands on Flow Regulation in a Tropical River Catchment
Reprinted from: *Land* 2023, 12, 2158, doi:10.3390/land12122158 1

Paulo Flores Ribeiro and José Lima Santos
Exploring the Effects of Climate Change on Farming System Choice:
A Farm-Level Space-for-Time Approach
Reprinted from: *Land* 2023, 12, 2113, doi:10.3390/land12122113 26

Yujin Park, Junga Lee, Se-Rin Park and Sang-Woo Lee
Assessing the Resilience of Stream Ecosystems to Rainfall Impact
Reprinted from: *Land* 2023, 12, 2072, doi:10.3390/land12112072 41

Haitao Ji, Xiaoshun Li, Yiwei Geng, Xin Chen, Yuexiang Wang, Jumei Cheng and Zhuang Chen
Delineation of Urban Development Boundary and Carbon Emission Effects in Xuzhou City, China
Reprinted from: *Land* 2023, 12, 1819, doi:10.3390/land12091819 55

Xiyu Li, Le Yu and Xin Chen
New Insights into Urbanization Based on Global Mapping and Analysis of Human Settlements
in the Rural–Urban Continuum
Reprinted from: *Land* 2023, 12, 1607, doi:10.3390/land12081607 71

Abebaw Andarge Gedefaw
Analysis of the Contribution of Land Registration to Sustainable Land Management in East
Gojjam Zone, Ethiopia
Reprinted from: *Land* 2023, 12, 1157, doi:10.3390/land12061157 94

Cynthia Simmons, Marta Astier, Robert Walker, Jaime Fernando Navia-Antezana, Yan Gao, Yankuic Galván-Miyoshi and Dan Klooster
Forest Transition and Fuzzy Environments in Neoliberal Mexico
Reprinted from: *Land* 2023, 12, 840, doi:10.3390/land12040840 115

Yuan Qi, Xin Chen, Jiaqing Zhang, Yaoyao Li and Daolin Zhu
How Do Rising Farmland Costs Affect Fertilizer Use Efficiency? Evidence from Gansu and
Jiangsu, China
Reprinted from: *Land* 2022, 11, 1730, doi:10.3390/land11101730 130

Mark Tilzey
Ill Fares the Land: Confronting Unsustainability in the U.K. Food System through Political
Agroecology and Degrowth
Reprinted from: *Land* 2024, 13, 594, doi:10.3390/land13050594 148

Haojun Xie, Quan Sun and Wei Song
Exploring the Ecological Effects of Rural Land Use Changes: A Bibliometric Overview
Reprinted from: *Land* 2024, 13, 303, doi:10.3390/land13030303 187

Article

The Effect of Papyrus Wetlands on Flow Regulation in a Tropical River Catchment

Alem Oyarmoi, Stephen Birkinshaw *, Caspar J. M. Hewett and Hayley J. Fowler

School of Engineering, Newcastle University, Newcastle upon Tyne NE1 7RU, UK; a.oyarmoi2@newcastle.ac.uk (A.O.); caspar.hewett@newcastle.ac.uk (C.J.M.H.); hayley.fowler@newcastle.ac.uk (H.J.F.)
* Correspondence: stephen.birkinshaw@newcastle.ac.uk

Abstract: Africa has the largest area of wetlands of international importance, and papyrus constitutes the most dominant species for many of these wetlands. This hydrological modelling study assesses and quantifies the impacts of these papyrus wetlands on historical baseflow and quickflow, as well as future flood and low flows in the Mpologoma catchment in Uganda. Assessment over the historic period shows that wetlands strongly attenuate quickflow while moderately enhancing baseflow. They play a moderating role in most months, except for the first dry season (June and July), due to the reversal of flows between wetlands and rivers that often occur during this period. Annual estimates show that wetlands are four times better at regulating quickflow than baseflow. Examination of changes at 2 and 4 °C global warming levels (GWLs) indicate that wetlands will play critical roles in mitigating flood risks, with a lesser role in supporting low flows. Wetlands are predicted to lower future mean flood magnitude by 5.2 and 7.8% at GWL2 and GWL4, respectively, as well as halving the average number of flood events in a year, irrespective of the warming level. This work shows that papyrus-dominated wetlands strongly influence catchment hydrology, with significant roles on quickflow, including floods, and highlights the need for their conservation and protection.

Keywords: papyrus wetlands; hydrological regulating services; Wetland-Specific Index; flood flows; low flows

1. Introduction

Wetland loss has been estimated at 64 to 71% since 1900 [1]. This is a major issue given their contribution to ecosystem services, including maintaining the hydrological cycle, regulating climate, and protecting biodiversity [2]. Among the continents, Africa has the most extensive wetlands of international importance [3], covering an impressive 7% of the continent [4]. Despite their vast scale, African wetlands are continually under pressure [5]. Up to 42% of wetlands were lost between 1970 and 2015 [6]. Threats to wetlands often stem from political and social debates on whether wetland areas are at their highest economic use, which fuels negative sentiments about them [7]. Threats to these wetlands include population pressure and agricultural expansion [1,8,9], river regulation [8], and climate change [9,10]. High population growth rates and food shortages often trigger wetland encroachment for crop production, given that they provide year-round moist soil conditions [11].

The giant sedge *Cyperus papyrus* L. (Cyperaceae), the largest of the 400 tropical sedge species within the genus, constitutes the most widespread dominant species for many of the African wetlands [9]. At 5–6 m in height, papyrus exists as rooted or floating marshes in riverine and lacustrine landscapes [12]. Their hydro-periods are characterised by permanent inundation (such as in rivers and lakes) or seasonal inundation (such as in floodplains and creeks) [12]. In Uganda, wetland coverage is 11 to 13%, with papyrus being the dominant or co-dominant plant [12]. Major papyrus wetlands include those in the shorelines of lakes such as Victoria, Albert, and Kyoga and riverine systems such as

Mpologoma, Nabajuzzi, and Namatala. Like wetland trends in Africa, papyrus wetlands in Uganda are under pressure. Loss rates are estimated at 0.5 to 5% per annum, although few studies have been undertaken [10].

The two primary ecosystem services of papyrus wetlands are provisioning and regulating services. Provisioning services include carbon storage, fisheries, papyrus biomass, and wetland reproduction [13–21]. Regulating services include storing floodwaters and maintaining surface water flow during dry periods [22–28]. Van Dam et al. [10] provided research and policy priorities for papyrus wetlands, emphasising the need for improved understanding and modelling of the often alluded to but rarely quantified regulating services. These regulating services hold a far more significant advantage than provisioning services but are difficult to evaluate [9], thus hindering the economic valuation and conservation of papyrus wetlands [10]. Studies in Uganda that quantified the microscale papyrus regulating services include Kayendeke et al. [25] and Kayendeke and French [24], who, respectively, assessed the response of papyrus root mats to changing water levels and the seasonal variations in papyrus wetland water balance. Elsewhere in Africa, Di Vittorio and Georgakakos [28], Howell et al. [27], Hurst [22], and Sutcliffe and Parks [23,26] described the regulation services of papyrus wetland complexes at the Sudd and Okavango wetlands in South Sudan and Botswana. Wetland ecosystem services emanate from the aggregated effects of wetlands [29]; however, no studies have attempted to quantify papyrus flow regulatory services at the catchment scale. Catchment scale studies incorporate the dynamic effects of wetlands, which can mitigate the risks of adverse policies, engineering, and management solutions emanating from single wetland studies [30].

As a transition zone between terrestrial and aquatic ecosystems, wetlands play an essential role in climate change. Although generally resistant to change [31], wetlands are susceptible to temporal dynamics, especially if thresholds (e.g., seasonal variations in hydrology) are exceeded [32]. As temperature increases, the rate of wetland evapotranspiration increases [31,33], which could lower the water levels, particularly during dry seasons [31]. Meanwhile, changes in the timing and amount of precipitation can result in flow regime changes. Extremes in temperature and flow regimes lead to changes in wetland biogeochemistry [31] and biodiversity [31,34]. Research that helps predict wetlands' role in mitigating future impacts of climate-induced changes in hydrology is valuable.

Models are crucial to developing insights into the dynamics of wetland ecosystem services [32]. Their selection, however, requires a conceptualisation of the critical processes to be modelled [35]. At the catchment scale, these processes include hydrologic and hydraulic fluxes [36,37] such as precipitation, evapotranspiration, surface water, and groundwater. 'Getting the water right' is a prerequisite to understanding the dynamics of wetland systems [38], which require the use of models that have an explicit representation of the various water transfer mechanisms [37], with the coupled surface-groundwater flows a critical component [36]. SHETRAN [39] is a suitable model, as it is a physically based, spatially distributed catchment modelling tool that simulates surface-groundwater flows and can quantify the effects of wetlands. It is freeware and has also successfully been applied globally in land use and climate change studies in predominantly rural catchments (e.g., [40–42]).

This paper investigates the impacts of papyrus-dominated wetlands on catchment hydrology under climate change, focusing on the climate data-limited Mpologoma catchment. It is worth noting that human comprehension of catchment systems is underpinned by the availability and level of confidence in observational data [43,44], of which atmospheric data are most critical [45]. This study, therefore, focuses on addressing the following research questions. (a) What are the limitations of freely available global precipitation datasets in developing a catchment model in the study region? (b) How do papyrus wetlands regulate baseflow and quickflow in the Mpologoma catchment? (c) What roles will papyrus wetlands play in regulating future extreme flows (flood and low flows) in the Mpologoma catchment?

2. Materials and Methods

2.1. Study Area

The Mpologoma catchment, located between longitude 33.4–33.6° E and latitude 0.3–1.3° N, is a transboundary watershed between Uganda and Kenya (Figure 1) with over 80% of its area in Uganda. Covering 8989 km^2 [46], its highest peak, Mount Elgon, is to the northeast and at approximately 4298 m above sea level (a.s.l). A more significant part of the catchment (over 80%) lies between 996 and 1150 m a.s.l, the former being the mean lake level at Lake Kyoga, Mpologoma's outlet. Key tributaries include Namatala, Manafwa, Malaba, and Naigombwa. The region's rainfall regime is bimodal (peak months of March to May and August to November) [47], with a mean annual rainfall of 1215 to 1660 mm [47,48]. Following Köppen–Geiger's climate classification [49], the catchment can be categorised as wet Equatorial Monsoonal. Subsistence farming is the main economic activity, with major crops being rice, maize, beans, groundnuts, cassava, sugarcane, etc. Seven unique land use and cover types were identified in 2019 [50]. These include woodland (6.2%), grassland (5.2%), built-up areas (11.6%), subsistence crops (53.2%), wetland (21.5%), open water (1.9%), and commercial farms (0.5%).

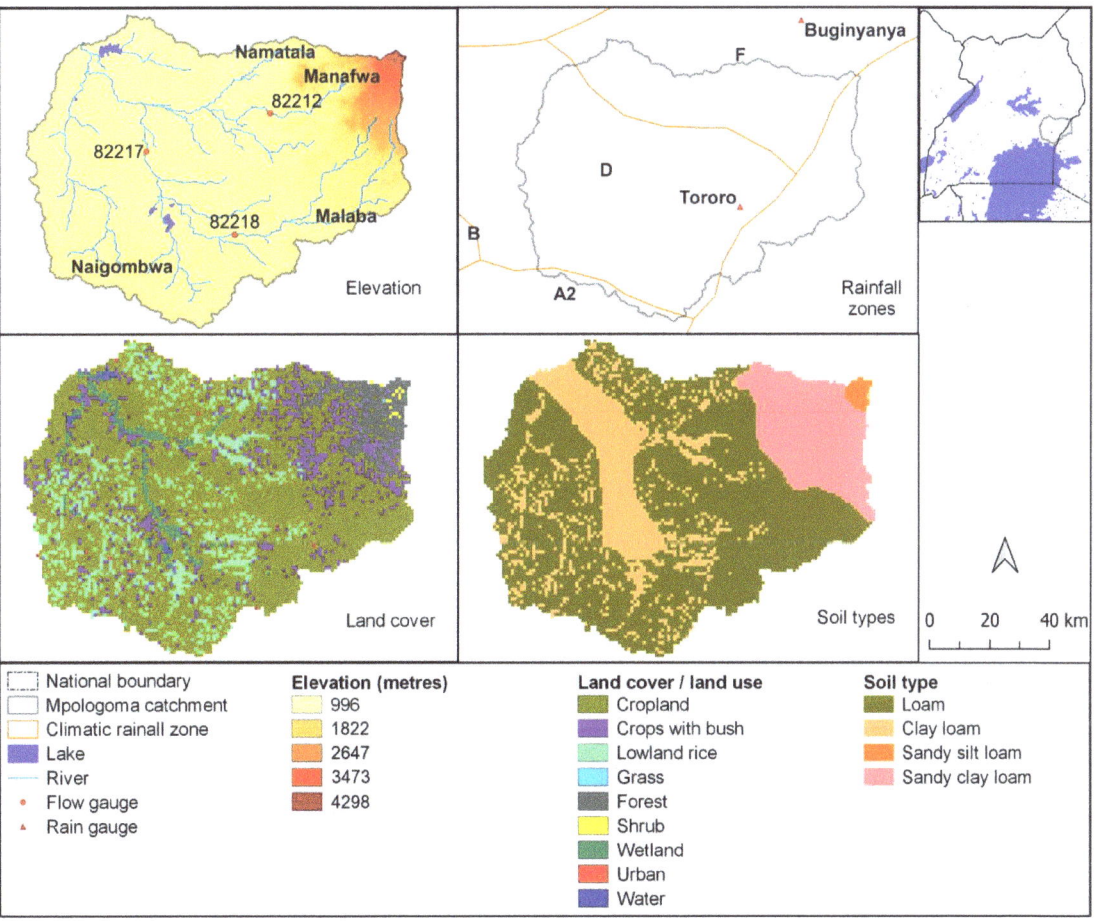

Figure 1. Mpologoma catchment elevation, rainfall zones, land cover, and soil types.

2.2. Data Sources and Processing

Hydrological modelling in SHETRAN requires spatially explicit input datasets of climate (precipitation and potential evapotranspiration), land surface topography, land use, land cover, and soil and lithological distribution. Measured surface flow and/or groundwater levels are also required for model calibration and validation. The following subsections describe the selection and preparation of these datasets. Figure S1 in the Supplementary Materials shows a schematisation of the steps followed in the study.

2.2.1. Selection and Evaluation of Rainfall Products

Satellite-based and reanalysis rainfall products have often been favoured for hydrological modelling in regions with low rain gauge networks, such as Sub-Saharan Africa [51,52]. However, the rainfall fields of reanalysis products are less accurate in the tropics compared to satellite estimates [53]. As the predominant rainfall over inland East Africa is convective [54], satellite products may provide more accurate data. The essential satellite-based precipitation products (SPPs) that have been applied over Africa include TAMSAT, CHIRPS, ARC, RFE, MSWEP, PERSIANN, CMORPH, and TRMM 3B42 (Table A1 in Appendix A).

This study evaluated the SPPs mentioned above by ground truthing (i.e., direct comparison with rain gauge data) followed by hydrological modelling. Point-to-pixel ground-truthing, with gauges at the point scale and SPPs at the grid scale, was carried out for climatological rainfall zones D and F (Figure 1) at daily and monthly time scales, which was preceded by quality control of gauge data, as outlined in Maidment et al. [53]. The point-to-pixel comparison was used to eliminate interpolation error, which is a concern for regions with sparse gauge networks [55]. Given that RFE2.0 and MSWEP2.2 start and end, respectively, in January 2001 and October 2017, evaluations were limited to January 2001 and December 2016. Overall, 0.79% and 12.25% of gauge records for Tororo and Buginyanya, respectively (Figure 1), were not included in the assessment due to missing data.

Following other studies on Africa (e.g., [56,57]), the point-to-pixel evaluation comprised assessment for accuracy in daily rainfall identification and daily and monthly rainfall totals. The extent of rainfall identification was assessed through measures of False Alarm Ratio (FAR), Probability of Detection (POD), Frequency Bias (FB), and Heidke skill score (HSS) (the equations are in Section S2 of the Supplementary Materials). In contrast, rainfall totals were assessed through correlation coefficient (R) and Nash–Sutcliffe efficiency (NSE). FAR represents the fraction of satellite-estimated rain days not captured by the gauge, and POD is the fraction of gauge rainfall days identified by the satellite product. FB is the ratio of total rain days in the satellite product to that of the gauge. It ranges from 0 to ∞, with values less (greater) than one indicating under (over) estimation of rain days. HSS, on the other hand, assesses the overall skill of a satellite product's rain-day detection while accounting for random chance. Its value ranges from $-\infty$ to 1, with zero indicating that the satellite product cannot detect rain days. A value less than zero implies that random chance is better than the product, whereas one denotes perfect skill.

MSWEP, TAMSAT, and CHIRPS performed best at the monthly timescale (see results in Appendix B). These SPPs were then hydrological assessed to select the best SPP for the study area.

2.2.2. Estimation of Potential Evapotranspiration (PET)

Duan et al. [58] evaluated the performance of Climate Forecast System Reanalysis (CFSR) daily temperature fields over the eastern African region of Ethiopia. They concluded they are as good as the observed temperature when employed in hydrological models. Hargreaves' method [59] uses temperature values for PET estimation, and it is recommended for regions with limited data [60]. Di Vittorio and Georgakakos [28] estimated monthly evapotranspiration rates over the Sudd wetlands in South Sudan using four different methods, including Hargreaves'. Hargreaves' procedure performed best, with results similar to those generated using the Penman equation. Thus, this study used the CFSR daily minimum and maximum air temperature fields to calculate PET.

2.2.3. Measured Flow Data

Water level time series recorded at 08:00 and 16:00 h for 17 years (2000 to 2016) were acquired from the Ministry of Water and Environment (MWE), Uganda. The datasets comprised measurements for rivers Manafwa (Station 82212), Malaba (Station 82218), and Mpologoma (Station 82217), Figure 1. Rating curves for each station, also supplied by MWE, were used in deriving daily average flows. Given that a quality check of flow measurements is essential in hydrological modelling, the minimum steps recommended by Crochemore [61] were adopted. These include analysis of data availability (in terms of length and spatial distribution of time series) and quality checks for outliers, homogeneity, and trend. The above steps were preceded by a visual inspection of hydrographs for suspicious records (e.g., negative values, wrongly recorded missing data, and out-of-the-ordinary hydrograph patterns of variation). Data preprocessing can lower hydrologic data uncertainty by up to 10%, minimising bias and incorrect conclusions [62].

2.2.4. Land Surface Representation

Coarse grids are often adopted in physically based, spatially distributed models to lower unknowns and execution time [63], which increases prediction uncertainty [64]. However, 0.5 to 4 km grid scales are generally acceptable for flow prediction in large catchments. For example, Sreedevi and Eldho [64] simulated discharge and sediment yield at 1 and 4 km grid scales based on effective parameters of a 2 km SHETRAN model. No significant difference was detected in the monthly flow predictions. Similarly, Zhang [65] compared the SHETRAN model performance at 0.5, 1, and 2 km grid scales. Performance generally improved with grid resolution. Nonetheless, the 1 and 2 km grids gave good results. Although DEM resampling techniques do not significantly influence streamflow modelling [66], resampling to coarser grids greater than 1 km could substantially affect the spatial distribution of land use, soil types, and river links, which can affect model performance in capturing peak flows and runoff volumes [65]. Thus, for this study, the Advanced Space-borne Thermal Emission and Reflection Radiometer (ASTER) Global Digital Elevation Model (DEM) Version 3 [67], initially at 30 m × 30 m grid resolution, was resampled to 1 km × 1 km grid size and used in defining catchment boundary, ground surface elevations, and generating river flow paths.

2.2.5. Land Cover and Land Use Layer

The European Space Agency (ESA) Climate Change Initiative (CCI) land cover map [68] was used in categorising land use/cover in the study area. ESA CCI provides annual global land cover layers at 300 m spatial resolution from 1992 to recent times, with accuracies varying from 71% to 97% depending on the land use/cover type [68]. No significant change in permanent (i.e., papyrus-dominated) wetland areas was detected from 2000 to 2020. Thus, model calibration and validation used the 2007 land cover layer (Figure 1). Detailed information on ESA CCI land cover layers is online at https://www.esa-landcover-cci.org/.

2.2.6. Soil and Lithology

The FAO/UNESCO Digital Soil Map of the World version 3.6 [69] was used in defining soil categories. However, soils in some wetland areas and small water bodies were not correctly categorised. The topsoil in these areas is mostly clay loam [46]. Thus, the land cover map was used in identifying these locations, and the soil category was reclassified accordingly. As with the DEM, the soil layer was resampled to a 1 km × 1 km grid. Four soil types (loam, clay loam, sandy silt loam, and sandy clay loam) were identified in the catchment (Figure 1).

Lithological depths were estimated from borehole log data for the study area (1986 in total). Analysis of the records indicates 1668, 195, and 123 logs in the regions dominated by loam, clay loam, and sandy clay loam, respectively. No log data were available for the area dominated by sandy silt loam. Thus, its lithological depths were assumed to be the same as the soil category closest to it, sandy clay loam. In addition, sandy silt loam covers less than

1% of the catchment. Overall, three critical lithological layers (topsoil, weathered rock, and base rock) were identified from the log data (Table 1).

Table 1. Lithological depths in Mpologoma catchment.

Topsoil Type	Mean Bottom Depth from the Ground Surface (m)		
	Topsoil	Weathered Rock	Base Rock
Loam	3.893	12.553	30
Clay loam	3.857	12.943	30
Sandy silt loam/sandy clay loam	5.226	14.16	30

2.3. Modelling Approach

SHETRAN [39] is a 3D integrated surface and subsurface finite difference modelling system for water flow, sediment transport, and contaminant transport in catchments. Its water flow components comprise interception and evapotranspiration, overland and channel flow, variably saturated lithology and aquifers, and channel–aquifer interactions. The model's xy grid was set at 1 km × 1 km, with 35 cells in each grid column. The lithological cells ranged from 0.1 to 0.2 m in the topsoils but were set to 5 m in the rock layers. Selection of the number of cells is a trade-off between accuracy and simulation time. Generally, the cells are smaller nearer the surface (as the flow is more variable) and larger as you go down the column. Although models were driven with daily data (precipitation and evapotranspiration), the simulation timestep was set to 1 h. Detailed information on SHETRAN can be accessed at https://research.ncl.ac.uk/shetran/.

Model calibration and validation, preceded by sensitivity analysis (see Appendix C). In Phase 1, the second step for precipitation product evaluation (i.e., hydrological evaluation of SPPs), SHETRAN was calibrated and validated on the sub-catchment gauged at station 82212 (Figure 2). The calibration and validation were carried out independently for each SPP (i.e., MSWEP, TAMSAT, and CHIRPS) that performed relatively well in the ground-truth evaluation. The sub-catchment gauged at 82212 has an insignificant wetland coverage. Thus, it enabled model parameterisation for catchments without wetlands in the region.

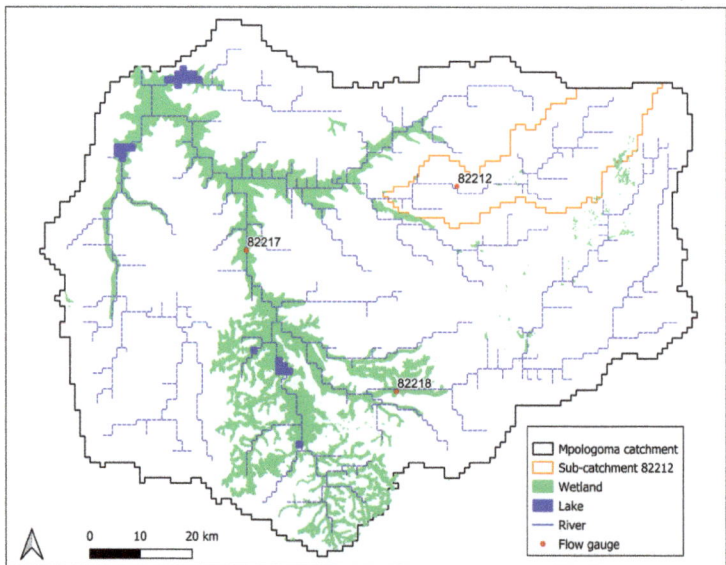

Figure 2. SHETRAN masks for Phase 1 (sub-catchment 82212) and Phase 2 (Mpologoma catchment) models, including the locations of river channels, lakes, and wetlands.

In Phase 2, the entire Mpologoma catchment was simulated, with parameters initially set to the effective values attained in Phase 1. The Phase 2 model was set up only for the precipitation product that performed best in Phase 1. Recalibration and revalidation on stations 82217 and 82218, respectively, were carried out with parameter adjustments limited to locations with wetlands.

River channels in the SHETRAN model flow along the edge of grid squares and are automatically created from the DEM. However, due to weaknesses in the SHETRAN automatic river generator in the flatter wetland areas, they were manually adjusted to follow the actual river network. River channels were removed where there are lakes, and the elevations of grid squares in lake locations were changed to depict average lake depths. The modified files were used in SHETRAN (version 4.4.5). Model performances were assessed statistically by quantifying the Nash–Sutcliffe efficiency (NSE), percent bias (Pbias), and the root mean square error standard deviation ratio (RSR). These metrics are generally recommended for model performance assessment in hydrology [70].

2.4. Impacts of Wetlands on Catchment Discharge

Wetland flow regulating functions can be assessed if a reference condition is defined, against which flow changes associated with the wetland are quantified [71]. Bullock and Acreman [72] documented the most used approaches for inferring wetland flow regulating functions. The 'with/without' approach, restricted to model-based studies, compares the same catchment with and without wetlands. The model is typically calibrated and validated for a 'with' or 'without' scenario. Simulations are then carried out with the alternate scenario, and any difference between the model outputs is attributed to the existence of wetlands. Papyrus plants are perennial herbaceous vegetation. Although communities around them harvest papyri in the dry season, they quickly regenerate at the onset of rains [12]. This study assumed that papyri stand remains the same all year round.

Furthermore, hydrological measures (flow indices) are employed in quantifying wetland flow regulating functions. These measures can be broadly classified into five groupings [72]: gross water balance, groundwater recharge, baseflow and low flows, flood response, and river flow variability. In this study, the historical impacts of wetlands on catchment flow were assessed using indices of baseflow and quickflow. In contrast, future effects were evaluated using flood and low flow indices. Quickflow is the part of precipitation that reaches the river fastest through surface runoff and interflow and is mainly responsible for floods. Thus, baseflow and quickflow are good indicators of how wetlands respond to slow and fast flow processes. However, a more holistic approach to catchment management requires the characterisation of flows in terms of flood/low flow indices (e.g., magnitude, duration, frequency, etc.) [73]; thus, it was adopted for the future period.

2.4.1. Impacts of Wetlands on Baseflow and Quickflow

The Wetland Specific Impact (WSI) metric [74] was employed in quantifying wetland flow regulating functions. It normalises the hydrological impact, which is an increase or decrease in flow metrics with respect to the wetland area (Equation (1)).

$$WSI_{BF/QF} = [R_{BFw/QFw} - R_{BFwo/QFwo}] \div A \quad (1)$$

where WSI ($m^3/s/km^2$) is the index of wetland impact expressed as a flow parameter (baseflow (BF) or quickflow (QF)) per unit area of wetland; R (m^3/s) is discharge (baseflow or quickflow) for the situation with (w) and without (wo) wetlands; and A (km^2) is the total wetland area.

The hydrological impacts of wetlands are highly dependent on climatic conditions, leading to seasonal and inter-annual variability [74]; thus, WSI values were computed at monthly and annual timescales. Boxplots showing the mean, median, and interquartile range and upper and lower limits were employed to illustrate this variability. However, only the mean WSI values and non-outlying WSI range (upper and lower limits) were used to describe wetlands' roles. WSI values were computed over 30 years (1984 to 2013) for the model outputs forced with 'observed' climatic data (i.e., MSWEP and CFSR, Table 2). Baseflow and quickflow components were extracted from total flows using WETSPRO, a flow filtering tool based on the extended Chapman filter [75] and available online at https://bwk.kuleuven.be/hydr/pwtools.htm#Wetspro (accessed on 10 August 2022).

2.4.2. Impacts of Wetlands on Future Flood and Low Flows

This study assessed climate change impacts using the first ensemble (r1i1p1f1) daily climate variables from the Coupled Model Intercomparison Project phase 6 (CMIP6). Given that suitable corrections can only be attained if the scale gap between the model and observational data is realistic [76], the CMIP6 models were restricted to 100 km nominal resolution. Thus, only four (GFDL-ESM4, CESM2-WACCM, MRI-ESM2-0, and NorESM2-MM) of the nine CMIP6 models (Table A2 in Appendix A) recommended for application over the East African region were selected. As of July 2022, CESM2-WACCM had no historical minimum and maximum temperature at the official CMIP6 web portal (https://esgf-node.llnl.gov/projects/cmip6/). Thus, GFDL-ESM4, MRI-ESM2-0 and NorESM2-MM were the only Global Climate Models (GCMs) used. These GCMs were bias-corrected using the quantile delta mapping method in RStudio's 'MBC' package [77], with daily MSWEP and CFSR as observed precipitation and temperature fields, respectively.

The focus was on the impacts of representative concentration pathway 8.5 (RCP8.5), given that current greenhouse gas emissions are closer to it than the other RCPs [78]. Although there is consensus to limit warming to 2 °C [79], Intergovernmental Panel on Climate Change (IPCC) predictions indicate end-of-century warming of up to 4 °C if the current emission trend continues [80]. Thus, this study assessed climate change impacts at 2 and 4 °C global warming levels (GWLs), with NOAA GISS Surface Temperature Analysis [81] global observational data as a reference. A collection of 30-year windows corresponding to 2 and 4 °C GWLs were extracted for each GCM, following the method in Vautard et al. [82].

Following Xu et al. [83], the mitigating effects of wetlands on future flood flows were assessed using flow duration, magnitude, and frequency indices by analysing model outputs with and without wetlands. These were similarly adapted for low flows. Table 2 shows the total number of simulations carried out with the various CMIP6 datasets at the baseline (BL), GWL2 and GWL4. A 2-year return period flood and low flow threshold of the baseline scenario were used to detect the occurrence of flood and low flows, assuming that river cross sections do not change at GWL2 and GWL4. The 2-year threshold is often used as an estimate of bank-full discharge [84], thus a proxy for flood flow (i.e., starting point of inundation). The same return period was chosen for low flow as a review of droughts over Africa showed that, in the worst case, droughts occur every two years in Uganda [85]. A flood event was considered one or more consecutive days when the daily flow was larger than the bank-full flow. Similarly, a low flow event is one or more consecutive days when daily flow is less than or equal to the low flow threshold. The estimated 2-year return period flood and low flow threshold of the baseline scenario are 33.2 and 11.5 m^3/s, respectively.

The various flow indices were calculated over the water year (March to February). Flow duration was calculated as the average of individual events in a water year. The average per event was calculated before averaging over a year for flow magnitude. Flow frequency is the number of flood or low flow events yearly. The analysis assumed the same wetland size for baseline and future scenarios.

Table 2. List of models (with and without wetlands) simulated/forced with 'observed' (i.e., MSWEP and CFSR) and CMIP6 climatic datasets.

S. No.	Climatic Datasets	Simulation Period (Model Warmup)	Wetlands Present or Not
1	MSWEP rainfall and CFSR PET	1979–2013 (1979–1983)	Yes
2			No
3	GFDL-ESM4 at BL	1979–2013 (1979–1983)	Yes
4			No
5	MRI-ESM2-0 at BL	1979–2013 (1979–1983)	Yes
6			No
7	NorESM2-MM at BL	1979–2013 (1979–1983)	Yes
8			No
9	GFDL-ESM4 at GWL2	2028–2063 (2028–2032)	Yes
10			No
11	MRI-ESM2-0 at GWL2	2030–2065 (2030–2034)	Yes
12			No
13	NorESM2-MM at GWL2	2025–2060 (2025–2029)	Yes
14			No
15	GFDL-ESM4 at GWL4	2059–2094 (2059–2063)	Yes
16			No
17	MRI-ESM2-0 at GWL4	2051–2086 (2051–2055)	Yes
18			No
19	NorESM2-MM at GWL4	2049–2084 (2049–2053)	Yes
20			No

3. Results

3.1. Model Calibration and Validation Results

Phase 1 model calibration and validation results for the best performing SPPs (MSWEP, TAMSAT, and CHIRPS) are shown in Figures 3 and 4 for daily and monthly timesteps, respectively. Overall, MSWEP performed best with a daily NSE (Pbias/RSR) and monthly NSE (Pbias/RSR) of 0.47 (−11.4%/0.73) and 0.73 (−5.15%/0.52) at calibration and 0.44 (−0.68%/0.75) and 0.87 (−3.39%/0.36) at validation. CHIRPS followed this at 0.24 (10.31%/0.87) and 0.55 (−6.95%/0.67) for calibration and 0.45 (−4.68%/0.74) and 0.86 (−7.17%/0.38) for validation. Meanwhile, TAMSAT attained −0.33 (1.28%/1.16) and 0.39 (1.34%/0.78) at calibration and −0.07 (−0.44%/1.03) and 0.18 (−4.02%/0.9) at validation. Model performance generally improves with timestep and concurs with studies over Africa (e.g., [43,86]) that recommend hydrological performance assessment of SPPs at longer timescales (e.g., monthly instead of daily). This is because validation errors in SPPs are often offset upon spatial and temporal integration at the catchment scale [87]. In addition, real-time control, design, and management of most catchment systems (such as irrigation and water supply, reservoirs, land use change, climate change, and environmental impact studies) can be achieved with monthly hydrological model estimates [88]. Mutenyo et al. [89] modelled the catchment gauged at 82212 using SWAT, a semi-distributed hydrological model, with daily gauge rainfall data from 1955 to 1961. The model performed best at the monthly timescale with NSE values (Pbias) of 0.72 (−0.49%) and 0.64 (20.5%) for calibration and validation, respectively, which are similar to the results obtained in this study.

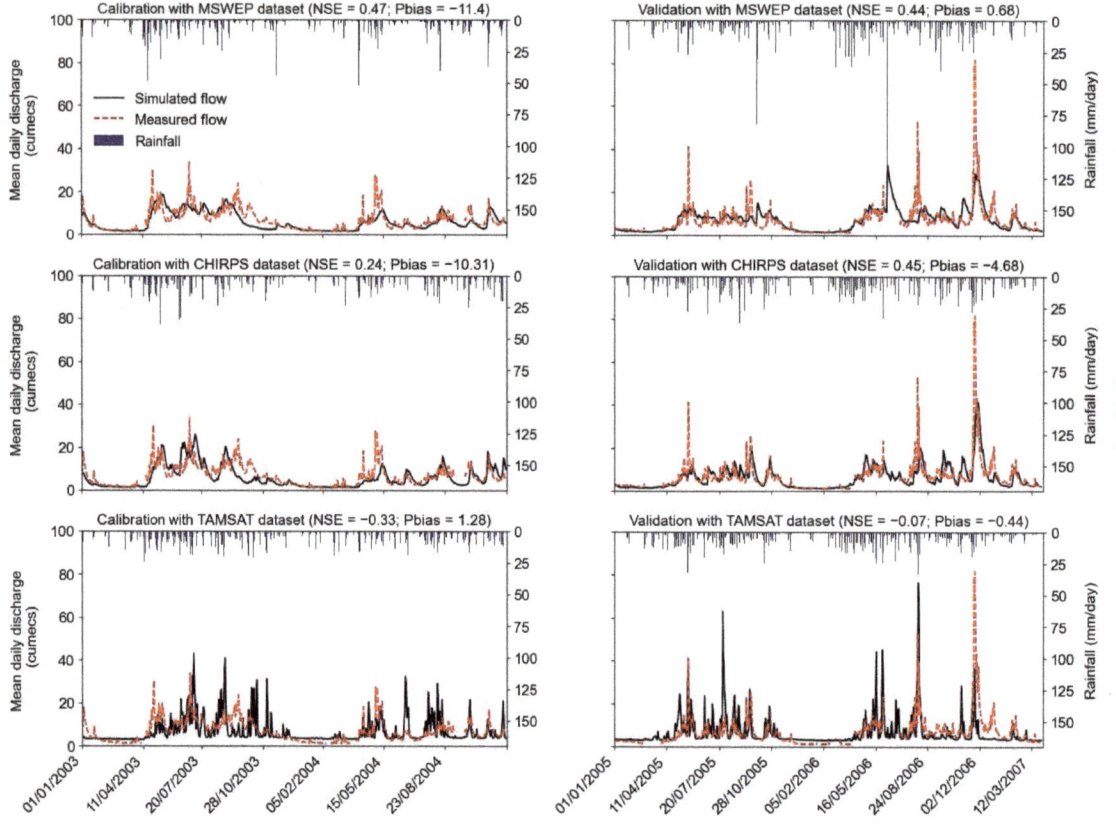

Figure 3. Daily timestep calibration and validation results at station 82212. Date format in Day Month Year.

Given its superiority in this study, MSWEP was further used in Phase 2 of model calibration and validation, focusing on the recalibration of channel Strickler coefficient (Krc) in locations with wetlands. Strickler coefficient is the inverse of Mannings coefficient An optimal Krc of 7 m$^{1/3}$/s, from an initial value of 30 m$^{1/3}$/s (Phase 1), denoting a 76.6% change in Strickler coefficient, was attained, with daily and monthly NSE (Pbias/RSR) of 0.52 (−2.46%/0.69) and 0.64 (−6.04%/0.6) at calibration and 0.55 (−13.67%/0.69) and 0.59 (−13.66%/0.64) at validation, respectively (Figure 5). To ensure the robustness of the Phase 2 model, calibration and validation were carried out at two different sites (82217 for calibration and 82218 for validation). Due to uneven data gaps in the records of the two gauging stations, calibration was carried out from January 2011 to November 2013, whereas validation was from January 2000 to December 2003.

3.2. Historical Impacts of Wetlands on Catchment Hydrology

3.2.1. Overall Impacts of Wetlands on Catchment Hydrology

The mean annual water balance components over the Mpologoma catchment are shown in Table 3. Precipitation (P), actual evapotranspiration (AET), and catchment outflow (Q) were extracted from the SHETRAN model output for the situation with and without wetlands over the period 1984 to 2013 (30 years). The models were driven with MSWEP and CFSR precipitation and PET datasets. Overall, wetlands modulate river flow in the Mpologoma catchment by decreasing annual discharge by 5.5% on average. In contrast, evapotranspiration is enhanced annually by 0.4% on average.

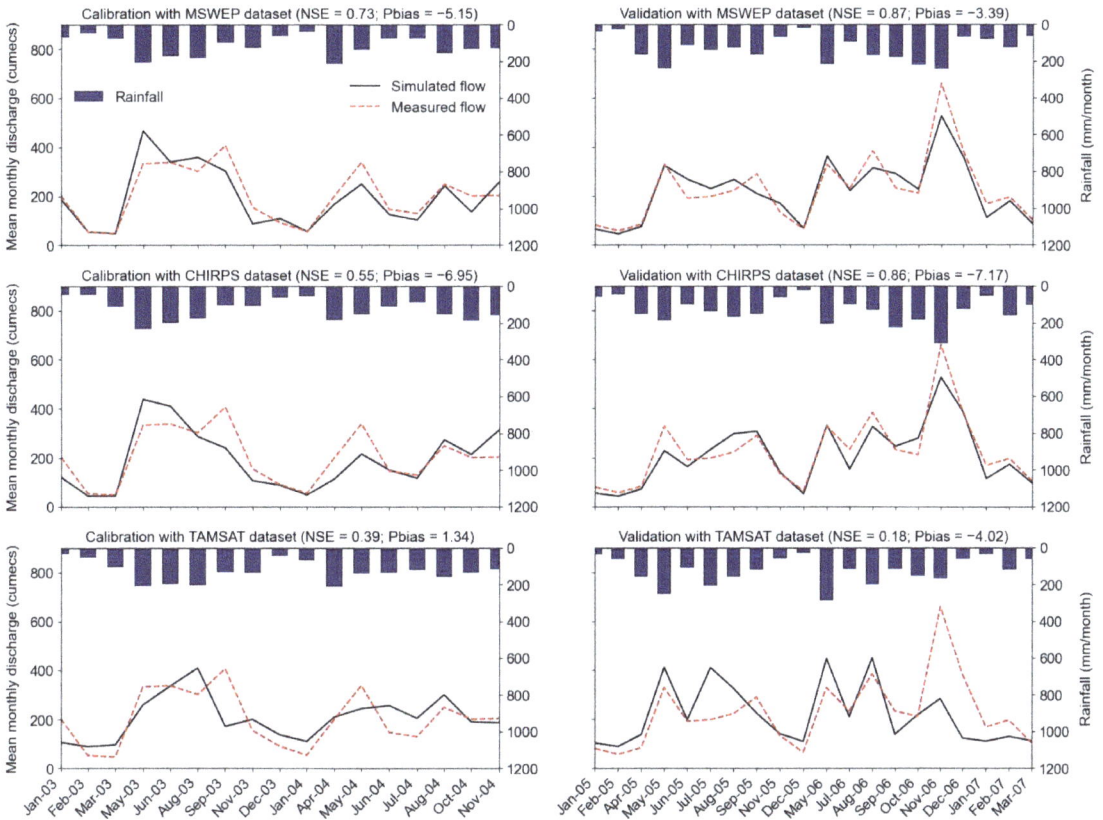

Figure 4. Monthly timestep calibration and validation results at station 82212. Months with data gaps in measured flow are not included. Date format in Month Year.

Table 3. Mean annual precipitation, actual evapotranspiration, and catchment discharge over 30 years (1984 to 2013) for the situation with and without wetlands. Models were driven with MSWEP and CFSR precipitation and PET datasets, respectively.

Water Balance Component	Without Wetlands	With Wetlands	Relative Change (%)
Precipitation (mm)	1299.3	1299.3	
Actual evapotranspiration (mm)	1218.2	1222.5	0.4
Catchment outflow (mm)	78.6	74.3	−5.5

3.2.2. Impacts of Wetlands on Baseflow and Quickflow

Based on mean WSI values (Figure 6), wetlands moderately support baseflow at the catchment scale in most months of the year except June (WSI of −0.02 m^3/s/km^2). Quickflow is suppressed in most months, except July and February (positive WSI of 0.16 and 0.01 m^3/s/km^2, respectively). Negative baseflow and positive quickflow WSI values indicate a reversal of the supportive role of wetlands towards these flow components, whereas zero implies a null response. The critical periods during which baseflow is often positively impacted are from March to May (WSI of 0.13, 0.12 and 0.05 m^3/s/km^2, respectively) and from October to February (WSI of 0.05, 0.05, 0.05, 0.08 and 0.1 m^3/s/km^2, respectively). Meanwhile, the corresponding periods for quickflow are from April to June (WSI of −0.48, −1.21 and −0.43 m^3/s/km^2, respectively) and September to November (WSI of −0.25, −0.16 and −0.36 m^3/s/km^2, respectively).

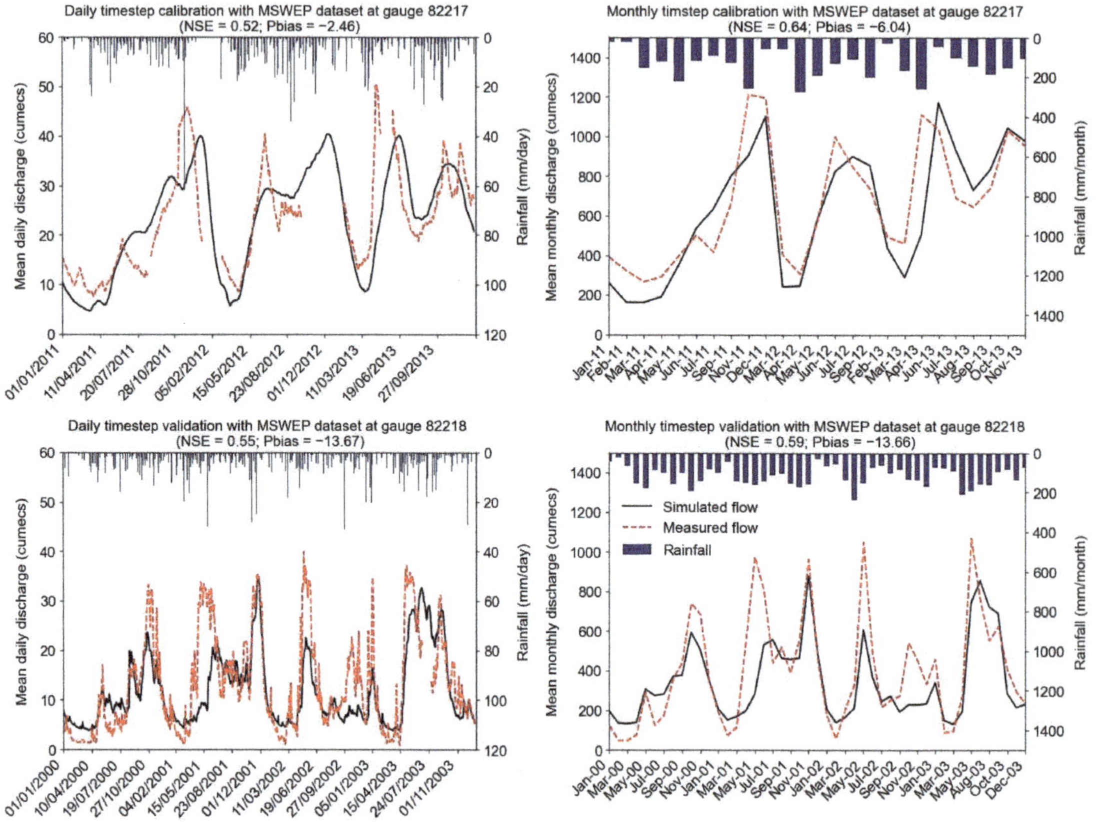

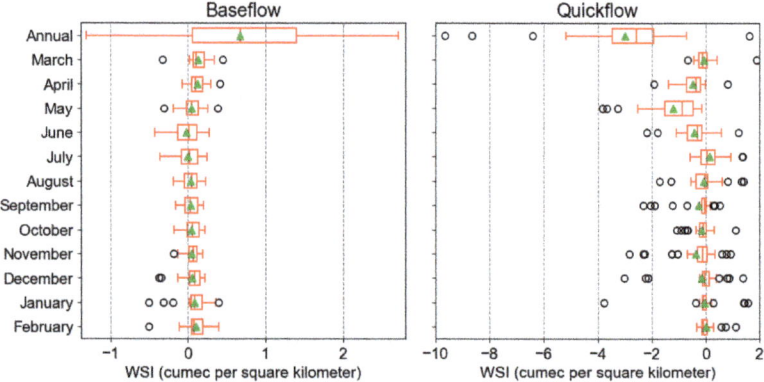

Figure 5. Daily and monthly timestep calibration (at station 82217) and validation (at station 82218) results. For the monthly plots, months with data gaps in measured flow are not included. Date format in Day Month Year for the daily plots and Month Year for the monthly plots.

Figure 6. Monthly and annual wetland-specific impact (WSI) on baseflow and quickflow at the catchment outlet. Computations are over 30 years (1984 to 2013). The boxplots show the mean (triangles), median, and interquartile range. The whiskers show the upper and lower limits, excluding any outliers (indicated as circles).

At the annual timescale, wetlands generally enhance baseflow (mean WSI = 0.68 m^3/s /km^2) and attenuate quickflow (mean WSI = -3 m^3/s/km^2). However, the yearly WSI non-outlying range shows that wetlands significantly reduced quickflow (-5.19 to -0.71 m^3/s/km^2) compared to enhanced baseflow (-1.31 to 2.7 m^3/s/km^2).

3.3. Impacts of Wetlands on Future Catchment Hydrology

3.3.1. The Overall Impact of Wetlands on Future Catchment Hydrology

The mean annual water balance components over the Mpologoma catchment for the ensemble baseline (BL), GWL2 and GWL4 are summarised in Table 4. Like the water balance for the models driven by MSWEP and CFSR precipitation and PET products (Table 3), water balance components were extracted over 30 years, with the baseline period set from 1984 to 2013. GWL2 and GWL4 water balance components are tied to a future 30-year period with a mid-point at 2 and 4 °C global warming levels, respectively. As the various individual GCM products were bias-corrected with MSWEP and CFSR datasets, model predictions for the BL are similar to those driven by MSWEP and CFSR. Wetlands in the BL modulate river flow by decreasing annual discharge (-6.6%) while enhancing evapotranspiration (0.4%) on average. These effects continue with warming but with a reduced impact on discharge.

Table 4. Mean annual precipitation, actual evapotranspiration, and catchment discharge over 30 years for the situation with and without wetlands. Models were forced with bias-corrected GCM precipitation and PET datasets.

Water Balance Component	Without Wetlands	With Wetlands	Relative Change (%)
	Model ensemble—baseline period		
Precipitation (mm)	1294.9	1294.9	
Actual evapotranspiration (mm)	1212.7	1217.0	0.4
Catchment discharge (mm)	82.4	77.0	-6.6
	Model ensemble—GWL2		
Precipitation (mm)	1403.4	1403.4	
Actual evapotranspiration (mm)	1298.8	1304.3	0.4
Catchment discharge (mm)	111.8	107.0	-4.3
	Model ensemble—GWL4		
Precipitation (mm)	1456.9	1456.9	
Actual evapotranspiration (mm)	1361.8	1367.4	0.4
Catchment discharge (mm)	105.7	100.5	-5.0

3.3.2. Impacts of Wetlands on Future Flood and Low Flows

Figure 7 shows violin plots of flood and low flow indices for the situation with and without wetlands. For flood flows and a particular catchment treatment (i.e., with or without wetlands), the 30-year mean value (Table 5) indicates that both duration and magnitude increase with warming. However, the difference between GWL2 and GWL4 indices is minimal. For the situation with wetlands, the mean flood duration per event increases from 85 days at BL to 128 and 110 days at GWL2 and GWL4, respectively. Similarly, for the situation without wetlands, the duration increases from 46 days at BL to 72 and 61 days at GWL2 and GWL4, respectively. The mean flood magnitude per event, for the situation with wetlands, increases from 37.65 m^3/s at BL to 42.87 and 41.21 m^3/s at GWL2 and GWL4, respectively. For the case without wetlands, the magnitude per event increases from 42.23 m^3/s at BL to 45.21 and 44.7 m^3/s at GWL2 and GWL4, respectively. In general, wetlands in the Mpologoma catchment are predicted to increase the mean flood duration by 77.8% (56 days) and 80.3% (49 days) at GWL2 and GWL4, respectively, and lower the mean flood magnitude by 5.2 and 7.8%.

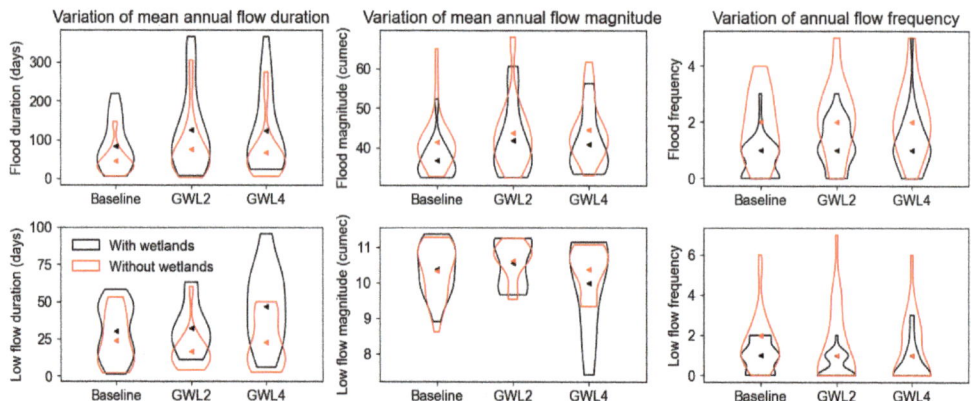

Figure 7. Violin plots of flood and low flow indices for the situation with and without wetlands. Flow duration and magnitude plots depict mean annual values, whereas frequency depicts the number of events in a year. The triangles indicate the average flow index over 30 years.

Table 5. Mean and range of flow indices (duration, magnitude, and frequency) for the situation with (w) and without (wo) wetlands over 30 years.

Scenario	Flow Duration (Days)		Flow Magnitude (m^3/s)		Event Frequency	
	Mean	Range	Mean	Range	Mean	Range
			Flood flows			
BLw	85	3–216	37.65	33.27–52.70	1	0–2
GWL2w	128	9–365	42.87	34.14–60.68	1	0–3
GWL4w	110	21–295	41.21	33.70–56.78	1	0–4
BLwo	46	3–144	42.23	33.37–65.43	2	0–4
GWL2wo	72	16–250	45.21	35.35–68.29	2	0–5
GWL4wo	61	8–272	44.70	33.92–59.92	2	0–5
			Low flows			
BLw	32	3–59	10.42	8.91–11.40	1	0–2
GWL2w	33	14–65	10.60	9.69–11.30	1	0–2
GWL4w	48	15–97	10.04	7.40–11.17	1	0–4
BLwo	22	2–53	10.47	8.62–11.54	2	0–5
GWL2wo	19	5–61	10.71	9.56–11.27	1	0–7
GWL4wo	27	3–88	10.45	9.39–11.32	1	0–6

From Table 5, the estimated mean flood frequency is constant, irrespective of warming level, at 1 and 2 flood events per year for the situation with and without wetlands, respectively. However, the potential for a larger number of flood events in a year (the range) increases with warming. For example, for the situation with wetlands, annual flood events increase from the BL range of 0–2 flood events to 0–3 and 0–4 events at GWL2 and GWL4, respectively. For the situation without wetlands, the range increases from 0–4 to 0–5 at both GWL2 and GWL4.

Unlike flood flows, the impacts of warming levels on low flows are generally heterogeneous. For example, for the situation with wetlands, the mean low flow duration at GWL2 (33 days) is the same as that at BL (32 days) but increases to 48 days at GWL4. However, the range of low flow duration per event increases with warming from 3–59 days at BL to 14–65 and 15–97 days at GWL2 and GWL4, respectively. Similarly, for the situation without wetlands, the mean low flow duration at GWL2 (19 days) is the same as that at BL (22 days) but increases to 27 days at GWL4. The range, however, increases with warming from 2–53 days at BL to 5–61 and 3–88 days at GWL2 and GWL4, respectively. In general, wetlands are projected to increase future mean low flow duration by 73.7% (14 days) and 77.8% (21 days) at GWL2 and GWL4, respectively.

For the case of low flow magnitude, there is no significant change in the 30-year mean flow value, irrespective of catchment treatment. For example, for the situation with wetlands, the mean flow magnitude per event increases and drops minimally from 10.42 m^3/s at BL to 10.6 and 10.04 m^3/s at GWL2 and GWL4, respectively. Similarly, for the situation without wetlands, it increases and drops minimally from 10.47 to 10.71 and 10.45 m^3/s, respectively. No significant change can be seen in the range of low flow values, except for GWL4 with wetlands. The range of low flow magnitude increases from 8.91–11.4 m^3/s at BL to 7.4–11.17 m^3/s at GWL4. In general, though not significant, projections suggest that wetlands will reduce future mean low flow magnitude by 1 and 3.9% at GWL2 and GWL4, respectively. However, a considerable decline of 21.2% is expected in the lower limit of low flow at GWL4.

The mean low flow frequency is generally constant, irrespective of the warming level. On average, one low flow event occurs per year regardless of catchment treatment. However, the range of low flow events is lower for the situation with wetlands. Events vary from 0–2 and 0–4 at GWL2 and GWL4, respectively, compared to 0–7 and 0–6 for the situation without wetlands. Overall, though wetlands are predicted to negatively impact future low flow magnitude and duration, they are expected to positively impact low flow frequencies by lowering the range of events in a year by 71.4 and 33.3% at GWL2 and GWL4, respectively.

4. Discussion

4.1. Historical Impacts of Wetlands on Catchment Hydrology

The catchment water balance (Table 3) shows that the Mpologoma floodplain wetlands reduce annual average river flow, increase evapotranspiration compared to most land types, and enhance groundwater recharge. The response is typical of wetlands, depending on the underlying lithology [72]. Additionally, the mean annual WSI, depicting enhanced baseflow and attenuated quickflow, show that wetlands regulate flow variability at the catchment scale [90]. Past studies in Africa, although restricted to wetland complexes (e.g., the Sudd wetland in South Sudan [22,23] and the Okavango Delta in Botswana [23]), have also shown that papyrus-based riparian wetlands play significant roles in dampening flow variability. The greater impact on quickflow compared to baseflow, as seen in this study's WSI non-outlying range, concurs with the work by Kadykalo and Findlay [91], who concluded that wetlands perform better in flood mitigation than low flow augmentation. Overall, the mean annual WSI values show that wetlands in the Mpologoma catchment are at least four times better at attenuating quickflow than enhancing baseflow. Like the mean annual WSI values, the mean monthly WSI results show that papyrus wetlands are most effective in curtailing quickflow. The periods in which quickflow is suppressed coincide with the region's wet seasons (March to May and September to November).

Although the mean monthly WSI shows baseflow enhancement in most months, reversal of the supportive role of wetlands occurs in the first dry season (June to July). One would expect baseflow enhancement and nearly zero quickflow WSI during the first dry season (June to July), as seen in the second dry season (December to February). The disparity between the two dry periods could be due to the different antecedent conditions. Both dry periods start immediately after a wet period; however, the first rainy season is much wetter than the second, and the second dry spell is much drier [47]. This affects storage and connectivity pathways between wetlands and rivers. Hydraulic gradients and hydrological connectivity in riparian wetlands, typical of those in the Mpologoma catchment, depend on adequate water levels in both the river and wetlands, thus influencing discharge to or recharge from rivers [74]. This exchange of fluxes between wetlands and rivers is driven by catchment scale factors such as rainfall (intensity and duration), landscape (topography, soil and geology, drainage area, land use, and land cover), and initial conditions [91–93]. Kayendeke and French [24] showed that, for a microsite of 0.18 km^2 in the Naigombwa sub-catchment of the Mpologoma catchment, papyrus wetland flows are dominated by surface inflows, with negligible contribution from groundwater (i.e., wetland soils are mostly heavy clay loam). The lack of baseflow support from wetlands in June–July may result from a reversal in hydrological dynamics between rivers and wetlands, as shown in Figure 8. The figure shows hydrographs of baseflow and total flow for water years 1985 and 2010 when wetlands reduced baseflow by the largest amount in June and July. For these two years, baseflow with wetlands was, respectively, 0.61 and 0.64 of the baseflow with no wetlands for June and 0.69 and 0.72 for July. Figure 8 indicates that the wetland's supportive role of providing baseflow discharge to rivers was reversed between May and September. This reversal was also observed by Kayendeke and French [24] for the site mentioned above between July 2015 and January 2016. Overall, wetland interconnections are complex at the catchment scale due to the aggregation effects of multiple wetlands on flow regimes and the influences of various flow paths [30].

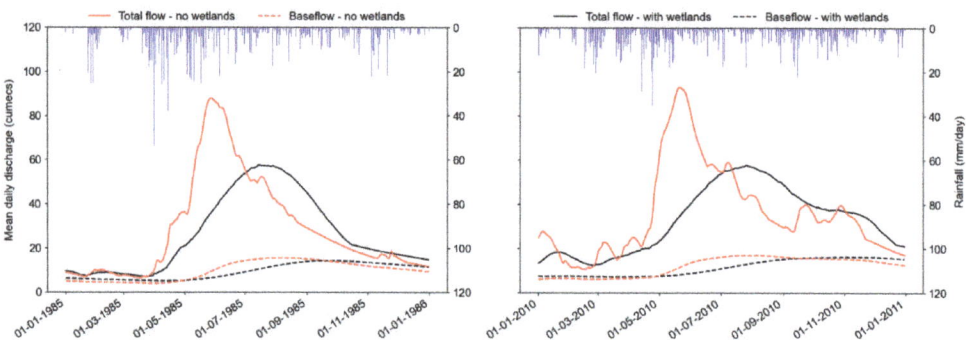

Figure 8. Hydrograph of baseflow and total flow for the water years 1985 and 2010.

4.2. Impacts of Wetlands on Future Catchment Hydrology

The projected catchment water balance (Table 4) shows that wetlands will decrease discharge at GWL2 and GWL4 but at rates lower than BL. CMIP6 models project increased precipitation over East Africa [94,95], as seen in the 30-year ensemble mean of the GCMs used in this study (Table 4 and Figure S7 in the Supplementary Materials). Thus, it explains the expected reduced effect of wetlands on the future annual discharge and the predicted increase, irrespective of catchment treatment (i.e., with or without wetlands), in the 30-year mean flood duration and magnitude and the range of flood frequencies in a year (Table 5).

Although the mean flood magnitude and the range of flood frequencies increase at GWL2 and GWL4, they are lower for the situation with wetlands. Meanwhile, the predicted mean flood duration is more significant in the case of wetlands. This agrees with other studies that show that floodplain wetlands lower flood magnitude, reduce the frequency of flooding, increase time to peak (flood duration), and delay floods [72,91].

Future predictions show that wetlands will worsen low flow's mean duration and magnitude. Warmer climates accelerate wetland water loss through evapotranspiration [31], leading to a decline in dry season flows, particularly during severe drought [74]. Floodplain wetlands in other parts of Africa, such as Sierra Leone and South Africa, have been shown to negatively impact dry season river flow due to high evapotranspiration rates [72]. Meanwhile, wetlands in the Mpologoma catchment will positively impact future low flow frequencies by lowering the range of frequencies in a year. The predicted reduction could be associated with water storage and slow water release wetland services. Acting like a 'sponge', wetlands store water during the wet season and release it in the dry period [96], thus minimising low flow episodes.

4.3. Implications and Limitations

Our findings show that wetlands greatly influence the hydrological footprint (i.e., the influence of watershed features on downstream discharge) of the Mpologoma catchment. However, increased wetland conversion for crop production, the most significant driver of wetland loss in rural catchments in Uganda [97], poses threats to sustainable catchment management. Changing climates characterised by increased precipitation, typical of the East African region [94,95], coupled with large-scale wetland loss, are predicted to aggravate flood risk [98]. Bunyangha [50] estimated a wetland loss rate of 0.6% per annum between 1986 and 2019 in the Mpologoma catchment, with a predicted loss rate of 0.2% between 2019 and 2039. This calls for restoration and conservation measures, given that most urban centres and rural communities in the Mpologoma catchment rely on natural systems (upstream wetlands) for flood management. The importance of wetland restoration for flow regulation services has been exemplified in several studies such as Acreman et al. [99], Mitsch and Day [100], Wu et al. [101], and Yang et al. [102]. Furthermore, Gulbin et al. [98] demonstrated that, even if a significant change to flow regulation is not achieved upon wetland restoration under the current climate, it could provide a future complementary measure for mitigating the negative impacts of existing flood management strategies. For example, current adequate flood protection structures such as dikes could fail due to increased precipitation. Thus, restored wetlands can play important supplementary roles.

Looking at the broader impact of climate change, irrespective of catchment treatment (i.e., with or without wetlands), our projections indicate increased flood and low flow risks, a trend similar to global projections [103]. The broader impact of flood risk in the Mpologoma catchment is projected to manifest through increased mean flood magnitude and duration and increased variability in flood frequency. Low flow risks are expected to manifest primarily in the form of increased variability in duration and number of events in a year. Although mean low flow magnitude may not significantly change with warming, extreme warming (i.e., at GWL4) is predicted to significantly reduce low flows and increase their variability. These findings on the broader impact of climate change are similar to other studies in Uganda, with indications of increased future flood magnitude [104–108] and occurrence of drought flows [104]. Thus, there is a need for the development of sustainable catchment flood and low flow management plans.

This paper provides useful information for wetland managers and policymakers on the supportive roles played by papyrus-based wetland systems, indicating a need for their restoration and conservation. However, our results should be treated with caution, given that modelling studies on flow regulation services of wetlands tend to predict larger impacts compared to empirical studies [91]. Future climate data generally contribute the largest uncertainty to hydrologic studies [109]. Lee et al. [110] assessed the impacts of uncertainties arising from climate change data on the streamflow of a catchment with wetlands. They concluded that the variability of GCM projections was the most significant contributor to flow prediction uncertainty. Nine CMIP6 GCMs were initially identified for this study, but only three were used due to limitations on resolution. We recommend using a larger GCM or regional climate model ensemble to lower prediction uncertainty.

5. Conclusions

This study investigated the effects of papyrus-dominated wetlands on baseflow and quickflow, including future flood and low flows within the Mpologoma catchment in Uganda. With the aid of the physically based, spatially distributed catchment modelling tool SHETRAN, the study quantified the catchment-scale flow regulating roles of papyrus wetlands. Findings show that papyrus wetlands wield a strong influence on catchment hydrology. They significantly affect quickflow (including floods), with a minor role on baseflow and most low flow indices. This indicates that wetland management is integral to water resources and flood management, particularly in tropical Africa, where papyrus wetlands naturally occur. These wetlands provide nature-based solutions against floods to several communities in the developing nations of Africa, thus providing solutions to challenges faced by flood engineers and water resources managers.

The modelled high wetland evapotranspiration rates with increased risks of low flows, especially in dry years, indicate an exaggeration of water scarcity. However, conservation decisions on whether papyrus wetlands are essential despite water scarcity risks should be taken in the broader context, depending on other wetland functions (e.g., biodiversity, human health and food, flood mitigation, recreation, etc.).

Supplementary Materials: The following supporting information can be downloaded at: https://www.mdpi.com/article/10.3390/land12122158/s1, Figure S1: Schematisation of steps followed in the study; Table S1: FAR, POD, FB and HSS at Tororo and Buginyanya rain gauge stations over January 2001 and December 2016; Figure S2: Scatter plots of daily rainfall total of SPPs and rain gauge data at Tororo and Buginyanya. The dashed line shows the perfect fit that could be attained if the gauge and SPP data were equal; Figure S3: Bar graphs of percent bias, mean absolute error and Nash-Sutcliffe efficiency at daily timescale for the various SPPs at Tororo and Buginyanya; Figure S4: Scatter plots of monthly rainfall totals (satellite products against gauge) at Tororo and Buginyanya. The dashed line shows the perfect fit that could be attained if the gauge and SPP data were equal; Figure S5: Bar graphs of percent bias, mean absolute error and Nash-Sutcliffe efficiency at monthly timescale for the various SPPs at Tororo and Buginyanya; Figure S6: Scatter plots of model response to changes in key parameters. Inset is the sensitivity index (SI); Figure S7: 30-year ensemble mean of 'observed' (MSWEP and CFSR) and bias-corrected CMIP6 models over the Mpologoma catchment. The plots show mean monthly rainfall and potential evapotranspiration (PET) at baseline (BL) and 2 and 4 °C warming levels.

Author Contributions: Conceptualisation, A.O., S.B., C.J.M.H. and H.J.F.; methodology, A.O., S.B. C.J.M.H. and H.J.F.; formal analysis, A.O. and S.B.; data curation, A.O.; writing—original draft preparation, A.O.; writing—review and editing, S.B., C.J.M.H. and H.J.F.; visualisation, A.O.; supervision, S.B., C.J.M.H. and H.J.F. All authors have read and agreed to the published version of the manuscript.

Funding: This research was funded by the Commonwealth Scholarship Commission and the Foreign Commonwealth and Development Office in the UK (Grant Number UGCS-2019-752). All views expressed here are those of the authors not the funding body.

Data Availability Statement: Due to restrictions, measured flow data and rain gauge data cannot be shared; however, they can be acquired at a cost from the Ministry of Water and Environment (https://www.mwe.go.ug/) and the Uganda National Meteorological Authority (https://www.unma.go.ug/). The data generated during the current study are available from the corresponding author on reasonable request.

Conflicts of Interest: The authors declare no conflict of interest.

Appendix A

Table A1. Essential satellite-based precipitation products (SPPs) applied over Africa.

S. No.	SPP	Description	Resolution (km)	Reference
1	TAMSATv3.1	Tropical Applications of Meteorology using SATellite (TAMSAT) and ground-based observations version 3.1; developed by the University of Reading, UK.	4	[111]
2	CHIRPSv2.0	Rainfall Estimates from Rain Gauge and Satellite Observations version 2.0; developed by the U.S. Geological Survey Earth Resources Observation and Science Centre, in collaboration with Santa Barbara Climate Hazards Group of the University of California.	6	[112]
3	ARC2	Africa Rainfall Climatology (ARC) version 2.0; developed by NOAA Climate Prediction Centre.	11	[113]
4	RFE2	African Rainfall Estimation Algorithm (RFE) version 2.0; developed by NOAA Climate Prediction Centre.	11	[114,115]
5	MSWEPv2.2	Multi-Source Weighted-Ensemble Precipitation (MSWEP) version 2.2; developed by.	11	[116]
6	PERSIANN-CDR	Precipitation Estimation from Remotely Sensed Information using Artificial Neural Networks (PERSIANN-CDR); developed by UCI Centre for Hydrometeorology & Remote Sensing.	28	[117]
7	CMORPHv1.0ADJ	Climate Prediction Centre (CPC) morphing technique (CMORPH) bias-corrected with gauge data (ADJ) version 1.0; developed by NOAA Climate Prediction Centre.	8	[118]
8	TRMM 3B42v7	Tropical Rainfall Measuring Mission (TRMM) Multi-satellite Precipitation Analysis (TMPA) version 7; developed by NASA and Japan's National Space Development Agency.	28	[119]

Table A2. List of best-performing CMIP6 GCM models over the region of Uganda. The list is based on the findings of Ayugi et al. [120] and Ngoma et al. [121].

S. No.	GCM Model	Institution	Resolution (km) for Ensemble Members r1i1p1f1
1	CanESM5	Canadian Centre for Climate Modelling and Analysis, Environment and Climate Change Canada, Victoria, Canada.	500
2	CESM2-WACCM	National Centre for Atmospheric Research, USA.	100
3	CNRM-CM6-1	Centre National de Recherches Météorologiques (CNRM); Centre Européen de Recherches et de Formation Avancéeen Calcul Scientifique, France.	157
4	GFDL-ESM4	Geophysical Fluid Dynamics Laboratory (GFDL), USA.	100
5	MPI-ESM1-2-LR	Max Planck Institute for Meteorology, Germany.	250
6	MRI-ESM2-0	Meteorological Research Institute, Japan.	100
7	NorESM2-LM	Norwegian Climate Centre, Norway.	250
8	NorESM2-MM	Norwegian Climate Centre, Norway.	100
9	UKESM1-0-LL	UK Met Office Hadley Centre, UK.	209×139

Appendix B

Evaluation of Satellite Precipitation Products (SPPS) with Gauge Data

Accuracy of SPPs in Daily Rainfall Identification

The statistics of daily rainfall identification for each SPP and rain gauge location are summarised in Table S1 in the Supplementary Materials. Overall, Probability of Detection (POD) statistics show that SPPs perform better in detecting rain at low altitudes (Tororo) in comparison to high altitudes (Buginyanya). This is because satellite rain detection heavily depends on ice at cloud tops [122], which both infrared and passive microwave algorithms weakly detect in the relatively warm orographic rains that dominate mountainous regions [123]. This weakness in rain detection could explain the relatively lower False Alarm Ratio (FAR) at a high altitude. Similar to findings by Diem et al. [124] for a study in the north-western region of Uganda, high POD and Frequency Bias (FB) are generally linked with larger FAR. In general, though all assessed products utilise gauge data for calibration, those that ingest gauge data at longer timescales (e.g., TRMM and PERSIANN-CDR at monthly scale) performed worst.

Accuracy of SPPs in Capturing Daily and Monthly Rainfall Totals

Most satellite precipitation products performed weakly in capturing daily rain totals, with correlation coefficient (R) and Nash-Sutcliffe efficiency (NSE) varying from 0.1 to 0.39 (Figure S2 in the Supplementary Materials) and −0.37 to 0.14 (Figure S3 in the Supplementary Materials), respectively. However, significant improvement was attained on aggregation to monthly totals (Figure S4 and Figure S5, respectively, in the Supplementary Materials), with MSWEP, TAMSAT and CHIRPS performing best. Their monthly R and NSE vary from 0.71 to 0.78 and 0.35 to 0.6, respectively. SPPs have inherent detection, systematic and random errors [125,126]. However, systematic errors increase slightly for aggregation windows greater than 15 days, while random errors have a negligible effect upon aggregation [127]. Some of the studies over Africa that have reported improvement of SPPs on aggregation include Bhatti et al. [127], Dembélé and Zwart [56], Dinku [57] and Gebrechorkos [128]. The above-average performance of TAMSAT and CHIRPS could be attributed to the assimilation of gauge datasets from local and regional meteorological authorities. Meanwhile, MSWEP's performance could be due to ingesting multiple datasets, including gauge and reanalysis products.

Appendix C

Model Sensitivity Analysis

Model sensitivity is usually assessed using local or global methods [129]. Though unreliable for high-dimensional and non-linear models [130], local approaches such as the 'one parameter at a time' are simple and computationally less expensive [131]. Further, they have been applied in numerous hydrological modelling studies [131], including SHETRAN [132]. Birkinshaw [133] gives a detailed description of SHETRAN parameters, including the applicable range. Some of these parameters are not very sensitive. For this study, the parameters assessed for sensitivity include overland (Kro) and channel (Krc) Strickler coefficients (the Strickler coefficient is the inverse of Manning's coefficient), canopy leaf area index (CLAI), topsoil saturated conductivity (Ksat) and the ratio of actual to potential evapotranspiration at field capacity (AET/PET ratio). The 'one parameter at a time' approach was employed in assessing the above parameters. Each parameter was evaluated over the range of ±90% by plotting the change in catchment discharge against the change in parameter base value. The sensitivity index (SI) for each parameter was also calculated using the equation below, where O0 is catchment discharge at base value, and O90 and O-90 are catchment discharge at +90 and -90% change in base value. This study's sensitivity analysis was carried out only for the Phase 1 model driven by the MSWEP precipitation product.

$$SI_{90} = |O_{90} - O_{-90}| \div O_0 \qquad (A1)$$

Results of Model Sensitivity Analysis

Sensitivity indices for overland (Kro) and channel (Krc) Strickler coefficients, canopy leaf area index (CLAI), topsoil saturated conductivity (Ksat), and the ratio of actual to potential evapotranspiration at field capacity (AET/PET) are shown in Figure S6 in the Supplementary Materials. The most sensitive parameters, in descending order, are Krc, AET/PET and Kro, with SI values of 1.1, 0.83 and 0.72, respectively. Ksat was the least sensitive, with SI = 0.06. As expected, larger Kro, Krc, and Ksat values increase catchment outflow. On the contrary, increments in CLAI and AET/PET lower outflows. The sensitivity results are similar to findings by Mutenyo et al. [89] for a SWAT model of the Manafwa sub-catchment in the Mpologoma catchment, in which parameters that control overland and channel flow, canopy cover, evapotranspiration, and flow in soil were found to be most sensitive.

References

1. Davidson, N.C. How Much Wetland Has the World Lost? Long-Term and Recent Trends in Global Wetland Area. *Mar. Freshw. Res.* **2014**, *65*, 934–941. [CrossRef]
2. Hu, S.; Niu, Z.; Chen, Y.; Li, L.; Zhang, H. Global Wetlands: Potential Distribution, Wetland Loss, and Status. *Sci. Total Environ.* **2017**, *586*, 319–327. [CrossRef] [PubMed]
3. Xu, T.; Weng, B.; Yan, D.; Wang, K.; Li, X.; Bi, W.; Li, M.; Cheng, X.; Liu, Y. Wetlands of International Importance: Status, Threats, and Future Protection. *Int. J. Environ. Res. Public Health* **2019**, *16*, 1818. [CrossRef] [PubMed]
4. Junk, W.; An, S.; Finlayson, M.; Gopal, B.; Květ, J.; Mitchell, S.; Mitsch, W.; Robarts, R. Current State of Knowledge Regarding the World's Wetlands and Their Future under Global Climate Change: A Synthesis. *Aquat. Sci.* **2013**, *75*, 151–167. [CrossRef]
5. Tockner, K. Riverine Flood Plains: Present State and Future Trends. *Environ. Conserv.* **2002**, *29*, 308–330. [CrossRef]
6. Darrah, S.E.; Shennan-Farpón, Y.; Loh, J.; Davidson, N.C.; Finlayson, C.M.; Gardner, R.C.; Walpole, M.J. Improvements to the Wetland Extent Trends (WET) Index as a Tool for Monitoring Natural and Human-Made Wetlands. *Ecol. Indic.* **2019**, *99*, 294–298. [CrossRef]
7. Woodward, R.T.; Wui, Y.S. The Economic Value of Wetland Services: A Meta-Analysis. *Ecol. Econ.* **2001**, *37*, 257–270. [CrossRef]
8. Kashaigili, J.J.; McCartney, M.P.; Mahoo, H.F.; Lankford, B.A.; Mbilinyi, B.P.; Yawson, D.K. *Use of a Hydrological Model for Environmental Management of the Usangu Wetlands, Tanzania*; International Water Management Institute: Colombo, Sri Lanka, 2009.
9. Pacini, N.; Hesslerová, P.; Pokorný, J.; Mwinami, T.; Morrison, E.H.J.; Cook, A.A.; Zhang, S.; Harper, D.M. Papyrus as an Ecohydrological Tool for Restoring Ecosystem Services in Afrotropical Wetlands. *Ecohydrol. Hydrobiol.* **2018**, *18*, 142–154. [CrossRef]
10. Van Dam, A.A.; Kipkemboi, J.; Mazvimavi, D.; Irvine, K. A Synthesis of Past, Current and Future Research for Protection and Management of Papyrus (*Cyperus papyrus* L.) Wetlands in Africa. *Wetl. Ecol. Manag.* **2014**, *22*, 99–114. [CrossRef]
11. Dixon, A.B.; Wood, A.P. Wetland Cultivation and Hydrological Management in Eastern Africa: Matching Community and Hydrological Needs through Sustainable Wetland Use. *Nat. Resour. Forum* **2003**, *27*, 117–129. [CrossRef]
12. Kipkemboi, J.; Van Dam, A.A. Papyrus Wetlands. In *The Wetland Book*; Finlayson, C., Milton, G., Prentice, R., Davidson, N., Eds.; Springer: Dordrecht, The Netherlands, 2018; pp. 183–197.
13. Gaudet, J.J. Mineral Concentrations in Papyrus in Various African Swamps. *J. Ecol.* **1975**, *63*, 483–491. [CrossRef]
14. Emerton, L.; Iyango, L.; Luwum, P.; Malinga, A. *The Present Economic Value of Nakivubo Urban Wetland, Uganda*; IUCN, Regional Office for Eastern Africa: Nairobi, Kenya, 1999.

5. Kansiime, F.; Nalubega, M. *Wastewater Treatment by a Natural Wetland: The Nakivubo Swamp, Uganda: Processes and Implications*; A.A. Balkema: Rotterdam, The Netherlands, 1999.
6. Jones, M.B.; Humphries, S.W. Impacts of the C4 Sedge *Cyperus papyrus* L. on Carbon and Water Fluxes in an African Wetland. *Hydrobiologia* **2002**, *488*, 107–113. [CrossRef]
7. Kiwango, Y.A.; Wolanski, E. Papyrus Wetlands, Nutrients Balance, Fisheries Collapse, Food Security, and Lake Victoria Level Decline in 2000–2006. *Wetl. Ecol. Manag.* **2008**, *16*, 89–96. [CrossRef]
8. Opio, A.; Jones, M.; Kansiime, F.; Otiti, T. Growth and Development of *Cyperus papyrus* in a Tropical Wetland. *Open J. Ecol.* **2014**, *04*, 113–123. [CrossRef]
9. Saunders, M.J.; Kansiime, F.; Jones, M.B. Reviewing the Carbon Cycle Dynamics and Carbon Sequestration Potential of *Cyperus papyrus* L. Wetlands in Tropical Africa. *Wetl. Ecol. Manag.* **2014**, *22*, 143–155. [CrossRef]
10. Ssanyu, G.A.; Kipkemboi, J.; Mathooko, J.M.; Balirwa, J. Land-Use Impacts on Small-Scale Mpologoma Wetland Fishery, Eastern Uganda: A Socio-Economic Perspective. *Lakes Reserv. Sci. Policy Manag. Sustain. Use* **2014**, *19*, 280–292. [CrossRef]
11. Terer, T.; Muasya, A.M.; Higgins, S.; Gaudet, J.J.; Triest, L. Importance of Seedling Recruitment for Regeneration and Maintaining Genetic Diversity of *Cyperus papyrus* during Drawdown in Lake Naivasha, Kenya. *Aquat. Bot.* **2014**, *116*, 93–102. [CrossRef]
12. Hurst, H.E. The Sudd Region of the Nile. *J. R. Soc. Arts* **1933**, *81*, 720–736.
13. Sutcliffe, J.V.; Parks, Y.P. Comparative Water Balances of Selected African Wetlands. *Hydrol. Sci. J.* **1989**, *34*, 49–62. [CrossRef]
14. Kayendeke, E.; French, H.K. Characterising the Hydrological Regime of a Tropical Papyrus Wetland in the Lake Kyoga Basin, Uganda. In *Agriculture and Ecosystem Resilience in Sub Saharan Africa: Livelihood Pathways under Changing Climate*; Bamutaze, Y., Kyamanywa, S., Singh, B.R., Nabanoga, G., Lal, R., Eds.; Springer International Publishing: Cham, Switzerland, 2019; pp. 213–236. ISBN 978-3-030-12974-3.
15. Kayendeke, E.J.; Kansiime, F.; French, H.K.; Bamutaze, Y. Spatial and Temporal Variation of Papyrus Root Mat Thickness and Water Storage in a Tropical Wetland System. *Sci. Total Environ.* **2018**, *642*, 925–936. [CrossRef] [PubMed]
16. Sutcliffe, J.V.; Parks, Y.P. Hydrological Modelling of the Sudd and Jonglei Canal. *Hydrol. Sci. J.* **1987**, *32*, 143–159. [CrossRef]
17. Howell, P.; Lock, M.; Cobb, S. *The Jonglei Canal: Impact and Opportunity*; Cambridge University Press: Cambridge, UK, 2009; ISBN 9780521105491.
18. Di Vittorio, C.A.; Georgakakos, A.P. Hydrologic Modeling of the Sudd Wetland Using Satellite-Based Data. *J. Hydrol. Reg. Stud.* **2021**, *37*, 100922. [CrossRef]
19. Cohen, M.J.; Creed, I.F.; Alexander, L.; Basu, N.B.; Calhoun, A.J.K.; Craft, C.; D'Amico, E.; DeKeyser, E.; Fowler, L.; Golden, H.E.; et al. Do Geographically Isolated Wetlands Influence Landscape Functions? *Proc. Natl. Acad. Sci. USA* **2016**, *113*, 1978–1986. [CrossRef] [PubMed]
20. Thorslund, J.; Jarsjo, J.; Jaramillo, F.; Jawitz, J.W.; Manzoni, S.; Basu, N.B.; Chalov, S.R.; Cohen, M.J.; Creed, I.F.; Goldenberg, R.; et al. Wetlands as Large-Scale Nature-Based Solutions: Status and Challenges for Research, Engineering and Management. *Ecol. Eng.* **2017**, *108*, 489–497. [CrossRef]
21. Salimi, S.; Almuktar, S.A.A.A.N.; Scholz, M. Impact of Climate Change on Wetland Ecosystems: A Critical Review of Experimental Wetlands. *J. Environ. Manag.* **2021**, *286*, 112160. [CrossRef]
22. Langan, C.; Farmer, J.; Rivington, M.; Smith, J.U. Tropical Wetland Ecosystem Service Assessments in East Africa; A Review of Approaches and Challenges. *Environ. Model. Softw.* **2018**, *102*, 260–273. [CrossRef]
23. Wang, H.; Xu, S.; Sun, L. Effects of Climatic Change on Evapotranspiration in Zhalong Wetland, Northeast China. *Chin. Geogr. Sci.* **2006**, *16*, 265–269. [CrossRef]
24. Döll, P.; Zhang, J. Impact of Climate Change on Freshwater Ecosystems: A Global-Scale Analysis of Ecologically Relevant River Flow Alterations. *Hydrol. Earth Syst. Sci.* **2010**, *14*, 783–799. [CrossRef]
25. Barnes, C.; Bonell, M. How to Choose an Appropriate Catchment Model. In *Forests, Water and People in the Humid Tropics: Past, Present and Future Hydrological Research for Integrated Land and Water Management*; Bruijnzeel, L.A., Bonell, M., Eds.; International Hydrology Series; Cambridge University Press: Cambridge, UK, 2005; pp. 717–741. ISBN 9780521829533.
26. Golden, H.E.; Lane, C.R.; Amatya, D.M.; Bandilla, K.W.; Raanan Kiperwas, H.; Knightes, C.D.; Ssegane, H. Hydrologic Connectivity between Geographically Isolated Wetlands and Surface Water Systems: A Review of Select Modeling Methods. *Environ. Model. Softw.* **2014**, *53*, 190–206. [CrossRef]
27. Acreman, M.C.; Miller, F. Hydrological Impact Assessment of Wetlands. In *The Global Importance of Groundwater in the 21st Century, Proceedings of the International Symposium on Groundwater Sustainability, Alicante, Spain, 24–27 January 2006*; Ragone, S., Hernandez-Mora, N., de la Hera, A., Berkamp, G., Mckay, J., Eds.; National Groundwater Association Press: Westerville, OH, USA, 2007; pp. 225–255.
28. Fitz, H.C.; Hughes, N. *Wetland Ecological Models*; SL257; University of Florida, Institute of Food and Agricultural Sciences: Gainesville, FL, USA, 2008.
29. Ewen, J.; Geoff, P.; O'Connell, E.P. SHETRAN: Distributed River Basin Flow and Transport Modeling System. *J. Hydrol. Eng.* **2000**, *5*, 250–258. [CrossRef]
30. Birkinshaw, S.; Guerreiro, S.; Nicholson, A.; Liang, Q.; Quinn, P.; Zhang, L.; He, B.; Yin, J.; Fowler, H. Climate Change Impacts on Yangtze River Discharge at the Three Gorges Dam. *Hydrol. Earth Syst. Sci.* **2017**, *21*, 1911–1927. [CrossRef]

41. Op de Hipt, F.; Diekkrüger, B.; Steup, G.; Yira, Y.; Hoffmann, T.; Rode, M.; Näschen, K. Modeling the Effect of Land Use and Climate Change on Water Resources and Soil Erosion in a Tropical West African Catch-Ment (Dano, Burkina Faso) Using SHETRAN. *Sci. Total Environ.* **2019**, *653*, 431–445. [CrossRef]
42. Zhang, R.; Corte-Real, J.; Moreira, M.; Kilsby, C.; Birkinshaw, S.; Burton, A.; Fowler, H.J.; Forsythe, N.; Nunes, J.P.; Sampaio, E.; et al. Downscaling Climate Change of Water Availability, Sediment Yield and Extreme Events: Application to a Mediterranean Climate Basin. *Int. J. Climatol.* **2019**, *39*, 2947–2963. [CrossRef]
43. Dembélé, M.; Schaefli, B.; van de Giesen, N.; Mariéthoz, G. Suitability of 17 Gridded Rainfall and Temperature Datasets for Large-Scale Hydrological Modelling in West Africa. *Hydrol. Earth Syst. Sci.* **2020**, *24*, 5379–5406. [CrossRef]
44. Hughes, D.A. Comparison of Satellite Rainfall Data with Observations from Gauging Station Networks. *J. Hydrol.* **2006**, *327*, 399–410. [CrossRef]
45. Wilby, R.L.; Clifford, N.J.; De Luca, P.; Harrigan, S.; Hillier, J.K.; Hodgkins, R.; Johnson, M.F.; Matthews, T.K.R.; Murphy, C.; Noone, S.J.; et al. The 'Dirty Dozen' of Freshwater Science: Detecting Then Reconciling Hydrological Data Biases and Errors. *WIREs Water* **2017**, *4*, e1209. [CrossRef]
46. MWE. *Mpologoma Catchment Management Plan*; Ministry of Water and Environment: Kampala, Uganda, 2018.
47. Basalirwa, C.P.K. Raingauge Network Designs for Uganda. Ph.D. Thesis, Nairobi University, Nairobi, Kenya, 1991.
48. Chombo, O.; Lwasa, S.; Makooma, T.M. Spatial Differentiation of Small Holder Farmers' Vulnerability to Climate Change in the Kyoga Plains of Uganda. *Am. J. Clim. Chang.* **2018**, *7*, 624. [CrossRef]
49. Kottek, M.; Grieser, J.; Beck, C.; Rudolf, B.; Rubel, F. World Map of the Köppen-Geiger Climate Classification Updated. *Meteorol. Z.* **2006**, *15*, 259–263. [CrossRef] [PubMed]
50. Bunyangha, J.; Majaliwa, M.J.G.; Muthumbi, A.W.; Gichuki, N.N.; Egeru, A. Past and Future Land Use/Land Cover Changes from Multi-Temporal Landsat Imagery in Mpologoma Catchment, Eastern Uganda. *Egypt. J. Remote Sens. Space Sci.* **2021**, *24*, 675–685. [CrossRef]
51. Bitew, M.M.; Gebremichael, M.; Ghebremichael, L.T.; Bayissa, Y.A. Evaluation of High-Resolution Satellite Rainfall Products through Streamflow Simulation in a Hydrological Modeling of a Small Mountainous Watershed in Ethiopia. *J. Hydrometeorol.* **2012**, *13*, 338–350. [CrossRef]
52. Fuka, D.R.; Walter, M.T.; Macalister, C.; Degaetano, A.T.; Steenhuis, T.S.; Easton, Z.M. Using the Climate Forecast System Reanalysis as Weather Input Data for Watershed Models. *Hydrol. Process.* **2013**, *28*, 5613–5623. [CrossRef]
53. Maidment, R.; Grimes, D.; Allan, R.; Greatrex, H.; Rojas, O.; Leo, O. Evaluation of Satellite-Based and Model Re-Analysis Rainfall Estimates for Uganda. *Meteorol. Appl.* **2013**, *20*, 308–317. [CrossRef]
54. Mutai, C.C.; Ward, M.N. East African Rainfall and the Tropical Circulation/Convection on Intraseasonal to Interannual Timescales. *J. Clim.* **2000**, *13*, 3915–3939. [CrossRef]
55. Gumindoga, W.; Rientjes, T.H.M.; Haile, A.T.; Makurira, H.; Reggiani, P. Performance of Bias-Correction Schemes for CMORPH Rainfall Estimates in the Zambezi River Basin. *Hydrol. Earth Syst. Sci.* **2019**, *23*, 2915–2938. [CrossRef]
56. Dembélé, M.; Zwart, S.J. Evaluation and Comparison of Satellite-Based Rainfall Products in Burkina Faso, West Africa. *Int. J. Remote Sens.* **2016**, *37*, 3995–4014. [CrossRef]
57. Dinku, T.; Funk, C.; Peterson, P.; Maidment, R.; Tadesse, T.; Gadain, H.; Ceccato, P. Validation of the CHIRPS Satellite Rainfall Estimates over Eastern Africa. *Q. J. R. Meteorol. Soc.* **2018**, *144*, 292–312. [CrossRef]
58. Duan, Z.; Tuo, Y.; Liu, J.; Gao, H.; Song, X.; Zhang, Z.; Yang, L.; Mekonnen, D.F. Hydrological Evaluation of Open-Access Precipitation and Air Temperature Datasets Using SWAT in a Poorly Gauged Basin in Ethiopia. *J. Hydrol.* **2019**, *569*, 612–626. [CrossRef]
59. Hargreaves, G.H.; Samani, Z.A. Reference Crop Evapotranspiration from Temperature. *Appl. Eng. Agric.* **1985**, *1*, 96–99. [CrossRef]
60. Allen, R.G.; Pereira, L.S.; Raes, D.; Smith, M. *Crop Evapotranspiration—Guidelines for Computing Crop Water Requirements*; FAO Irrigation and Drainage Paper No. 56.; Food and Agriculture Organization of the United Nations: Rome, Italy, 1998; ISBN 9251042195.
61. Crochemore, L.; Isberg, K.; Pimentel, R.; Pineda, L.; Hasan, A.; Arheimer, B. Lessons Learnt from Checking the Quality of Openly Accessible River Flow Data Worldwide. *Hydrol. Sci. J.* **2020**, *65*, 699–711. [CrossRef]
62. McMillan, H.K.; Westerberg, I.K.; Krueger, T. Hydrological Data Uncertainty and Its Implications. *WIREs Water* **2018**, *5*, e1319. [CrossRef]
63. Wildemeersch, S.; Goderniaux, P.; Orban, P.; Brouyère, S.; Dassargues, A. Assessing the Effects of Spatial Discretization on Large-Scale Flow Model Performance and Prediction Uncertainty. *J. Hydrol.* **2014**, *510*, 10–25. [CrossRef]
64. Sreedevi, S.; Eldho, T.I. Effects of Grid-Size on Effective Parameters and Model Performance of SHETRAN for Estimation of Streamflow and Sediment Yield. *Int. J. River Basin Manag.* **2021**, *19*, 535–551. [CrossRef]
65. Zhang, R. Integrated Modelling for Evaluation of Climate Change Impacts on Agricultural Dominated Basin. Ph.D. Thesis, University of Évora, Évora, Portugal, 2015.
66. Tan, M.L.; Ficklin, D.L.; Dixon, B.; Ibrahim, A.L.; Yusop, Z.; Chaplot, V. Impacts of DEM Resolution, Source, and Resampling Technique on SWAT-Simulated Streamflow. *Appl. Geogr.* **2015**, *63*, 357–368. [CrossRef]
67. NASA/Japan Space Systems. ASTER Global Digital Elevation Model V003. Available online: https://asterweb.jpl.nasa.gov/gdem.asp (accessed on 5 March 2020).

68. Defourny, P.; Bontemps, S.; Lamarche, C.; Brockmann, C.; Boettcher, M.; Wevers, J.; Kirches, G. *Land Cover CCI. Product User Guide Version 2*; UCL–Geomatics: Louvain-la-Neuve, Belgium, 2017.
69. FAO/UNESCO. The Digital Soil Map of The World—Version 3.6. Available online: https://data.apps.fao.org/map/catalog/srv/eng/catalog.search#/metadata/446ed430-8383-11db-b9b2-000d939bc5d8 (accessed on 5 March 2020).
70. Moriasi, D.N.; Arnold, J.G.; Van Liew, M.W.; Bingner, R.L.; Harmel, R.D.; Veith, T.L. Model Evaluation Guidelines for Systematic Quantification of Accuracy in Watershed Simulations. *Am. Soc. Agric. Biol. Eng.* **2007**, *50*, 885–900. [CrossRef]
71. Smakhtin, V.U.; Batchelor, A.L. Evaluating Wetland Flow Regulating Functions Using Discharge Time-Series. *Hydrol. Process.* **2005**, *19*, 1293–1305. [CrossRef]
72. Bullock, A.; Acreman, M. The Role of Wetlands in the Hydrological Cycle. *Hydrol. Earth Syst. Sci.* **2003**, *7*, 358–389. [CrossRef]
73. Postel, S.; Richter, B.D. *Rivers for Life—Managing Water for People and Nature*; Island Press: Washington, DC, USA, 2003; ISBN 155963443X.
74. Fossey, M.; Rousseau, A.N.; Savary, S. Assessment of the Impact of Spatio-Temporal Attributes of Wetlands on Stream Flows Using a Hydrological Modelling Framework: A Theoretical Case Study of a Watershed under Temperate Climatic Conditions. *Hydrol. Process.* **2016**, *30*, 1768–1781. [CrossRef]
75. Willems, P. A Time Series Tool to Support the Multi-Criteria Performance Evaluation of Rainfall-Runoff Models. *Environ. Model. Softw.* **2009**, *24*, 311–321. [CrossRef]
76. Maraun, D.; Widmann, M. *Statistical Downscaling and Bias Correction for Climate Research*; Cambridge University Press: Cambridge, UK, 2018; ISBN 9781107066052.
77. Cannon, A.J. Package 'MBC' User Guide. Multivariate Bias Correction of Climate Model Outputs. Available online: https://cran.r-project.org/web/packages/MBC/MBC.pdf (accessed on 1 May 2022).
78. Friedlingstein, P.; Jones, M.W.; O'Sullivan, M.; Andrew, R.M.; Bakker, D.C.E.; Hauck, J.; Le Quéré, C.; Peters, G.P.; Peters, W.; Pongratz, J.; et al. Global Carbon Budget 2021. *Earth Syst. Sci. Data* **2022**, *14*, 1917–2005. [CrossRef]
79. UNFCCC. *Decision 1/CP.18 Report of the Conference of the Parties on Its Eighteenth Session, Held in Doha from 26 November to 8 December 2012*; UNFCCC: New York, NY, USA, 2013.
80. Arias, P.; Bellouin, N.; Coppola, E.; Jones, C.; Krinner, G.; Marotzke, J.; Naik, V.; Plattner, G.-K.; Rojas, M.; Sillmann, J.; et al. *Climate Change 2021: The Physical Science Basis. Contribution of Working Group I to the Sixth Assessment Report of the Intergovernmental Panel on Climate Change; Technical Summary*; Masson-Delmotte, V., Zhai, P., Pirani, A., Connors, S.L., Péan, C., Berger, S., Caud, N., Chen, Y., Goldfarb, L., Gomis, M.I., et al., Eds.; Cambridge University Press: Cambridge, UK; New York, NY, USA, 2021.
81. Gistemp, T. GISS Surface Temperature Analysis (GISTEMP), Version 4. NASA Goddard Institute for Space Studies. Available online: https://data.giss.nasa.gov/gistemp/ (accessed on 1 June 2022).
82. Vautard, R.; Gobiet, A.; Sobolowski, S.; Kjellström, E.; Stegehuis, A.; Watkiss, P.; Mendlik, T.; Landgren, O.; Nikulin, G.; Teichmann, C.; et al. The European Climate under a 2 °C Global Warming. *Environ. Res. Lett.* **2014**, *9*, 34006. [CrossRef]
83. Xu, X.; Wang, Y.-C.; Kalcic, M.; Muenich, R.L.; Yang, Y.C.E.; Scavia, D. Evaluating the Impact of Climate Change on Fluvial Flood Risk in a Mixed-Use Watershed. *Environ. Model. Softw.* **2019**, *122*, 104031. [CrossRef]
84. Wu, Y.; Sun, J.; Jun Xu, Y.; Zhang, G.; Liu, T. Projection of Future Hydrometeorological Extremes and Wetland Flood Mitigation Services with Different Global Warming Levels: A Case Study in the Nenjiang River Basin. *Ecol. Indic.* **2022**, *140*, 108987. [CrossRef]
85. Masih, I.; Maskey, S.; Mussá, F.E.F.; Trambauer, P. A Review of Droughts on the African Continent: A Geospatial and Long-Term Perspective. *Hydrol. Earth Syst. Sci.* **2014**, *18*, 3635–3649. [CrossRef]
86. Bisselink, B.; Zambrano-Bigiarini, M.; Burek, P.; de Roo, A. Assessing the Role of Uncertain Precipitation Estimates on the Robustness of Hydrological Model Parameters under Highly Variable Climate Conditions. *J. Hydrol. Reg. Stud.* **2016**, *8*, 112–129. [CrossRef]
87. Stisen, S.; Sandholt, I. Evaluation of Remote-Sensing-Based Rainfall Products through Predictive Capability in Hydrological Runoff Modeling. *Hydrol. Process.* **2010**, *24*, 879–891. [CrossRef]
88. Blöschl, G.; Sivapalan, M. Scale Issues in Hydrological Modelling: A Review. *Hydrol. Process.* **1995**, *9*, 251–290. [CrossRef]
89. Mutenyo, I.; Nejadhashemi, A.P.; Woznicki, S.A.; Giri, S. Evaluation of SWAT Performance on a Mountainous Watershed in Tropical Africa. *Hydrol. Curr. Res.* **2013**, *6*, 7. [CrossRef]
90. Quin, A.; Destouni, G. Large-Scale Comparison of Flow-Variability Dampening by Lakes and Wetlands in the Landscape. *Land Degrad. Dev.* **2018**, *29*, 3617–3627. [CrossRef]
91. Kadykalo, A.N.; Findlay, C.S. The Flow Regulation Services of Wetlands. *Ecosyst. Serv.* **2016**, *20*, 91–103. [CrossRef]
92. Acreman, M.; Holden, J. How Wetlands Affect Floods. *Wetlands* **2013**, *33*, 773–786. [CrossRef]
93. Rains, M.C.; Leibowitz, S.G.; Cohen, M.J.; Creed, I.F.; Golden, H.E.; Jawitz, J.W.; Kalla, P.; Lane, C.R.; Lang, M.W.; McLaughlin, D.L. Geographically Isolated Wetlands Are Part of the Hydrological Landscape. *Hydrol. Process.* **2016**, *30*, 153–160. [CrossRef]
94. Makula, E.K.; Zhou, B. Coupled Model Intercomparison Project Phase 6 Evaluation and Projection of East African Precipitation. *Int. J. Climatol.* **2022**, *42*, 2398–2412. [CrossRef]
95. Ayugi, B.; Shilenje, Z.W.; Babaousmail, H.; Lim Kam Sian, K.T.C.; Mumo, R.; Dike, V.N.; Iyakaremye, V.; Chehbouni, A.; Ongoma, V. Projected Changes in Meteorological Drought over East Africa Inferred from Bias-Adjusted CMIP6 Models. *Nat. Hazards* **2022**, *113*, 1151–1176. [CrossRef] [PubMed]

96. Bucher, E.; Bonetto, A.; Boyle, T.; Canevari, P.; Castro, G.; Huszar, P.; Stone, T. *Hidrovia: An Initial Environmental Examination of the Paraguay-Parana Waterway*; Humedades para las Americas. Publicacao; Publicatio.; Wetlands for the Americas: Manomet, MA, USA, 1993.
97. Turyahabwe, N.; Tumusiime, D.; Kakuru, W.; Barasa, B. Wetland Use/Cover Changes and Local Perceptions in Uganda. *Sustain. Agric. Res.* **2013**, *2*, 95–105. [CrossRef]
98. Gulbin, S.; Kirilenko, A.P.; Kharel, G.; Zhang, X. Wetland Loss Impact on Long Term Flood Risks in a Closed Watershed. *Environ. Sci. Policy* **2019**, *94*, 112–122. [CrossRef]
99. Acreman, M.C.; Riddington, R.; Booker, D.J. Hydrological Impacts of Floodplain Restoration: A Case Study of the River Cherwell, UK. *Hydrol. Earth Syst. Sci.* **2003**, *7*, 75–85. [CrossRef]
100. Mitsch, W.J.; Day, J.W. Restoration of Wetlands in the Mississippi–Ohio–Missouri (MOM) River Basin: Experience and Needed Research. *Ecol. Eng.* **2006**, *26*, 55–69. [CrossRef]
101. Wu, Y.; Zhang, G.; Rousseau, A.N.; Xu, Y.J.; Foulon, É. On How Wetlands Can Provide Flood Resilience in a Large River Basin: A Case Study in Nenjiang River Basin, China. *J. Hydrol.* **2020**, *587*, 125012. [CrossRef]
102. Yang, W.; Wang, X.; Liu, Y.; Gabor, S.; Boychuk, L.; Badiou, P. Simulated Environmental Effects of Wetland Restoration Scenarios in a Typical Canadian Prairie Watershed. *Wetl. Ecol. Manag.* **2010**, *18*, 269–279. [CrossRef]
103. Thompson, J.R.; Gosling, S.N.; Zaherpour, J.; Laizé, C.L.R. Increasing Risk of Ecological Change to Major Rivers of the World With Global Warming. *Earths Future* **2021**, *9*, e2021EF002048. [CrossRef]
104. Nyenje, P.M.; Batelaan, O. Estimating the Effects of Climate Change on Groundwater Recharge and Baseflow in the Upper Ssezibwa Catchment, Uganda. *Hydrol. Sci. J.* **2009**, *54*, 713–726. [CrossRef]
105. Bahati, H.K.; Ogenrwoth, A.; Sempewo, J.I. Quantifying the Potential Impacts of Land-Use and Climate Change on Hydropower Reliability of Muzizi Hydropower Plant, Uganda. *J. Water Clim. Chang.* **2021**, *12*, 2526–2554. [CrossRef]
106. Gabiri, G.; Diekkrüger, B.; Näschen, K.; Leemhuis, C.; van der Linden, R.; Majaliwa, J.-G.M.; Obando, J.A. Impact of Climate and Land Use/Land Cover Change on the Water Resources of a Tropical Inland Valley Catchment in Uganda, East Africa. *Climate* **2020**, *8*, 83. [CrossRef]
107. Mehdi, B.; Dekens, J.; Herrnegger, M. Climatic Impacts on Water Resources in a Tropical Catchment in Uganda and Adaptation Measures Proposed by Resident Stakeholders. *Clim. Chang.* **2021**, *164*, 10. [CrossRef]
108. Mileham, L.; Taylor, R.G.; Todd, M.; Tindimugaya, C.; Thompson, J. The Impact of Climate Change on Groundwater Recharge and Runoff in a Humid, Equatorial Catchment: Sensitivity of Projections to Rainfall Intensity. *Hydrol. Sci. J.* **2009**, *54*, 727–738. [CrossRef]
109. Prudhomme, C.; Jakob, D.; Svensson, C. Uncertainty and Climate Change Impact on the Flood Regime of Small UK Catchments. *J. Hydrol.* **2003**, *277*, 1–23. [CrossRef]
110. Lee, S.; Qi, J.; McCarty, G.W.; Yeo, I.-Y.; Zhang, X.; Moglen, G.E.; Du, L. Uncertainty Assessment of Multi-Parameter, Multi-GCM, and Multi-RCP Simulations for Streamflow and Non-Floodplain Wetland (NFW) Water Storage. *J. Hydrol.* **2021**, *600*, 126564. [CrossRef]
111. Maidment, R.I.; Grimes, D.; Black, E.; Tarnavsky, E.; Young, M.; Greatrex, H.; Allan, R.P.; Stein, T.; Nkonde, E.; Senkunda, S.; et al. A New, Long-Term Daily Satellite-Based Rainfall Dataset for Operational Monitoring in Africa. *Sci. Data* **2017**, *4*, 170063. [CrossRef]
112. Funk, C.; Peterson, P.; Landsfeld, M.; Pedreros, D.; Verdin, J.; Shukla, S.; Husak, G.; Rowland, J.; Harrison, L.; Hoell, A.; et al. The Climate Hazards Infrared Precipitation with Stations—A New Environmental Record for Monitoring Extremes. *Sci. Data* **2015**, *2*, 150066. [CrossRef]
113. Novella, N.S.; Thiaw, W.M. African Rainfall Climatology Version 2 for Famine Early Warning Systems. *J. Appl. Meteorol. Climatol.* **2013**, *52*, 588–606. [CrossRef]
114. NOAA-CPC. *RFE 2.0 Technical Description Summary*; NOAA Climate Prediction Center: College Park, MD, USA, 2001.
115. Xie, P.; Arkin, P.A. Analyses of Global Monthly Precipitation Using Gauge Observations, Satellite Estimates, and Numerical Model Predictions. *J. Clim.* **1996**, *9*, 840–858. [CrossRef]
116. Beck, H.E.; Wood, E.F.; Pan, M.; Fisher, C.K.; Miralles, D.G.; van Dijk, A.I.J.M.; McVicar, T.R.; Adler, R.F. MSWEP V2 Global 3-Hourly 0.1° Precipitation: Methodology and Quantitative Assessment. *Bull. Am. Meteorol. Soc.* **2019**, *100*, 473–500. [CrossRef]
117. Ashouri, H.; Hsu, K.-L.; Sorooshian, S.; Braithwaite, D.K.; Knapp, K.R.; Cecil, L.D.; Nelson, B.R.; Prat, O.P. PERSIANN-CDR: Daily Precipitation Climate Data Record from Multisatellite Observations for Hydrological and Climate Studies. *Bull. Am. Meteorol. Soc.* **2015**, *96*, 69–83. [CrossRef]
118. Xie, P.; Joyce, R.; Wu, S. *Bias-Corrected CMORPH—Climate Algorithm Theoretical Basis Document*. NOAA Climate Data Record Program. CDRP-ATBD-0812, Rev. 0; NOAA: Washington, DC, USA, 2018.
119. Huffman, J.G.; Adler, F.R.; Bolvin, T.D.; Nelkin, J.E. The TRMM Multi-Satellite Precipitation Analysis (TMPA). In *Satellite Rainfall Applications for Surface Hydrology*; Gebremichael, M., Hossain, F., Eds.; Springer: Dordrecht, The Netherlands, 2011; pp. 3–22. ISBN 978-90-481-2915-7.
120. Ayugi, B.; Zhihong, J.; Zhu, H.; Ngoma, H.; Babaousmail, H.; Rizwan, K.; Dike, V. Comparison of CMIP6 and CMIP5 Models in Simulating Mean and Extreme Precipitation over East Africa. *Int. J. Climatol.* **2021**, *41*, 6474–6496. [CrossRef]
121. Ngoma, H.; Wen, W.; Ayugi, B.; Babaousmail, H.; Karim, R.; Ongoma, V. Evaluation of Precipitation Simulations in CMIP6 Models over Uganda. *Int. J. Climatol.* **2021**, *41*, 4743–4768. [CrossRef]

22. Asadullah, A.; Mcintyre, N.; Kigobe, M. Evaluation of Five Satellite Products for Estimation of Rainfall over Uganda. *Hydrol. Sci. J.* **2008**, *53*, 1137–1150. [CrossRef]
23. Dinku, T.; Ceccato, P.; Grover-Kopec, E.; Lemma, M.; Connor, S.J.; Ropelewski, C.F. Validation of Satellite Rainfall Products over East Africa's Complex Topography. *Int. J. Remote Sens.* **2007**, *28*, 1503–1526. [CrossRef]
24. Diem, J.E.; Hartter, J.; Ryan, S.J.; Palace, M.W. Validation of Satellite Rainfall Products for Western Uganda. *J. Hydrometeorol.* **2014**, *15*, 2030–2038. [CrossRef]
25. AghaKouchak, A.; Mehran, A.; Norouzi, H.; Behrangi, A. Systematic and Random Error Components in Satellite Precipitation Data Sets. *Geophys. Res. Lett.* **2012**, *39*, 4. [CrossRef]
26. Maggioni, V.; Massari, C. On the Performance of Satellite Precipitation Products in Riverine Flood Modeling: A Review. *J. Hydrol.* **2018**, *558*, 214–224. [CrossRef]
27. Bhatti, H.A.; Rientjes, T.; Haile, A.T.; Habib, E.; Verhoef, W. Evaluation of Bias Correction Method for Satellite-Based Rainfall Data. *Sensors* **2016**, *16*, 884. [CrossRef]
28. Gebrechorkos, S.H.; Hülsmann, S.; Bernhofer, C. Evaluation of Multiple Climate Data Sources for Managing Environmental Resources in East Africa. *Hydrol. Earth Syst. Sci.* **2018**, *22*, 4547–4564. [CrossRef]
29. Van Griensven, A.; Meixner, T.; Grunwald, S.; Bishop, T.; Diluzio, M.; Srinivasan, R. A Global Sensitivity Analysis Tool for the Parameters of Multi-Variable Catchment Models. *J. Hydrol.* **2006**, *324*, 10–23. [CrossRef]
30. Wagener, T.; Kollat, J. Numerical and Visual Evaluation of Hydrological and Environmental Models Using the Monte Carlo Analysis Toolbox. *Environ. Model. Softw.* **2007**, *22*, 1021–1033. [CrossRef]
31. Sreedevi, S.; Eldho, T.I. A Two-Stage Sensitivity Analysis for Parameter Identification and Calibration of a Physically-Based Distributed Model in a River Basin. *Hydrol. Sci. J.* **2019**, *64*, 701–719. [CrossRef]
32. Op de Hipt, F.; Diekkrüger, B.; Steup, G.; Yira, Y.; Hoffmann, T.; Rode, M. Applying SHETRAN in a Tropical West African Catchment (Dano, Burkina Faso)—Calibration, Validation, Uncertainty Assessment. *Water* **2017**, *9*, 101. [CrossRef]
33. Birkinshaw, S. *SHETRAN Version 4: Data Requirements, Data Processing and Parameter Values*; Newcastle University: Newcastle Upon Tyne, UK, 2008.

Disclaimer/Publisher's Note: The statements, opinions and data contained in all publications are solely those of the individual author(s) and contributor(s) and not of MDPI and/or the editor(s). MDPI and/or the editor(s) disclaim responsibility for any injury to people or property resulting from any ideas, methods, instructions or products referred to in the content.

Article

Exploring the Effects of Climate Change on Farming System Choice: A Farm-Level Space-for-Time Approach

Paulo Flores Ribeiro * and José Lima Santos

Forest Research Centre, Associate Laboratory TERRA, School of Agriculture, University of Lisbon, Tapada da Ajuda, 1349-017 Lisbon, Portugal; jlsantos@isa.ulisboa.pt
* Correspondence: pfribeiro@isa.ulisboa.pt

Abstract: Climate change is expected to affect the agricultural sector in ways that are often unclear to predict. If in the short- and medium-terms farmers may adapt to climate change by adjusting their agricultural practices, in the long-term, these adjustments may become insufficient, forcing farmers to change their farming systems. The extent and direction in which these farming system transitions will occur is still a subject that is underexplored in the literature. We propose a new framework to explore the effect of climate change on the choice of farming system while controlling the effect of other drivers that are also known to influence the farming system choice. Using a spatially explicit farming system choice model developed by a previous study in an extensive agricultural region of southern Portugal, we applied a space-for-time approach to simulate the effect of climate change on the future dynamics of the farming systems in the study area. The results suggest that climate change will force many farmers to change the farming system in a foreseeable future. The extent of the projected changes in farming systems is likely to trigger significant social, economic, and environmental impacts, which should require early attention from policy makers.

Keywords: climate change; farming systems; space-for-time; choice modelling; climate scenarios

Citation: Ribeiro, P.F.; Santos, J.L. Exploring the Effects of Climate Change on Farming System Choice: A Farm-Level Space-for-Time Approach. *Land* **2023**, *12*, 2113. https://doi.org/10.3390/land12122113

Academic Editors: Le Yu and Pengyu Hao

Received: 14 September 2023
Revised: 21 November 2023
Accepted: 24 November 2023
Published: 27 November 2023

Copyright: © 2023 by the authors. Licensee MDPI, Basel, Switzerland. This article is an open access article distributed under the terms and conditions of the Creative Commons Attribution (CC BY) license (https://creativecommons.org/licenses/by/4.0/).

1. Introduction

Climate change is expected to bring substantial impacts on the agricultural sector [1,2]. Plenty of research has been published on the subject, mostly focusing on assessing the effects of climate change on crop and livestock productivity [3–5], on the shifting of agroclimatic zones [6,7], or on global food security [4,8,9]. However, the literature focused on anticipating farmers' decisions in their climate change adaptation strategies is surprisingly less abundant.

Farm management decisions are typically made at the whole-farm level, in a comprehensive and internally coherent approach that accounts for the interdependencies between crops, livestock rearing, and other activities, rather than independently for each crop or activity [10]. So, to understand and anticipate farmers' decisions in their adaptation strategies to climate change, an integrated farming systems (FSs) approach should be adopted rather than resorting to a simple crop modelling analysis focused on understanding how climate change will affect a particular crop or activity [11–13]. In this sense, an FS acts as a classification scheme that is useful for identifying groups of farms that carry out roughly the same activities, with similar land use and livestock patterns, employing identical technologies and production methods and which, therefore, can be expected to react in a similar way to external stimuli [10,14]. This is the FS concept adopted in this study. We believe that this FS approach to anticipate farmers' response to climate change is new in the literature where adaptation studies focused on adjustments to agricultural practices within the same FS have prevailed.

The farming system choice is a farmer's decision that involves the joint consideration of several factors acting alone or in interaction, expanding or narrowing the farmer's set of possible choices, which include both socioeconomic factors (e.g., policy or market context farm size, water or labour availability, or farmer's idiosyncrasies) and biophysical factors (e.g., soil, slope, or climate) [10,15,16].

The process of modelling the FS choice at the farm level is often faced with data constraints. On the one hand, information is needed to establish the typology that will represent the available FS portfolio for farmers, which may require collecting data on crops, land uses, livestock, and production methods for each farm. On the other hand, it requires characterising farms in terms of the above-mentioned drivers of farming system choice, which can be carried out by resorting to a GIS analysis, if appropriate data are available, or by carrying out expensive and time-consuming surveys. Additional challenges may stem from the fact that some relevant drivers of FS choice are subject to temporal variations, such as prices, or to space–time variations such as climate. In the latter case, the difficulty is further inflated by the fact that in most cases, it is not practicable to collect the required data for a period of time that is long enough to capture the effect of climate change on farmers' choice of FS.

Such difficulties have led researchers who are interested in modelling the effect of climate change on FS choice to resort to proxy approaches, often relaxing the farming system concept to a farm-type approach, such as the typologies built from official statistical databases (e.g., the EU Farm Accountancy Data Network database—FADN, used to classify farms based on production orientation and economic size) [17,18], or the use of farm types derived from farm surveys [5,15]. Such farm-type approaches are not suitable for representing farmers' choices, as they typically include components that are exogenous to that choice, such as the physical or economic size of the farm.

Other works on farming resilience to climate change focus on short- to medium-term adaptation strategies, such as crop diversification, the adoption of more drought-resistant varieties, adjustments in planting and harvest dates, or increasing irrigation [19,20], and do not focus on the long-term impacts where, by hypothesis, the magnitude of climate change may become incompatible with adjustments within the same FS, and may force farmers to undertake deeper changes, eventually leading them to switching into a different FS.

To test this hypothesis, we depart from a recent study that presented an FS choice model developed to explain the spatial distribution of 22 farming systems in the Alentejo region (southern Portugal), based on a diversity of socioeconomic and biophysical drivers, including climate variables [16]. Taking advantage of the considerable extent of the study area used to estimate this model and its internal climate variability, in this study, we used this model to develop a space-for-time approach to simulate climate change scenarios and assess how these will influence the FS choice in the future. The use of a space-for-time substitution procedure is based on the assumption that when the drivers of spatial variation are the same of variation over time, then time can be replaced by space to model future change patterns [21]. Space-for-time has been used before in studies on the effects of climate change on the agricultural sector, for example, to assess how regional crop suitability will shift with climate change [22]. Still, to our knowledge, this coupling of an FS discrete choice model with a space-for-time substitution approach is a pioneer in empirical studies on the effects of climate change applied to the agricultural sector.

The proposed modelling layout allows for the exploration of the effects of climate change in the choice of the FS while ensuring high control over the remaining drivers that are intended to be kept constant, which is a premise that is often assumed in the literature, albeit in an implicit way, and therefore not usually discussed. Thus, the objective of the present study is not to predict which FS farmers will effectively choose in the future, but to understand how climate change alone will influence this decision-making process.

The specific objectives of the present study are, therefore, as follows: (1) to simulate the effects of different climate scenarios on FS choice; (2) compare the results achieved with our farm-level FS choice approach with those from crop-modelling approaches; and (3) assess the limits of a space-for-time approach to explore the effects of climate change on farming system choice. Finally, insights on the usefulness of the proposed approach to assess the impacts of farming system dynamics induced by climate change are discussed.

2. Materials and Methods

2.1. Study Area

The Alentejo region (EU NUT II), in the southern part of mainland Portugal, was selected as the study area for this work for 3 main reasons: (1) it is large enough to have significant climate variation, which is a crucial requirement for the space-for-time modelling approach described below; (2) the land use/cover is largely dominated by agriculture; and (3) a recently developed farm-level mapping of farming systems is available for the entire region (see Section 2.2).

The Alentejo region extends ca. 31,550 km^2, covering about 1/3 of Portugal (Figure 1). Its climate is typically Mediterranean, characterised by warm and dry summers that contrast with cool, rainy winters. Average monthly temperatures range from 9.9 °C in January to 23.4 °C in August (annual average 16.3 °C), and total annual precipitation sums 619 mm, mostly concentrated from October to March. Climate gradients across the region are significant, with annual average temperatures increasing from 13.2 °C to 17.8 °C and precipitation decreasing from 1272 mm to 379 mm, while progressing from northwest to southeast.

Figure 1. Location of the Alentejo study area, Portugal.

Agriculture covers about 70% of the territory, with the remaining area corresponding mostly to forest areas (25%), water bodies (3%), and artificial areas (2%). Agriculture is dominated by permanent pastures, annual crops (mostly cereals and forages), and permanent crops (primarily olive groves), in descending order of their weights in farmland. Approximately 40% of the agricultural land is under the canopy of scattered trees, mainly cork and holm oaks (*Quercus suber* and *Q. rotundifolia*, respectively), an agroforestry system locally called "montado".

2.2. Baseline Data

The baseline data for the present research consisted primarily of an FS map for the Alentejo region, a machine learning model predicting these FSs from a set of socioeconomic and biophysical variables, including climate variables, and data on future climate scenarios.

2.2.1. Farming Systems Information

Information on FS was extracted from a recent paper [16] where a typology of 22 FSs was derived from a cluster analysis applied on land use and livestock data for virtually all farms in the Alentejo region in 2017 (Figure 2 and Table S1 in Supplementary Information).

The data were taken from the Integrated Administration and Control System (IACS) and the Land Parcel Identification System (LPIS) provided by the national agency responsible for Common Agricultural Policy (CAP) payments.

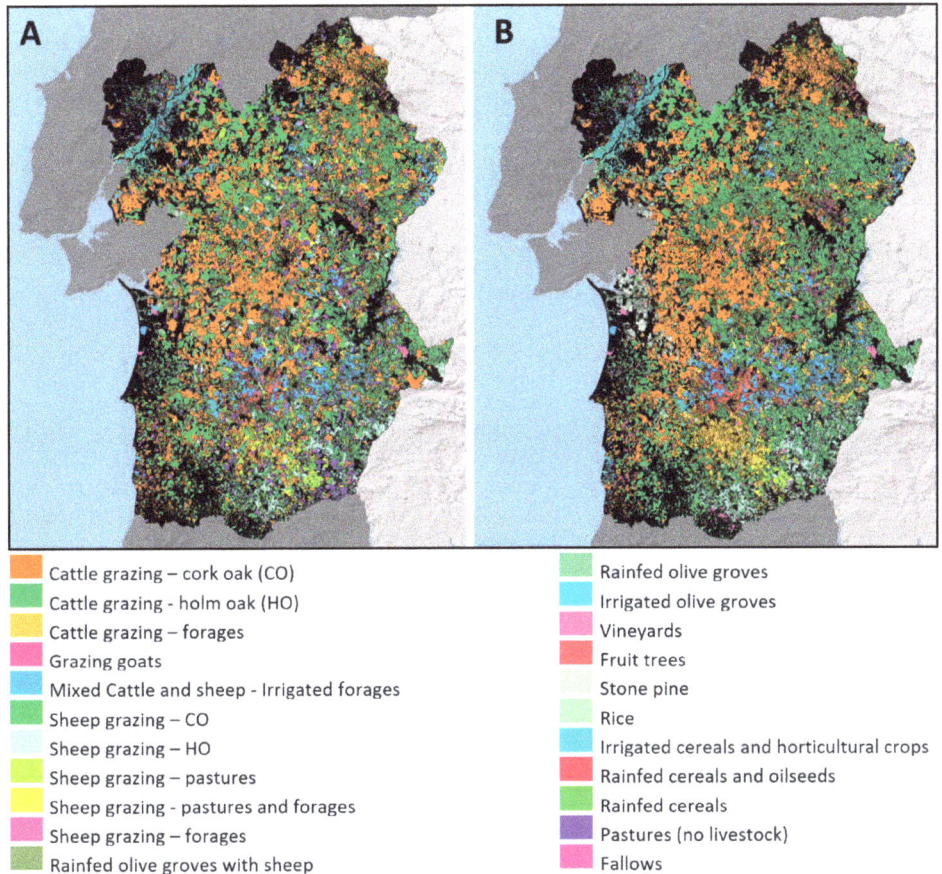

Figure 2. Observed (**A**) and predicted (**B**) spatial distribution of the farming systems of Alentejo, Portugal in 2017. Non-coloured areas (in black) refer to areas that did not apply for CAP payments in 2017, most of which are assumed to be non-agricultural areas, mostly forests (adapted from Ribeiro et al. [16]); see Supplementary Information in the same reference for detailed spatial distribution of each of the 22 farming systems.

All parcels declared by the same CAP beneficiary in the reference year (2017) were aggregated and taken as a single farm. The polygons of these parcels were spatially identified under the LPIS, so farm mapping was made possible (Figure 2). A total of 24,313 farms were thus identified and mapped, roughly covering 2×10^6 ha of utilised agricultural area (about 87% of total utilised agricultural area in Alentejo, according to the most recent 2019 agricultural census in Portugal, and 65% of the total Alentejo territorial area).

2.2.2. Farming Systems Predictive Model

The FS predictive model was also taken from Ribeiro et al. [16], who used a random forest modelling approach to estimate an FS choice model for the Alentejo region. Random forest is a popular machine learning technique that has been extensively used for modelling spatial and spatiotemporal data [23]. Random forest can be applied to both regression and classification problems, depending on the nature of the dependent variable, using bootstrap and aggregation (bagging) to build multiple decision trees based on random subsets of the data and using a random subset of the candidate predictor variables for each node in each decision tree [24]. On a classification problem, each observation is assigned to a class according to the majority of votes from all trees. In this case, the random forest model was used in a classificatory approach to predict the FS (categorical variable) characterising each farm from a broad set of 27 predictors describing farm structure (e.g., farm size, farm spatial fragmentation, and irrigation use), socioeconomic context (e.g., population density, use of hired labour, and main sources of income), and biophysical variables describing soil quality (pH and texture), topography (slope classes), and climate. These climate variables, which are keys for the present study, included the maximum and minimum mean temperatures of the warmest and coldest months, respectively, and annual precipitation, averaged between 1971 and 2000, which were used in model estimation in [16] (at the time, the most recent publicly available 30-year climate averages for mainland Portugal).

All farms were therefore characterised under these variables by overlaying the farm map with raster layers representing the spatial distribution of each variable to extract the average of the pixel values covered by the polygons of each farm, using a GIS zonal statistics tool.

The random forest model showed a good global predictive accuracy (63.7% error rate, which was evaluated positively considering the high number of 22 classes in the dependent variable, for which the random error rate would be about 95.4%).

In this study, the same model was replicated to predict FS choice under climate change scenarios by replacing the values of the climate variables used to estimate the model with new values referring to those climate scenarios, following a space-for-time substitution procedure [21].

2.2.3. Climate Scenarios

Data for climate change scenarios were extracted from the WorldClim database (https://www.worldclim.org, assessed on 19 February 2023), which provides climate projections based on a range of pathways for emission scenarios, known as "Shared Socioeconomic Pathways" (SSPs) [25]. These have been used as inputs for the latest climate models, under the Intergovernmental Panel on Climate Change (IPCC) and the six-assessment report of the Coupled Model Intercomparison Projects (CMIP6) [2,25]. These climate projections consist of monthly values of minimum temperature, maximum temperature, and precipitation, derived from 23 global climate models (GCMs), for four SSPs (1–mm 2.6, 2–4.5, 3–7.0, and 5–8.5) related to increasing levels of anthropogenic greenhouse gas (GHG) emissions, and for four 20-year periods (2021–2040, 2041–2060, 2061–2080, and 2081–2100).

In the present study, long-term climate scenarios for the 2081–2100 period were selected to heighten the effect of climate change on FS choice, since we were interested in capturing the effects on FS change, and not just changes in agricultural practices, like moving to more drought-resistant crop varieties. Three SSP scenarios were used to simulate low (SSP 1–2.6), moderate (SSP 3–7.0), and high GHG emission (SSP 5–8.5) scenarios. For each of these, we extracted the median of the projections provided by 9 of the 23 climate models, which were those with complete information on all climate variables and all SSP scenarios, to extract predictions of maximum and minimum temperatures of the warmest and coldest month, respectively, and annual average precipitation, so they can be comparable to the baseline climate variables used in the estimation of the random forest FS predictive model.

2.3. Scenario Assessment

The effects of climate change on FS choice were primarily assessed based on the changes in the land shares of the different FSs in the study area in each scenario. Sankey diagrams [26] were used to graphically visualise the predicted FS areal transitions. Transition matrices with the relative values of the areal transfers between FSs are also shown in Supplementary Information.

3. Results

3.1. Climate Change Predictions

The climate scenarios for the study area in the 2081–2100 period predict significant changes in the climate variables under study, even in the low emission scenario (SSP 1-2.6) (Table 1). The climate anomaly in the average value of the minimum temperature of the coldest month is expected to range from an increase of 1.0 °C in the low emission scenario to 3.3 °C in the high emission scenario. The average maximum temperature in the warmest month is expected to rise by 2.0 °C in the low emission scenario by and 6.6 °C in the high emission scenario. The average annual precipitation is anticipated to decrease by between −19.9 mm in the low emission scenario and −96.5 mm in the high emission scenario, which corresponds to a decrease of about 16% in the total annual precipitation of the baseline period (1971–2000). In the moderate emission scenario (SSP 3-7.0), the forecasts point to anomaly values between the limits of the two extreme scenarios.

Table 1. Distribution of the three climate variables used in the models in the baseline situation (average 1971–2000) in the study area and their predicted anomalies for the long-term scenario of the 2081–2100 period. Figures embedded in the graphics depict the corresponding mean values of the anomaly.

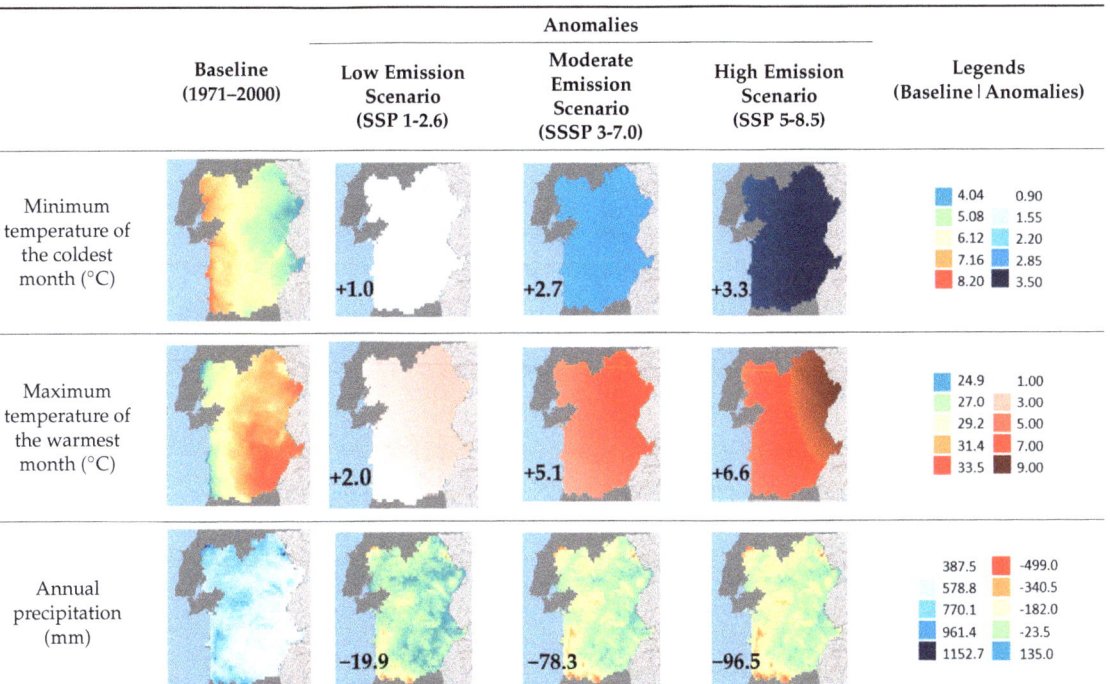

As expected, these climatic anomalies are not uniform across the study area, showing variations that are particularly evident in the case of the maximum temperature, whose anomalies are about 4 °C higher in the interior, compared to the coastal areas. In the case of precipitation, the effect of relief on climate models is apparent, with more significant decreases expected with an increasing altitude. In the lower areas of the interior, where the current precipitation is quite low (400 mm or less), the forecasts even predict an increase in the average annual precipitation.

3.2. Effects of Climate Change on Farming System Choice

The model predictions show that climate change will substantially impact the FS choice in the study area. Altogether, 23%, 37%, or 40% of the total agricultural area of the study area is expected to change the FS, respectively, in the low, moderate, or high GHG emission scenarios (Figure 3). The impact of climate change, however, is not expected to be the same across all FSs; while some will barely be affected, others will undergo significant changes, either gaining or losing area (Table 2).

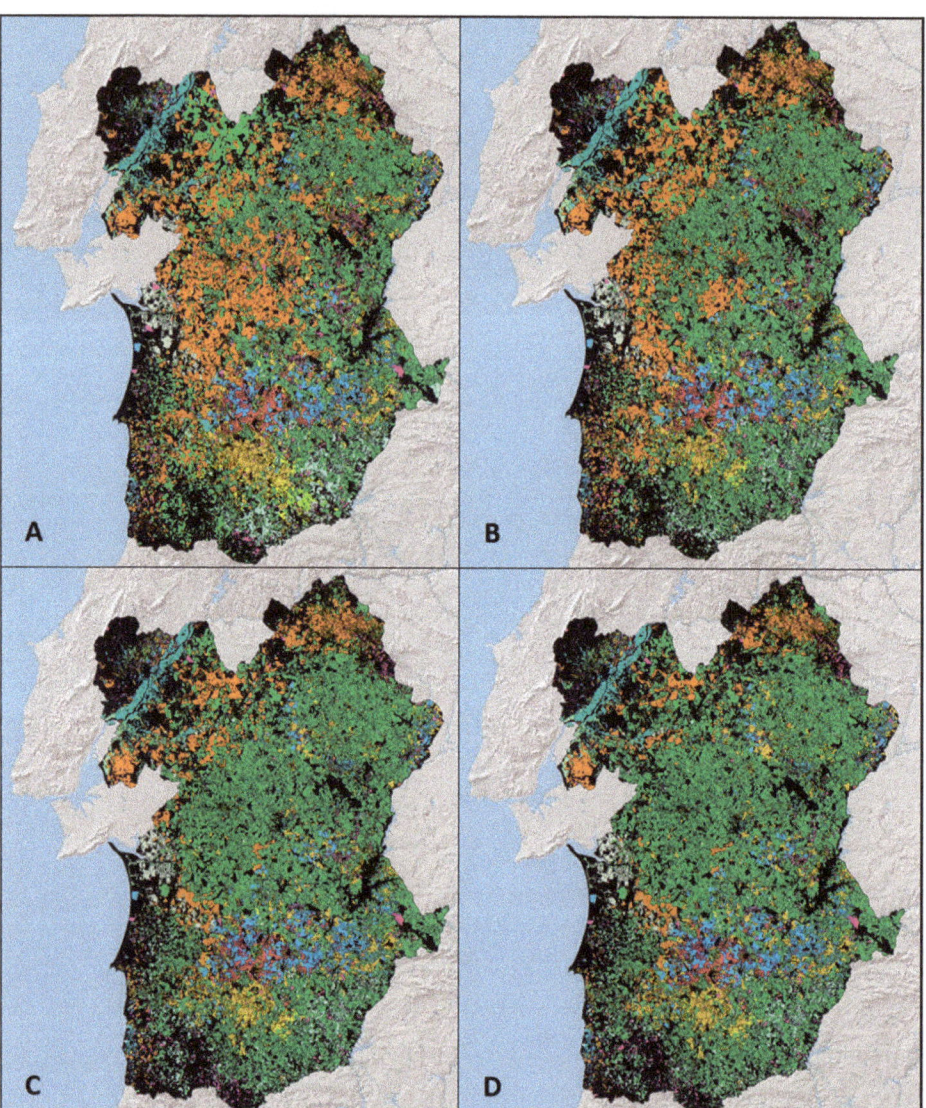

Figure 3. Predicted farming system distribution in the study area in 2017 ((**A**)—baseline scenario) and for the 2081–2100 period in low ((**B**)—SSP 1-2.6), moderate ((**C**)—SSP 3-7.0), and high emission ((**D**)—SSP 5-8.5) climate scenarios (see Figure 2 for colour legend).

Table 2. Expected effects of climate change on current area of farming systems in the long-term (2081–2100) scenarios.

Farming System	Area in 2017 (Predicted)		Expected Relative Changes in Area in Each 2081–2100 Scenario		
	ha	%	Low (SSP 1-2.6)	Moderate (SSP 3-7.0)	High (SSP 5-8.5)
Cattle grazing–CO	495,655	25.2	−22%	−62%	−72%
Cattle grazing–HO	580,656	29.5	44%	80%	89%
Cattle grazing–forages	168,219	8.5	−3%	17%	23%
Grazing goats	21,075	1.1	−12%	8%	6%
Mixed cattle and sheep–irrigated forages	20,090	1.0	−1%	−13%	−18%
Sheep grazing–CO	132,621	6.7	−55%	−89%	−93%
Sheep grazing–HO	100,952	5.1	−11%	4%	5%
Sheep grazing–pastures	32,734	1.7	−48%	−62%	−66%
Sheep grazing–pastures and forages	36,330	1.8	−45%	−61%	−63%
Sheep grazing–forages	18,602	0.9	−23%	−48%	−59%
Rainfed olive groves with sheep	13,396	0.7	3%	−22%	−28%
Rainfed olive groves	15,472	0.8	−15%	−12%	−7%
Irrigated olive groves	89,647	4.6	11%	42%	46%
Vineyards	21,947	1.1	−45%	−58%	−64%
Fruit trees	10,256	0.5	−8%	−33%	−37%
Stone pine	54,665	2.8	−11%	−26%	−29%
Rice	22,220	1.1	7%	−9%	−12%
Irrigated cereals and horticultural crops	50,042	2.5	−9%	−22%	−21%
Rainfed cereals and oilseeds	39,694	2.0	9%	−9%	−12%
Rainfed cereals	19,182	1.0	7%	14%	22%
Pastures without livestock	12,775	0.6	−75%	−84%	−84%
Fallows	12,700	0.6	−13%	−5%	−1%
Total	1,968,929	100.0	–	–	–

The cattle grazing–HO system, which is currently the dominant system covering ca. 30% of the total agricultural area, is expected to be the one experiencing the greatest area increase in any of the three climate scenarios, almost doubling the area in the high emission scenario (89% increase). Most of this growth will come from area currently under the cattle grazing–CO system (Figure 4), which is one of the top area losers, expected to drop between 22% and 72% of its current area, respectively, in the low and high emission scenarios (Table 2). This extensive replacement of cork oak with holm oak agroforestry systems is probably related to the distinct agroecological preferences of both oak species, which currently makes the cork oak dominant to the west and north of Alentejo, where the climate tends to be less hot and dry due to the proximity of the Atlantic Ocean [27], and makes the holm oak more frequent in the warmer and drier southern interior of the Alentejo region [16,28]. Thereby, the widespread decrease in precipitation and increase in summer heat will likely prompt holm oak to expand to areas that are currently dominated by cork oak. However, this expansion of the cattle grazing–HO system towards the more northwestern parts of Alentejo seems to suffer some resistance, only overthrown under the higher GHG emission scenarios, which may be related to the predominance of light texture soils (sandy soils) in this region, which tend to favour cork oak and not holm oak [16].

In addition to the Cattle grazing–CO system, other systems are expected to suffer significant declines in area. Except for the sheep grazing–HO system, all other sheep-specialised systems (sheep grazing–CO, sheep grazing–forages, sheep grazing–pastures, and sheep grazing–pastures and forages) are predicted to lose area, particularly the sheep grazing–CO system, which will likely lose more than half of its current area (55%) in the low emission scenario, and will almost disappear in the high emission scenario (93% drop). Although the available data do not allow for clarification, this may be related to a preference of the cattle systems for larger farms, which are predominant in the Alentejo region [16], fostering its expansion over sheep systems.

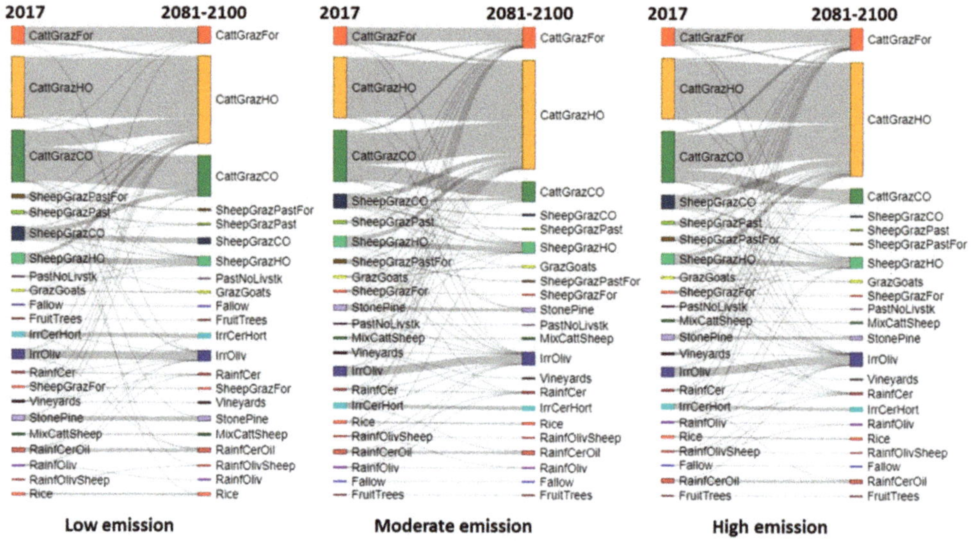

Figure 4. Areal transitions between farming systems from the baseline period (2017) to the 2081–2100 period in each of the three climate scenarios (to avoid over-cluttering the figure, transitions below 1000 ha—ca. 0.05% of the study area—are not shown; farming system names are abbreviated from names in Figure 2).

The grazing goats FS will predictably be one of the livestock-specialised FSs that will be the least affected, probably due to being a low-demanding FS in terms of the socioeconomic and biophysical contexts, only associated with a sloping terrain [16], which will not be altered by climate change.

The vineyards and the stone pine FS are likely to lose considerable area irrespective of the climate scenario. The former will lose area especially to the irrigated olive grove FS, and the latter will lose area especially to the cattle grazing–HO system.

As for FSs that are strongly dependent on irrigation, the irrigated cereal and horticultural crops FS is expected to lose area, while the irrigated olive groves FS is projected to expand significantly in response to climate change, which may expand by up to 46% in the high emission scenario, mainly at the expense of vineyards and rainfed olive groves. The rice FS is expected to remain mostly unchanged, since its choice is strongly influenced by the availability of suitable areas next to water courses, which is a driver of low relevance for most other FSs.

Rainfed cereals are expected to expand in all three scenarios, at the expense of areas from different systems (see transition matrices in Supplementary Information).

Other systems show inconsistent trends across the three scenarios. Cattle grazing–forages and sheep grazing–HO are predicted to lose area in the low emission scenario, but they are expected to expand in the moderate and high emission scenarios. Conversely, the rainfed olive groves with sheep system is anticipated to expand in the low emission scenario, but expected to lose area in the moderate and high emission scenarios.

4. Discussion

4.1. Climate Change and Farming System Choice

The proposed modelling framework proved its suitability to explore the effects of climate change on FS choice. The use of a space-for-time substitution approach proved to be a wise choice to overcome the limitation of having only stationary data on the spatial distribution of FS (relative to a single year), which we believe to be a novelty in the research on the effects of climate change applied to the agricultural sector.

As hypothesised, our findings suggest that the magnitude of climate change expected in the long run scenarios will force many farmers to adjust well beyond agricultural practices, pushing them to undertake an effective FS change. Moreover, such effects are expected in both high and low GHG emission scenarios, although, unsurprisingly, they are clearly more substantial in high emission scenarios. This supports the claims of previous authors on the need for a farming systems approach when investigating farmers' adaptation strategies to climate change, to the detriment of the conventional crop-modelling approaches that populate most of the literature [13].

The FS choice model used in the simulations included a variety of drivers of FS choice that were intentionally kept constant except for the climate variables, making it possible to attribute the observed FS dynamics predominantly to the isolated effect of climate change, as was the objective of the present study. This does not mean that the social, economic, technological, policy, or demographic contexts, for example, will remain unchanged in the future, but only that we sought to detach the effects of our variables of interest (climate), to investigate their single contributions in the choice of future FSs, in a ceteris paribus approach. This use of multivariate choice models to study the effect of climate change on FS choice based on a multitude of socioeconomic and biophysical factors is also new in the literature, as far as we know.

The scrutiny of the effects of climate change on FS choice must consider that the simulations were carried out using a predictive model estimated from the data observed on the decisions of more than 23 thousand farmers in their FS choices, based on real farms, and mediated by a very high number of independent variables that included 13 socioeconomic drivers and 14 biophysical drivers, conjointly influencing the decision-making process. Therefore, the effect of simulations on climate variables must be assessed in the context of the joint action of all of these drivers, whose influences on the decision will act differently in each farm, since each one is unique in its characteristics. Also, climate change will not affect all farms equally, since the extent of those changes are differentiated in space, showing an increasing gradient of the average temperatures of the coldest and warmest months from the coast towards the inland, and a drop in precipitation marked by altitude. Under all of these premises, it is not surprising that climate change will not affect all FSs in the same way; while some will be severely affected by the gain or loss of area, others will persist mostly unaffected.

The primary FS shift expected to be induced by climate change until the end of the century refers to the significant westward expansion of the cattle grazing–HO system, towards areas that are currently dominated by the cattle grazing–CO system. These shifts, induced by climate change, are also probably being mediated by other drivers and constraints of FS choice, such as the agroecological preferences of both oak species [16,27,28] expressed by other variables in the model that, although kept constant, are also influencing the FS choice, together with the climate variables. The predicted sharp reduction in the cork oak area is likely to imply significant social and economic impacts at the national level, given the high importance of the cork cluster in Portugal [29]. Other livestock-specialised FSs are also expected to suffer substantial areal shifts, including most sheep grazing systems, which are expected to change to cattle systems, encouraged by the farmland structure of the region, which is marked by large farms [16].

Among the crop-specialised FSs, most will lose area in response to climate change, especially in the moderate and high emission scenarios. In the case of the vineyards and the stone pine FSs, both have strong regional identities in Alentejo agriculture, being pillars of important agro-industrial value chains on regional and national scales, so the prediction of its reduction can lead to important social and economic impacts.

The anticipated increase in the irrigated olive groves FS must be considered with caution, as it is highly dependent on large irrigation systems, whose future sustainability in the study area has been questioned due to the predicted drop in precipitation and its impacts on both the quantity and quality of irrigation water [30,31].

The rainfed cereals FS is also expected to expand in response to climate change, not so much because it is favoured by future climate conditions, but because they are likely to be less affected than most of the competing FSs, as they are typically rainfed extensive systems with a strong dependence on autumn–winter crops (cereals).

4.2. Comparing Farming System and Crop-Modelling Approaches

The proposed FS approach to explore the effects of climate change relies on a discrete-choice modelling framework, where farmers are set to choose the FS—a categorical variable—among a range of possible choices, based on socioeconomic and biophysical drivers. The crop-modelling approaches, or other species-specific approaches often found in the literature, typically assume that farmers operating a given farming system will adapt to climate change by adjusting their farming practices, such as increasing irrigation water or adjusting the sowing/harvesting dates, without shifting to other FSs. In this section, we discuss our results, with reference to those of previous studies focused on key crops or activities in the study area and carried out in comparable agroecological contexts, to explore the similarities and divergences between the results of both approaches.

Despite little current knowledge about the dynamics of Mediterranean oak woodlands in response to multiple drivers [28], previous studies have reported the likelihood of a decline in Mediterranean agroforestry systems resulting from climate change. Research on cork and holm oak canopy cover loss carried out in the same region as the present study (Alentejo) found a likely decline trend for cork oak associated with the increase in the mean temperature, while the decline in holm oak would be mostly associated with increasing cattle numbers [32]. Rising temperatures in recent decades have also been related with increased crown defoliation and tree mortality in both cork and holm oak [32]. Other studies carried out in Spanish Andalusia, a region that borders Alentejo to the west, report that a significant part of the cork oak plantations made in this region in the 1990s of the last century, largely driven by CAP policies encouraging the afforestation of less productive farmland, are probably doomed to succumb due to the deterioration of weather conditions in the future [33]. Our findings are in line with these previous studies, with the novelty of foreseeing an extensive replacement of the cork oak with holm oak.

As for the apparent substitution of sheep with cattle, pushed by the substantial expansion of the cattle grazing–HO system, although research references relating these effects to climate change are scarce, there is evidence mentioning that cattle grazing is prone to reduce grassland heterogeneity in Mediterranean regions, which may decrease the ability to adapt to climate change [34]. Further evidence suggests that climate change will negatively influence perennial grasslands and forage yields in Mediterranean ecosystems [35,36], which may raise doubts about the expansion of cattle grazing systems predicted by our simulations.

The patent increase in the irrigated olive groves FS suggested by our results seems to contrast with the findings of previous studies focused on the impacts of climate change on olive groves, which have questioned the future suitability of this crop in the Mediterranean basin, unless appropriate adaptation measures are implemented [37]. One such possible measure is irrigation [38,39], which is in line with our forecasts of an expansion of the irrigated olive groves FS and a contraction of the rainfed olive groves FS. Warming and drought trends expected for southern Europe in the coming decades, however, are likely to bring major challenges to irrigation expansion due to excessive heat and water stress [30,31]. Additionally, other works have warned of the possible increase in the risk of pest outbreaks in Mediterranean olive groves, as a result of climate change [40]. Therefore, doubts remain about the sustainability of our prediction of an increase in the irrigated olive groves FS, which is particularly important since this crop has been the target of large investments in new plantations in recent years in the study area, cultivated in intensive and super-intensive regimes with irrigation, already being one of the main irrigation crops in the Alentejo region [41].

Regarding our prediction of an expansion of the rainfed cereals FS, it contrasts with the results of previous studies that used crop modelling to assess the impact of climate change on wheat production in southern Portugal, foreseeing significant production losses depending on the climate scenario used [42]. As adaptation measures to reverse yield reductions, these authors propose the use of early flowering wheat varieties, or the anticipation of the sowing date. Studies that used the CERES-Wheat crop model to simulate yields under climate change in Mediterranean regions also identified a trend towards reduced yields, recommending the development of adaptation strategies and measures such as the use of adapted genotypes to counteract the negative impact of climate change [43].

We conclude that the two approaches can be complementary, assuming that the short- to medium-term farmers will be able to make this type of adjustment without changing the FS, while our results suggest that in the long term, this may no longer be possible, and many farmers will effectively be forced to change the FS. Whether or not, when the time comes, they will have the means, the knowledge, or the ability to carry out this change remains a critical issue that should concern policymakers, but which is beyond the scope of the present study.

4.3. Strengths and Weaknesses of the Proposed Approach

This work shows that climate change will eventually subject many farmers to an adaptation effort that goes far beyond simple adjustments in agricultural practices, varieties used, or sowing or harvesting dates, forcing them to abandon the current farming system and switch to new farming systems that are potentially very different from those they currently practice. For example, switching from a farming system based on annual crops to a cattle grazing system, which our results show could affect many farmers in this study area, could require large investments in the acquisition of herds, installation of pastures, fencing of grazing plots, and technical training, among others. It is not guaranteed that most farmers are prepared to embark on this process of change, or have the means to do so, whether financial, technological, or know-how. Studies such as the present one can help policy makers to anticipate the support needs that these farmers will require to undergo this change process, while also helping them to foresee its wider impacts on food supply, the environment, or nature and landscape conservation, enabling early action to alleviate the effects of climate change.

An important asset of the proposed framework is that it relies on very detailed and spatially explicit farm-level data, describing livestock and land use/cover at the plot scale, which was made possible by the opportunity to access IACS/LPIS data. Therefore, it is built on observed data from management decisions made by actual farmers, framed by the characteristics of their farms and their biophysical and socioeconomic contexts. This entails a significant advantage when compared to previous studies, mainly based on crop models or on declared data collected in surveys in response to hypothetical scenarios (e.g., [15]). Also, the spatial explicit feature provided by the connection to the LPIS enables areal trade-offs between FSs to be explored and to explicitly map where the changes are expected to take place, which may be valuable to, e.g., inform land planning assessment.

The random forest approach used in model estimation made it possible to work with a high dimensional categorical dependent variable, representing the 22 FS choice-sets available to farmers in adapting to climate change scenarios, which is unprecedented in the literature.

The potential limitation of working with agricultural data for a single year (2017), which, at the outset, would prevent the exploration of temporal dynamics, was overcome by resorting to an approach of substituting time for space, taking advantage of the considerable extent of the study area. Despite being a longstanding approach (see [44] and references therein), space-for-time substitution remains a widely used approach in several fields, especially in ecology, where it emerged (e.g., [21,45]), whenever only stationary data are available.

With minor adjustments, the same basic approach could be used to explore, for example, how public policies could be implemented to encourage farmers to adopt particular FSs, aimed at ensuring desired levels of food security or the sufficient provision of socially valued public goods, provided that the random forest choice model includes some comparative profitability variable discriminating the FS. Such possibilities stand as proposals for future research.

Regarding the shortcomings of the framework, it should be noted that it deals solely with changes in the averages of climate variables, and not with changes in their variability. Indeed, climate change pressures on farmers' decisions will likely be felt earlier—if not already felt—due to the increasing frequency of extreme events, such as droughts, heat waves, or floods, which may significantly anticipate the need for farmers to adapt to climate change. This drawback, however, is hardly avoidable because current climate models do not provide scenarios of climate variability change, but only changes in their average values.

The framework's implementation is also quite demanding in baseline data, both to derive the FS typology and to estimate the choice model. In fact, the approach is only feasible when data comparable to that in the IACS/LPIS are available, which is often not the case, particularly in developing regions where such research could be of particular interest, e.g., in the context of food security issues.

The fact that the framework deals only with existing FSs, observed in the reference year (2017), may also entail some weakness, as it hinders the emergence of new FSs that are potentially better suited to cope with climate change [10]. Nevertheless, the high number of categories in the FS typology must have contributed to minimise this possible problem.

Finally, it must be recognised that some of the results achieved are hard to explain based on the available information, such as the fact that the cattle grazing–forages FS loses area in the low emission scenario and increases area in the higher emission scenario.

5. Conclusions

Not underestimating that the relationship between climate and agriculture goes both ways, since agriculture is also a driver of climate change, the present study focused on investigating the effect of climate change on the choice of FS. The results indicate that climate change alone is prone to lead many farmers to change their FSs as an adaptation strategy. Such changes are likely to modify the pattern of ecosystem services that is currently provided by agriculture, including at the provisioning, regulating, supporting, or cultural levels. This calls for further research on the exploration of these effects, opening the way for climate change impact assessments and the consideration of policy options. Indeed, previous work has suggested that changes in policy, as well as technology or prices, may have stronger impacts on farmers' decisions than climate change [13,46], meaning that there will be room for policy to help ease the adaptation effort that farmers will have to endure.

Supplementary Materials: The following supporting information can be downloaded at https://www.mdpi.com/article/10.3390/land12122113/s1, Transition matrices for areal changes of farming systems from 2017 to 2081–2100 in three climate change scenarios in the Alentejo region, Portugal.

Author Contributions: Conceptualisation, methodology, validation, and writing—review and editing, P.F.R. and J.L.S.; formal analysis, data curation, and writing—original draft preparation, P.F.R. All authors have read and agreed to the published version of the manuscript.

Funding: This work was financed by national funds through FCT—Portuguese Foundation for Science and Technology, I.P., under Project UIDB/00239/2020 of the Forest Research Centre (CEF) and the Associate Laboratory TERRA.

Data Availability Statement: Data are contained within the article and Supplementary Materials.

Conflicts of Interest: The authors declare no conflict of interest.

References

1. European Environment Agency. Climate Change Adaptation in the Agriculture Sector in Europe—Publications Office of the EU, EEA Rep. No 04/2019, Publ. Off. 2019. Available online: https://op.europa.eu/en/publication-detail/-/publication/fb9bf9af-0117-11ea-8c1f-01aa75ed71a1/language-en/format-PDF/source-265745439 (accessed on 10 October 2022).
2. IPCC. *Climate Change 2022: Impacts, Adaptation and Vulnerability. Contribution of Working Group II to the Sixth Assessment Report of the Intergovernmental Panel on Climate Change*; Cambridge University Press: Cambridge, UK; New York, NY, USA, 2022. [CrossRef]
3. Asare-Nuamah, P.; Botchway, E. Understanding climate variability and change: Analysis of temperature and rainfall across agroecological zones in Ghana. *Heliyon* **2019**, *5*, e02654. [CrossRef] [PubMed]
4. Ayinu, Y.T.; Ayal, D.Y.; Zeleke, T.T.; Beketie, K.T. Impact of climate variability on household food security in Godere District, Gambella Region, Ethiopia. *Clim. Serv.* **2022**, *27*, 100307. [CrossRef]
5. Habtemariam, L.T.; Kassa, G.A.; Gandorfer, M. Impact of climate change on farms in smallholder farming systems: Yield impacts, economic implications and distributional effects. *Agric. Syst.* **2017**, *152*, 58–66. [CrossRef]
6. Adnan, S.; Ullah, K.; Gao, S.; Khosa, A.H.; Wang, Z. Shifting of agro-climatic zones, their drought vulnerability, and precipitation and temperature trends in Pakistan. *Int. J. Climatol.* **2017**, *37*, 529–543. [CrossRef]
7. Ceglar, A.; Zampieri, M.; Toreti, A.; Dentener, F. Observed Northward Migration of Agro-Climate Zones in Europe Will Further Accelerate Under Climate Change. *Earth's Future* **2019**, *7*, 1088–1101. [CrossRef]
8. Molotoks, A.; Smith, P.; Dawson, T.P. Impacts of land use, population, and climate change on global food security. *Food Energy Secur.* **2021**, *10*, e261. [CrossRef]
9. Ofori, S.A.; Cobbina, S.J.; Obiri, S. Climate Change, Land, Water, and Food Security: Perspectives From Sub-Saharan Africa. *Front. Sustain. Food Syst.* **2021**, *5*, 680924. [CrossRef]
10. Santos, J.L.; Moreira, F.; Ribeiro, P.F.; Canadas, M.J.; Novais, A.; Lomba, A. A farming systems approach to linking agricultural policies with biodiversity and ecosystem services. *Front. Ecol. Environ.* **2021**, *19*, 168–175. [CrossRef]
11. Hayman, P.; Rickards, L.; Eckard, R.; Lemerle, D. Climate change through the farming systems lens: Challenges and opportunities for farming in Australia. *Crop Pasture Sci.* **2012**, *63*, 203–214. [CrossRef]
12. Reidsma, P.; Ewert, F.; Lansink, A.O.; Leemans, R. Adaptation to climate change and climate variability in European agriculture: The importance of farm level responses. *Eur. J. Agron.* **2010**, *32*, 91–102. [CrossRef]
13. Reidsma, P.; Wolf, J.; Kanellopoulos, A.; Schaap, B.F.; Mandryk, M.; Verhagen, J.; Van Ittersum, M.K. Climate change impact and adaptation research requires integrated assessment and farming systems analysis: A case study in the Netherlands. *Environ. Res. Lett.* **2015**, *10*, 045004. [CrossRef]
14. Dixon, J. Concept and Classifications of Farming Systems. *Encycl. Food Secur. Sustain.* **2019**, *3*, 71–80. [CrossRef]
15. Etwire, P.M. The impact of climate change on farming system selection in Ghana. *Agric. Syst.* **2020**, *179*, 102773. [CrossRef]
16. PRibeiro, F.; Santos, J.L.; Canadas, M.J.; Novais, A.M.; Moreira, F.; Lomba, Â. Explaining farming systems spatial patterns: A farm-level choice model based on socioeconomic and biophysical drivers. *Agric. Syst.* **2021**, *191*, 103140. [CrossRef]
17. Iakovidis, D.; Gadanakis, Y.; Park, J. Farm-level sustainability assessment in Mediterranean environments: Enhancing decision-making to improve business sustainability. *Environ. Sustain. Indic.* **2022**, *15*, 100187. [CrossRef]
18. Leclère, D.; Jayet, P.A.; de Noblet-Ducoudré, N. Farm-level Autonomous Adaptation of European Agricultural Supply to Climate Change. *Ecol. Econ.* **2013**, *87*, 1–14. [CrossRef]
19. Meuwissen, M.P.M.; Feindt, P.H.; Spiegel, A.; Termeer, C.J.A.M.; Mathijs, E.; de Mey, Y.; Finger, R.; Balmann, A.; Wauters, E.; Urquhart, J.; et al. A framework to assess the resilience of farming systems. *Agric. Syst.* **2019**, *176*, 102656. [CrossRef]
20. van Zonneveld, M.; Turmel, M.S.; Hellin, J. Decision-Making to Diversify Farm Systems for Climate Change Adaptation. *Front. Sustain. Food Syst.* **2020**, *4*, 32. [CrossRef]
21. Wogan, G.O.U.; Wang, I.J. The value of space-for-time substitution for studying fine-scale microevolutionary processes. *Ecography* **2018**, *41*, 1456–1468. [CrossRef]
22. Holzkämper, A.; Calanca, P.; Fuhrer, J. Analyzing climate effects on agriculture in time and space. *Procedia Environ. Sci.* **2011**, *3*, 58–62. [CrossRef]
23. Hengl, T.; Nussbaum, M.; Wright, M.N.; Heuvelink, G.B.M.; Gräler, B. Random forest as a generic framework for predictive modeling of spatial and spatio-temporal variables. *PeerJ* **2018**, *6*, e5518. [CrossRef] [PubMed]
24. Liaw, A.; Wiener, M. Classification and Regression by randomForest. *R News* **2002**, *2*, 18–22.
25. Riahi, K.; van Vuuren, D.P.; Kriegler, E.; Edmonds, J.; O'Neill, B.C.; Fujimori, S.; Bauer, N.; Calvin, K.; Dellink, R.; Fricko, O.; et al. The Shared Socioeconomic Pathways and their energy, land use, and greenhouse gas emissions implications: An overview. *Glob. Environ. Chang.* **2017**, *42*, 153–168. [CrossRef]
26. Cuba, N. Research note: Sankey diagrams for visualizing land cover dynamics. *Landsc. Urban Plan.* **2015**, *139*, 163–167. [CrossRef]
27. Pérez-Girón, J.C.; Díaz-Varela, E.R.; Álvarez-Álvarez, P. Climate-driven variations in productivity reveal adaptive strategies in Iberian cork oak agroforestry systems. *For. Ecosyst.* **2022**, *9*, 100008. [CrossRef]
28. Acácio, V.; Dias, F.S.; Catry, F.X.; Rocha, M.; Moreira, F. Landscape dynamics in Mediterranean oak forests under global change: Understanding the role of anthropogenic and environmental drivers across forest types. *Glob. Chang. Biol.* **2017**, *23*, 1199–1217. [CrossRef]
29. Sørensen, I.H.; Torralba, M.; Quintas-Soriano, C.; Muñoz-Rojas, J.; Plieninger, T. Linking Cork to Cork Oak Landscapes: Mapping the Value Chain of Cork Production in Portugal. *Front. Sustain. Food Syst.* **2021**, *5*, 787045. [CrossRef]

30. Rocha, J.; Carvalho-Santos, C.; Diogo, P.; Beça, P.; Keizer, J.J.; Nunes, J.P. Impacts of climate change on reservoir water availability, quality and irrigation needs in a water scarce Mediterranean region (southern Portugal). *Sci. Total Environ.* **2020**, *736*, 139477. [CrossRef]
31. Tomaz, A.; Palma, P.; Fialho, S.; Lima, A.; Alvarenga, P.; Potes, M.; Salgado, R. Spatial and temporal dynamics of irrigation water quality under drought conditions in a large reservoir in Southern Portugal. *Environ. Monit. Assess.* **2020**, *192*, 93. [CrossRef]
32. Acácio, V.; Dias, F.S.; Catry, F.X.; Bugalho, M.N.; Moreira, F. Canopy Cover Loss of Mediterranean Oak Woodlands: Long-term Effects of Management and Climate. *Ecosystems* **2021**, *24*, 1775–1791. [CrossRef]
33. Duque-Lazo, J.; Navarro-Cerrillo, R.M.; Ruíz-Gómez, F.J. Assessment of the future stability of cork oak (*Quercus suber* L.) afforestation under climate change scenarios in Southwest Spain, For. *Ecol. Manag.* **2018**, *409*, 444–456. [CrossRef]
34. Faria, N. Predicting agronomical and ecological effects of shifting from sheep to cattle grazing in highly dynamic Mediterranean dry grasslands. *Land Degrad. Dev.* **2019**, *30*, 300–314. [CrossRef]
35. Dumont, B.; Andueza, D.; Niderkorn, V.; Lüscher, A.; Porqueddu, C.; Picon-Cochard, C. A meta-analysis of climate change effects on forage quality in grasslands: Specificities of mountain and mediterranean areas. *Grass Forage Sci.* **2015**, *70*, 239–254. [CrossRef]
36. Yang, C.; Fraga, H.; Van Ieperen, W.; Santos, J.A. Modelling climate change impacts on early and late harvest grassland systems in Portugal. *Crop Pasture Sci.* **2018**, *69*, 821–836. [CrossRef]
37. Fraga, H.; Moriondo, M.; Leolini, L.; Santos, J.A. Mediterranean olive orchards under climate change: A review of future impacts and adaptation strategies. *Agronomy* **2021**, *11*, 56. [CrossRef]
38. Fraga, H.; Pinto, J.G.; Santos, J.A. Olive tree irrigation as a climate change adaptation measure in Alentejo, Portugal. *Agric. Water Manag.* **2020**, *237*, 106193. [CrossRef]
39. Tanasijevic, L.; Todorovic, M.; Pereira, L.S.; Pizzigalli, C.; Lionello, P. Impacts of climate change on olive crop evapotranspiration and irrigation requirements in the Mediterranean region. *Agric. Water Manag.* **2014**, *144*, 54–68. [CrossRef]
40. Caselli, A.; Petacchi, R. Climate Change and Major Pests of Mediterranean Olive Orchards: Are We Ready to Face the Global Heating? *Insects* **2021**, *12*, 802. [CrossRef] [PubMed]
41. Branquinho, S.; Rolim, J.; Teixeira, J.L. Climate change adaptation measures in the irrigation of a super-intensive olive orchard in the south of portugal. *Agronomy* **2021**, *11*, 1658. [CrossRef]
42. Yang, C.; Fraga, H.; van Ieperen, W.; Trindade, H.; Santos, J.A. Effects of climate change and adaptation options on winter wheat yield under rainfed Mediterranean conditions in southern Portugal. *Clim. Change* **2019**, *154*, 159–178. [CrossRef]
43. Dettori, M.; Cesaraccio, C.; Duce, P. Simulation of climate change impacts on production and phenology of durum wheat in Mediterranean environments using CERES-Wheat model. *Field Crops Res.* **2017**, *206*, 43–53. [CrossRef]
44. Pickett, S.T.A. Space-for-Time Substitution as an Alternative to Long-Term Studies. In *Long-Term Studies in Ecology*; Springer: New York, NY, USA, 1989; pp. 110–135. [CrossRef]
45. Lovell, R.S.L.; Collins, S.; Martin, S.H.; Pigot, A.L.; Phillimore, A.B. Space-for-time substitutions in climate change ecology and evolution. *Biol. Rev.* **2023**, *98*, 2243–2270. [CrossRef] [PubMed]
46. Wolf, J.; Kanellopoulos, A.; Kros, J.; Webber, H.; Zhao, G.; Britz, W.; Reinds, G.J.; Ewert, F.; De Vries, W. Combined analysis of climate, technological and price changes on future arable farming systems in Europe. *AGSY* **2015**, *140*, 56–73. [CrossRef]

Disclaimer/Publisher's Note: The statements, opinions and data contained in all publications are solely those of the individual author(s) and contributor(s) and not of MDPI and/or the editor(s). MDPI and/or the editor(s) disclaim responsibility for any injury to people or property resulting from any ideas, methods, instructions or products referred to in the content.

Article

Assessing the Resilience of Stream Ecosystems to Rainfall Impact

Yujin Park [1], Junga Lee [2], Se-Rin Park [1] and Sang-Woo Lee [1,*]

[1] Department of Forestry and Landscape Architecture, Konkuk University, Seoul 05029, Republic of Korea; qkrdbwls333@konkuk.ac.kr (Y.P.); serin87@konkuk.ac.kr (S.-R.P.)
[2] Division of Environmental Science & Ecological Engineering, Korea University, Seoul 02841, Republic of Korea; archjung@korea.ac.kr
* Correspondence: swl7311@konkuk.ac.kr; Tel.: +81-2-450-3684

Abstract: In Republic of Korea, pronounced seasonal precipitation variability poses substantial challenges for stream water quality management and the effective utilization of water resources. Ecologically degraded streams are particularly vulnerable to these fluctuations, which can exacerbate their already fragile condition. We assessed the resilience of reference and impaired streams in response to rainfall through water quality system performance (WQSP). The WQSP is quantified as the concentration of BOD, T-N, and T-P, which represent streams' eutrophication and anaerobic conditions and respond quickly to disturbances. Reference and impaired streams are classified according to the biological condition and habitat environment of the streams in the Han River watershed of Republic of Korea. The resilience of the stream ecosystem was estimated using WQSP, the linear multiple regression model, and the generalized additive model for rainfall and WQSP. The WQSP reference streams have a lower sensitivity to disturbance and recover more quickly from the influence of rainfall; therefore, they have higher resilience than impaired streams to rainfall events. This study facilitates understanding changes in stream ecosystems of varying conditions in response to rainfall for ensuring long-term stability and adaptability.

Keywords: water quality; robustness; rapidity; climate change; system performance; generalized additive model

1. Introduction

In Republic of Korea, seasonal precipitation patterns are predominantly shaped by heavy rainfall during summer months and arid winters, influenced by monsoonal activities [1]. Streams within this region, characterized by their limited extent and steep gradients, are particularly sensitive to these seasonal precipitations, which is evident from their high riverbed coefficients [2]. These seasonal fluctuations directly and indirectly affect water quality and stream life by lowering the flow rate of streams and increasing the load of pollutants when rainfall is low [3–5]. In contrast, periods of intense rainfall can lead to the dilution of pollutants yet also risk the introduction of land-sourced contaminants into streams, thus degrading water quality and disrupting habitats [6,7]. Streams already impaired by poor water quality, habitat conditions, or ecological disturbances are especially susceptible to the adverse effects of rainfall variability, often facing challenges in restoring their pre-disturbance state [8]. Therefore, it is necessary to understand streams' resilience and rainfall's impact on stream resilience.

Resilience is the capacity of a system to return to its state before a disturbance. In stream ecosystems, resilience can be considered the ability to maintain the stream ecosystem by quickly restoring a degraded system's performance, such as its water quality and living organisms, even when disturbances and disasters occur [9–11]. Resilience can be characterized by four "Rs", namely, robustness (minimum value of the remaining system performance after the disturbance), rapidity (ability to restore original function within a short time), redundancy (ability to replace system function), and resourcefulness (ability to recover after a disaster). The four Rs constituting resilience help identify vulnerabilities in the stream ecosystem and set up disaster prevention measures according to the characteristics of the stream [12,13].

However, as it is difficult to measure resilience quantitatively, resilience can be delineated by estimating changes in system performance in response to various disasters [14]. Particularly, resilience can be estimated in stream ecosystems through an index of water quality system performance (WQSP), which provides a time-varying measure of how well stream ecosystems achieve a desired water quality criterion at a given time (t) [15]. Water quality in the stream ecosystem plays a pivotal role in determining the condition of the stream environment and the habitat for organisms, as well as having an important effect on humans' healthy use of water [16]. In addition, stream water quality responds more quickly and sensitively to the watershed environment and meteorological dynamics than biological indicators, facilitating an understanding of the effects of various watershed factors on the stream [17]. The resilience of stream ecosystems through WQSP can differ depending on the condition of the ecosystem. Accordingly, it is necessary to understand the relationship between rainfall and stream resilience to respond to rainfall variability and manage water quality appropriately according to stream conditions.

Therefore, this study aimed to examine the resilience of reference and impaired streams to rainfall through the performance of water quality systems. The overall objectives of this study were as follows: (a) to quantify the WQSP of reference and impaired streams for the water quality indicators, and (b) to estimate the sensitivity and stability of the stream ecosystem by identifying the robustness and rapidity of resilience through the WQSP values of the reference and impaired streams and the relationship between rainfall and WQSP. The findings of this study can provide profound insights into stream ecosystem resilience for stream management.

2. Materials and Methods

2.1. Study Area and Selecting Reference and Impaired Streams

The Han River watershed, the focus of this study, is the largest in Republic of Korea contains the largest river, and is home to more than half of the country's population. It is concentrated in land development pressure and population growth [18]. The Han River watershed consists of 913 streams, including 907 sampling sites from the National Aquatic Ecological Monitoring Program (NAEMP). Through the National Aquatic Ecological Monitoring Program (NAEMP), the Ministry of Environment (MOE) [19] is evaluating the condition of stream ecosystems using biological indicators, such as tropic diatom communities (TDI), benthic macroinvertebrate (BMI), the fish assessment index (FAI), and habitat condition. The NAMEP conducts biannual assessments of biological indices at a nationwide scale. These biological indices are quantified on a scale from 0 to 100 and categorized into five classes, ranging from class A ("very good") to class E ("very poor") to assess their ecological condition. Biological-grade results for each stream are obtained from the Water Environment Information System (http://211.114.21.27/web, accessed on 16 November 2023).

In this study, reference streams were defined as those with measured values exceeding 80, while impaired streams were identified by values falling below 35. Of the 907 sampling sites, 158 monitoring sites maintained continuous data records from 2013 to 2019. Reference and impaired stream classifications were determined using data from these 158 monitoring sites. Within these monitoring sites, 22 monitoring sites were designated as reference streams, while 17 were classified as impaired streams (Figure 1).

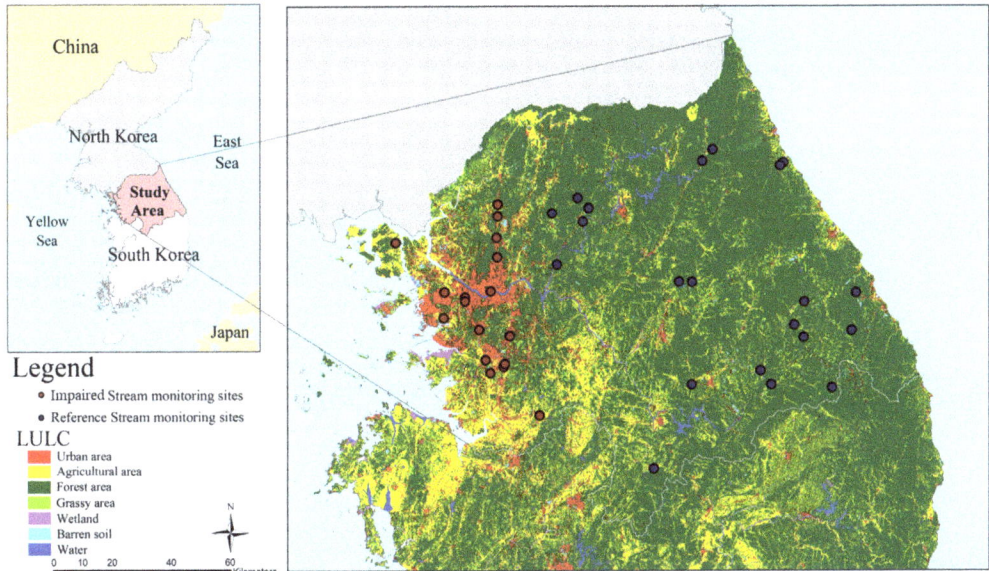

Figure 1. Han River watershed, land use classification, monitoring sites of reference, and impaired streams of the National Aquatic Ecological Monitoring Program in Republic of Korea.

2.2. System Performance as a Concept of Quantifying Resilience

Resilience is widely used to evaluate a system's performance and condition. In the field of aquatic ecology, it has started to be quantified through system performance in water resources [20]. Simonovic and Peck [21] developed a system performance framework to measure and quantify changes in the dynamic resilience of a system after disturbances due to climate change (Figure 2).

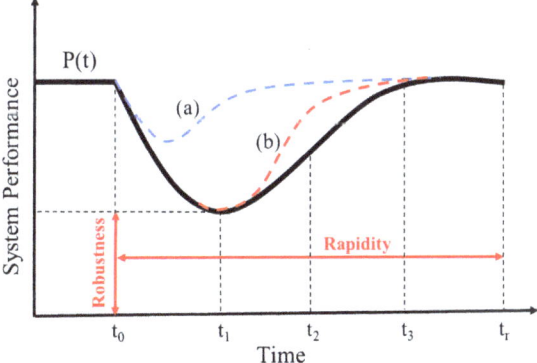

Figure 2. Change in the system performance after a disturbance. (a), the blue line represents a system performance with high robustness, characterized by a short recovery period and high resilience; (b), the red line represents a system performance with low robustness, featuring a long recovery period and low resilience.

Figure 2 shows the system's performance change after a disturbance. Line P(t) indicates the loss of system performance, t_0 signifies when the disturbance occurs, t1 indicates when the disturbance is finished, and tr signifies when recovery from the disturbance is complete. The system performance response to disturbance can be divided into three categories: a blue line (a), a black line (P(t)), and a red line (b). If P(t) represents the general system performance degradation and recovery, (a) has higher robustness, so the degradation of system performance with disturbance is not severe, and the performance recovery rapidity is short, resulting in high resilience. (b) shows a large drop in system performance with disturbance due to low robustness, and the recovery rapidity of system performance is longer, resulting in low resilience.

$$p(t) = \int_{t_0}^{t} [P_0 - P(t)]dt \text{ were } t \in [t_0, t_r] \quad (1)$$

System performance can be obtained through the cumulative value of the loss value compared to the cumulative values of the system's optimal value. System performance loss and optimal values can be calculated through Equation (1). To obtain the optimal and loss values of water quality system performance, 2013–2019 monthly average values of Biochemical Oxygen Demand (BOD), Total Nitrogen (T-N), and Total Phosphors (T-P) for the reference and impaired streams were used.

2.3. Analytic Framework

The analytical framework employed to assess the resilience of both reference and impaired streams is systematically presented in Figure 3. The determination of resilience robustness was based on assessing WQSP fluctuation ranges, utilizing WQSP values, and the time lag of WQSP changes associated with rainfall, which were derived through LM and GAM analyses. Resilience rapidity was quantified through the evaluation of the recovery period of WQSP after rainfall, also using WQSP values, and the rainfall period showed a significant relationship between WQSP and rainfall, determined through LM and GAM analyses. Evaluating the stream ecosystem's stability and sensitivity was founded upon the results obtained from robustness and rapidity assessments. The rainfall used in the analysis was averaged as the sum of monthly precipitation values in the sub-watershed containing each stream. Additionally, the following rainfall used to analyze the relationship between WQSP and rainfall was obtained by moving the sum of the previously calculated monthly precipitation to 1 to 5 months later.

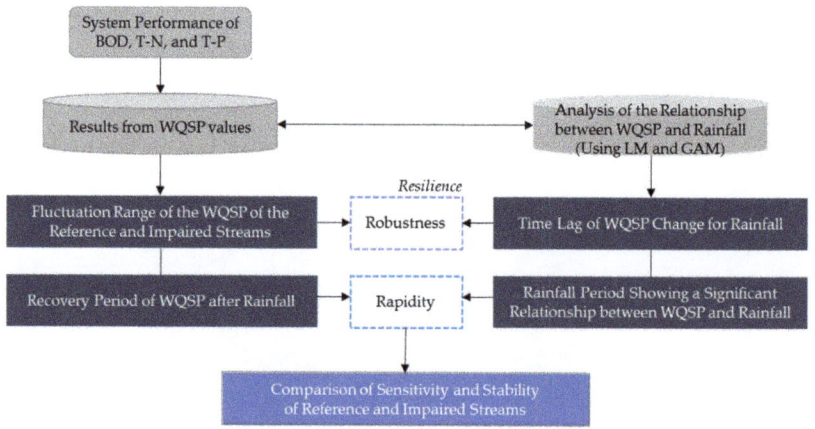

Figure 3. Flow diagram of the analytic framework.

2.4. Statistical Approach

A common approach to investigate the relationships between rainfall and water quality indices in streams is to use linear regression analyses. The linear multiple regression model

(LM) is an analysis method that explains the contributions of several causes to a result through several independent variables ($x_1, x_2, \ldots, x_n$) that explain the dependent variable (y):

$$LM_{wqsp} = \alpha + B_1 x_1 + B_2 x_2 + \cdots + B_n x_n + \varepsilon \quad (2)$$

Both linear correlation and regression analyses are useful for quantifying the direction, magnitude, and significance of the relationship between variables, but if the two variables are not linear, the relationship between the variables may not be accurately identified. To consider the nonlinear relationship between variables, in this study, a flexible regression model, the generalized additive model (GAM), was used along with linear correlation and regression analyses. The GAM can express a nonlinear relationship between the dependent variable and the independent variable while maintaining additivity and can be expressed as Equation (3):

$$g(GAM_{wqsp}) = \alpha + f_1 x_1 + f_2 x_2 + \cdots + f_n x_n + \varepsilon \quad (3)$$

We performed LM and GAM analyses to investigate the rapidity and robustness between streams through the relationship between rainfall and streams. LM and GAM analyses were performed using the R package, and Akaike's information criterion (AIC), Bayesian information criterion (BIC), coefficient of determination (R^2), and expected default frequency (EDF) of the LM and GAM analysis results were compared. The AIC and BIC are criteria for comparing the model's suitability; the smaller the AIC and BIC values are, the better the model. R^2 is a statistic representing the model's explanatory power; the closer it is to 1, the higher the explanatory power. EDF is a value that indicates whether the relationship between the explanatory variable and the independent variables is linear or nonlinear. The closer it is to 1, the closer the relationship is to a linear one.

3. Results

3.1. WQSP Variability Recovery Period of Reference and Impaired Streams

As shown in Figure 4, the range of fluctuations in WQSP and the recovery period of WQSP from rainfall were identified through the BOD, T-N, and T-P changes in the WQSP of the reference and impaired streams. The most precipitation occurred in July and August, and the least occurred in October, March, and January. The fluctuation ranges of the WQSP of the reference and impaired streams for BOD were 0.473 and 1.046, respectively. For T-N, the WQSP fluctuation ranges of the reference and impaired streams were 0.391 and 0.676, respectively, and similar to the BOD, the impaired streams showed a higher fluctuation range. However, the fluctuation range of WQSP for T-P was 1.223 for the reference stream and 0.842 for the impaired stream, indicating higher variability in the reference stream.

The recovery period of the WQSP of BOD and T-N took 11.14 months and 11.13 months on average, respectively, in the reference streams and 13.5 and 13.17 months, respectively, in the impaired streams. The recovery period of the reference and impaired streams differed by about 1–2 months but did not appear to be a significant difference. For T-P, the reference and impaired streams showed a rapidity of 11 and 11.43 months, respectively, indicating similar recovery periods.

3.2. Analysis of the Impact of Rainfall on WQSP

3.2.1. LM and GAM Analyses of Reference Streams for BOD, T-N, and T-P

To examine the relationship between the various rainfall values and WQSP of reference streams, LM and GAM were determined for all rainfall events (Table 1 and Figure 5). In LM, no rainfall variable was significantly associated with WQSP for BOD. The GAM analysis revealed a significant relationship between the four months following rainfall and WQSP. The EDF value was 4.69, which indicated a nonlinear relationship. The GAM better explained the relationship between rainfall and the WQSP of BOD. The findings indicate a four-month delay in the reference streams' BOD levels. Additionally, the GAM revealed that the rainfall period that showed a significant difference was limited to one month.

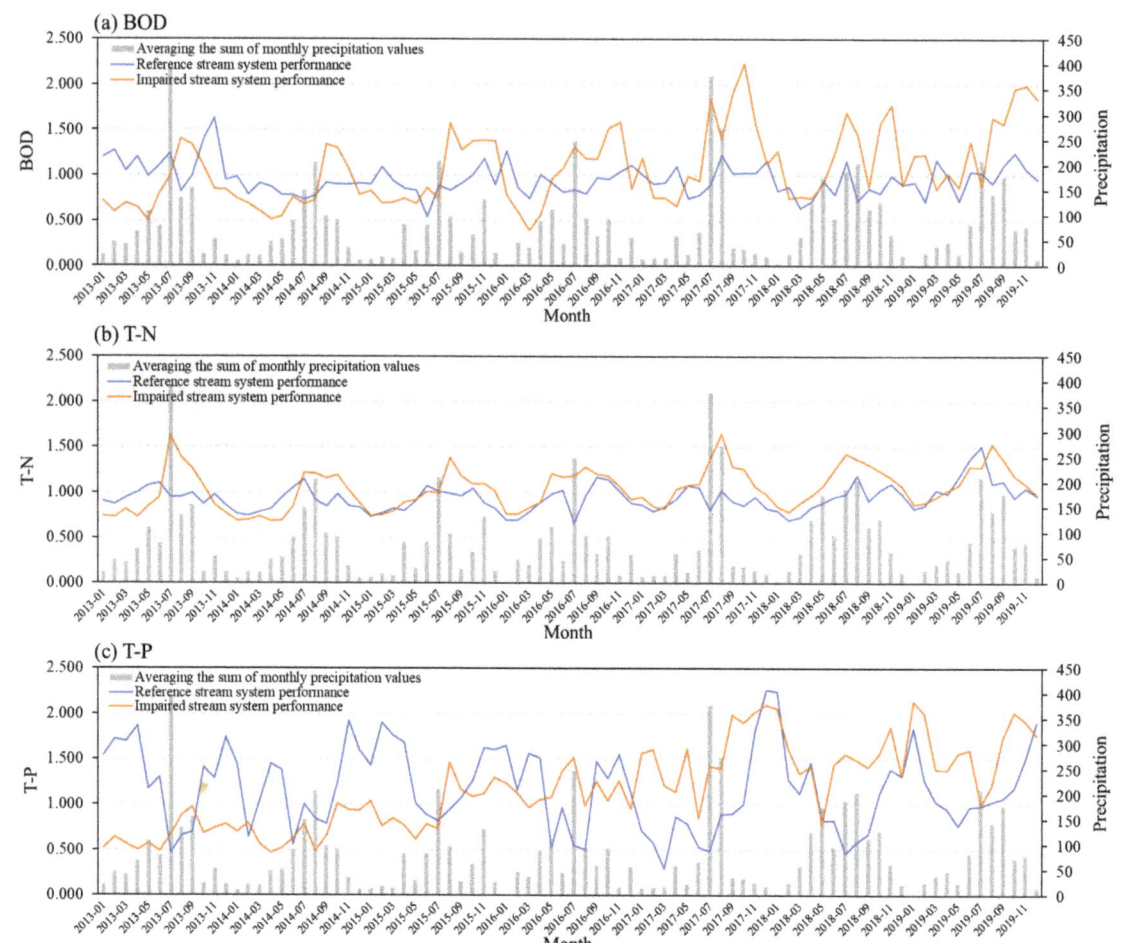

Figure 4. Graphs averaging the sum of monthly precipitation value and the WQSP change for reference and impaired streams. (**a**) BOD; (**b**) T-N; and (**c**) T-P.

Table 1. LM and GAM analysis results of the WQSP for BOD, T-N, and T-P in the reference streams The independent variables satisfied LM's low variance inflation factor (VIF) condition.

	Analysis Results of Reference Streams			Variables							Performance		
				P	P1	P2	P3	P4	P5	Constant	R^2	AIC	BIC
BOD	LM	Coefficients	b	0.0003	−0.0002	0.0003	−0.0001	0.0004	0.0003	0.841	0.12	−67.1	−47.9
			β	0.19	−0.09	0.16	−0.01	0.17	0.19	-			
		T-value		1.52	−0.7	1.2	−0.04	1.21	1.39	20.39 **			
	GAM	F-value		2.51	1.47	2.81	0.23	3.18 **	3.5	0.94 **	0.23	−78.3	−47.1
		EDF		1.18	1	1	1	4.69	1				

Table 1. Cont.

Analysis Results of Reference Streams			Variables							Performance		
			P	P1	P2	P3	P4	P5	Constant	R^2	AIC	BIC
T-N	LM	Coefficients b	0.0007	0.00003	0.0001	0.0001	0.0001	−0.0003	0.893	0.12	−67.1	−47.9
		β	0.34	0.02	0.07	0.35	0.06	−0.21	-			
		T-value	2.82 **	0.14	0.59	0.29	0.49	−1.75	24.62 **			
	GAM	F-value	3.05 *	2.59 *	0.1	0.001	0.29	1.57	0.94 **	0.23	−78.3	−47.1
		EDF	2.45	3.47	1	1	1	1				
T-P	LM	Coefficients b	−0.001	−0.001	−0.001	0.0001	0.001	0.002	1.13	0.41	67.9	87.3
		β	−0.27	−0.09	−0.11	0.03	0.23	0.32	-			
		T-value	−2.79 **	−0.94	−1.14	0.28	2.26 *	3.27 **	12.10 **			
	GAM	F-value	5.72 *	0.7	0.61	1.12	6.43 *	13.12 **	1.17 **	0.44	66.6	91.1
		EDF	1	1	1	2.48	1	1				

* $p < 0.05$, ** $p < 0.01$. AIC, Akaike's information criterion; BIC, Bayesian information criterion; EDF, expected default frequency. P means original rainfall, and P1 to P5 means 1 to 5 months following rainfall.

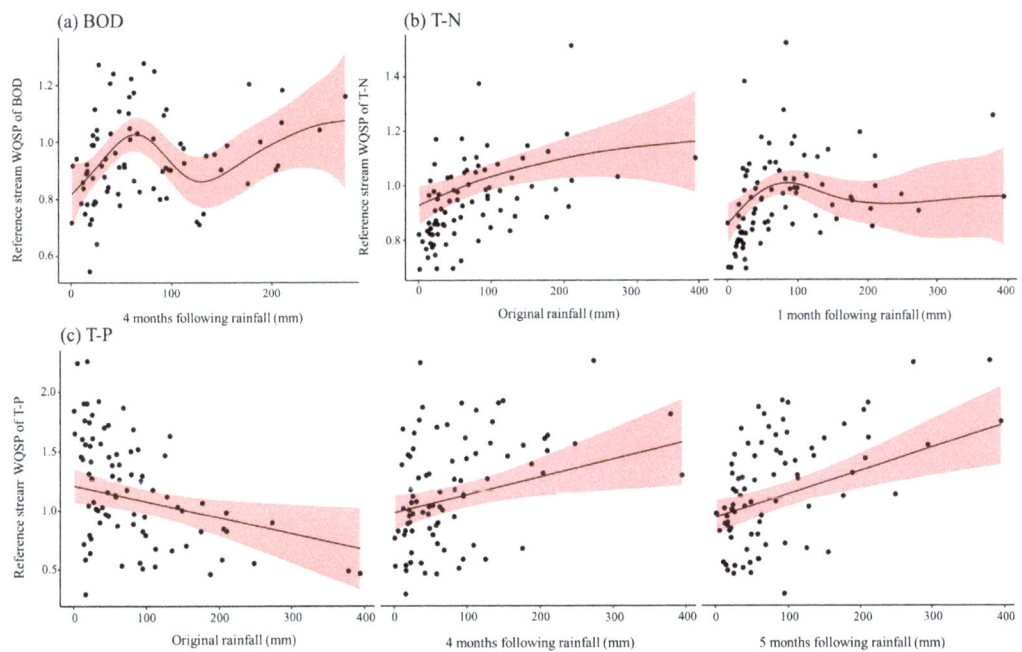

Figure 5. Smooth functions of the relationship between WQSP of reference streams and rainfalls which have a significant relationship: (**a**) WQSP of reference streams for BOD in relation to four months following rainfall; (**b**) WQSP of reference streams for T-N in relation to original rainfall and one month following rainfall; and (**c**) WQSP of reference streams for T-P in relation to original rainfall and four to five months following rainfall. The dot represents the water quality system performance value, and the red area represents the confidence interval.

A significant relationship was found in LM between the original rainfall and WQSP for T-N. In the GAM, the original rainfall and the one month following the rainfall had a significant relationship. The EDF values were 2.45 and 3.47, respectively, indicating that the relationship between rainfall and WQSP of T-N was nonlinear. Based on these results, the reference streams of T-N showed no time lag in either LM or GAM, and the significant rainfall period was one month in LM and two months in GAM.

In the reference stream of T-P, a significant relationship was found between the original rainfall and four to five months following the rainfall in both LM and GAM, and both EDF values were 1, indicating a linear relationship. Excluding the original rainfall, which has a low continuity with other significant rainfall events, the reference streams of T-P showed a time lag of four months, and the rainfall period with a significant relationship was two months in both LM and GAM.

3.2.2. LM and GAM Analyses of Impaired Streams for BOD, T-N, and T-P

To examine the relationship between the various rainfall values and WQSP of impaired streams, LM and GAM were determined for all rainfall events (Table 2 and Figure 6). In WQSP for BOD, a significant relationship was found between one and three months following rainfall in both LM and GAM, and both EDF values were 1, indicating a linear relationship. The impaired streams of BOD showed a time lag of one month, and the rainfall period with a significant relationship was three months for both LM and GAM.

Table 2. LM and GAM analysis results of the WQSP for BOD, T-N, and T-P in the impaired streams. The independent variables satisfied LM's low variance inflation factor (VIF) condition.

	Analysis Results of Impaired Streams			Variables							Performance		
				P	P1	P2	P3	P4	P5	Constant	R^2	AIC	BIC
BOD	LM	Coefficients	b	0.0005	0.001	0.001	0.003	0.0005	−0.0001	0.841	0.53	35.4	54.5
			β	0.09	0.25	0.25	0.44	0.08	−0.04	-			
		T-value		1.52	1.05	2.69 **	2.67 **	4.87 **	0.85	−0.37			
	GAM	F-value		2.51	1.16	9.65 **	8.07 **	27.23 **	0.84	0.21	0.56	34.5	60.5
		EDF		1.18	1.81	1	1	1	2.05	1			
T-N	LM	Coefficients	b	0.001	0.001	0.0005	0.0003	0.0001	−0.0002	0.893	0.74	−10.49	−85.4
			β	0.44	0.39	0.2	0.11	0.05	−0.1	-			
		T-value		2.82 **	6.30 **	5.41 **	2.94 **	1.66	0.67	−1.48			
	GAM	F-value		3.05 *	37.15 **	28.40 **	5.01 **	2.13	0.42	1.23	0.73	−104.3	−80.9
		EDF		2.45	1	1	1.59	1	1	1.45			
T-P	LM	Coefficients	b	−0.0004	0.0002	0.001	0.001	0.001	0.0004	1.13	0.04	113.4	132.9
			β	−0.08	0.04	0.17	0.12	0.1	0.07	-			
		T-value		−2.79 **	−0.63	0.33	1.32	0.97	0.82	0.57			
	GAM	F-value		5.72 *	0.43	0.2	1.94	0.89	0.66	0.17	0.05	113.7	135.4
		EDF		1	1.55	1	1	1	1	1			

* $p < 0.05$, ** $p < 0.01$. AIC, Akaike's information criterion; BIC, Bayesian information criterion; EDF, expected default frequency. P means original rainfall, and P1 to P5 means 1 to 5 months following rainfall.

In WQSP for T-N, a significant relationship was established between the original rainfall and one to two months following rainfall in both LM and GAM, while the three to five months following rainfall did not significantly contribute to the WQSP of T-N. Based on these results, the impaired streams of T-N showed no time lag and a significant rainfall period of three months in both LM and GAM. GAM and LM had similar explanatory power in the WQSP of BOD and T-N.

In WQSP for T-P, no rainfall had a significant relationship with WQSP in either the LM or GAM, so both models lacked explanatory power for the relationship between WQSP and rainfall.

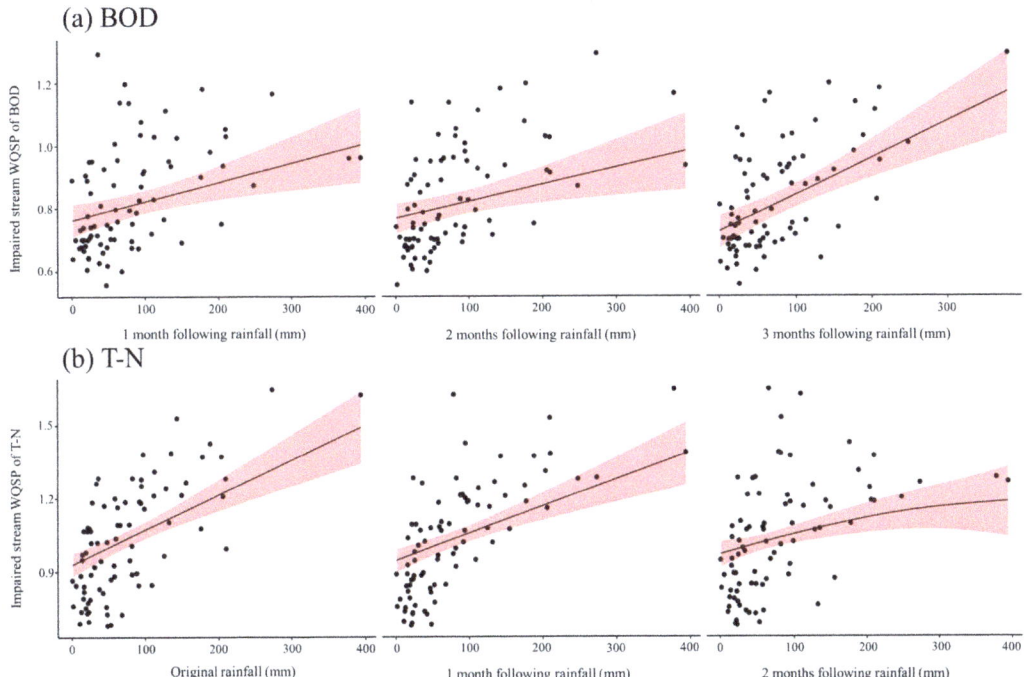

Figure 6. Smooth functions of the relationship between WQSP of impaired streams and rainfalls which have a significant relationship: (**a**) WQSP of impaired streams for BOD in relation to one to three months following rainfall; and (**b**) WQSP of reference streams for T-N in relation to original rainfall and one to two months following rainfall. The dot represents the water quality system performance value, and the red area represents the confidence interval.

3.3. Comparison of Sensitivity and Stability of Resilience
3.3.1. Comparison of Sensitivity through Robustness

To understand the sensitivity of stream ecosystem resilience, the fluctuation range of WQSP and the time lag of WQSP change with rainfall were compared (Table 3). The range of WQSP fluctuation for the reference streams for BOD and T-N was not larger than that of the impaired streams. The low fluctuation of the reference streams indicated no abrupt change in the WQSP due to rainfall. Therefore, it was concluded that the reference streams had a stream environment that responded less sensitively to rainfall and had higher robustness than the impaired streams. However, in the case of T-P, the range of the WQSP fluctuation for the reference streams was larger than that of the impaired streams, but the range of the fluctuations in impaired streams has been increasing since 2017, as shown in Figure 4. Therefore, T-P management must provide a stable habitat for the reference and impaired streams.

The time lag values derived through the LM and GAM indicated a one-month delay in the change in WQSP for BOD in impaired streams following rainfall. For T-N, the WQSP changed with the original rainfall event and reacted quite quickly to the rainfall. On the other hand, in the reference stream, the BOD and T-P showed a change in WQSP after three to four months, indicating that the time lag effect of rainfall was relatively long. The reference streams were considered to have a longer time lag since their tolerance to external environmental changes was not significant compared to that of the impaired streams, so it was concluded that the reference streams had higher robustness and lower sensitivity than the impaired streams. In the case of T-N, unlike other water quality indicators, the reference and impaired streams responded quickly to rainfall because the time lag was short. Unlike BOD, which is indirectly measured through dissolved oxygen and changes its concentration in decomposing organic matter that flows into the stream by microorganisms, T-N does not have a large time lag since the nitrogen component is flowed into and the concentration changes [22]. Furthermore, given that nitrogen readily dissolves in water, this phenomenon is attributed to the swift fluctuations in nitrogen concentration resulting from shifts in precipitation patterns [23,24].

Table 3. Robustness results include the WQSP fluctuation range and time lag for sensitivity comparison.

WQSP Classification		Fluctuation Range	Time Lag Results	
			LM	GAM
BOD	Reference	0.473	-	4 months
	Impaired	1.046	1 month	1 month
T-N	Reference	0.391	No time lag	No time lag
	Impaired	0.676	No time lag	No time lag
T-P	Reference	1.223	4 months	4 months
	Impaired	0.842	-	-

3.3.2. Comparison of Stability through Rapidity

To understand the stability of stream ecosystem resilience, the recovery period of the WQSP according to rainfall and the rainfall period that showed a continuous significant relationship with rainfall were compared (Table 4).

Table 4. Rapidity results include the recovery period and results of the rainfall period, which show a significant relationship for stability comparison.

WQSP Classification		Recovery Period	Rainfall Period Showing a Significant Relationship in Succession	
			LM	GAM
BOD	Reference	11.14	-	1 month
	Impaired	13.5	3 months	3 months
T-N	Reference	11.43	1 month	2 months
	Impaired	13.17	3 months	3 months
T-P	Reference	11.0	2 months	2 months
	Impaired	11.43	-	-

The period between WQSP degradation and recovery was approximately two months faster in the reference streams for BOD and T-N, confirming that the reference stream has better rapidity. However, for T-P, the recovery periods of the reference and impaired streams were similar, so it was concluded that management of the T-P is important.

The rainfall period that showed a significant relationship in LM and GAM in the reference streams was one to two months, and that in the impaired streams was three to four months. Therefore, the reference streams were affected by rainfall for a shorter period than the impaired streams, which indicated that the reference streams recovered more quickly from the effects of rainfall than the impaired streams and provided a more stable stream environment. Therefore, the reference streams have higher stability.

4. Discussion

4.1. Nonlinearity of WQSP for Reference and Impaired Streams

In the BOD and T-N of the reference streams and the T-N of the impaired streams, the relationship between the following rainfall and WQSP is generally nonlinear, and the smooth function for nonlinearity can be divided into three regions (Figure 7). Regions 1 and 3 showed a positive relationship with rainfall in this study, and Region 2 showed a negative relationship. The positive relationship between rainfall and WQSP in Region 1 seemed to be because the concentrations of BOD and T-N in the dry season remained high, but the effect of pollutant dilution due to rainfall was shown [25]. Kang et al. [26] showed similar results: the high water quality during the dry season decreased due to precipitation and runoff. However, for the BOD and T-N of the streams in Region 2, rainfall and WQSP showed a negative relationship, which is thought to be because nonpoint pollution from the watershed flows into the stream along with runoff due to the increased rainfall. According to Won et al. [27], since forests have a high soil penetration ability, runoff does not occur with low rainfall levels, and rainfall runoff increases as precipitation increases. Lee and Lee [28] also confirmed that T-N can be absorbed into the soil; thus, runoff containing T-N does not occur until a rainfall of 50 mm is reached. Therefore, in Region 2, where rainfall increased due to the absorption characteristics of T-N and the permeability of the forest area of the reference streams, the inflow of nonpoint pollution increased, showing a negative relationship.

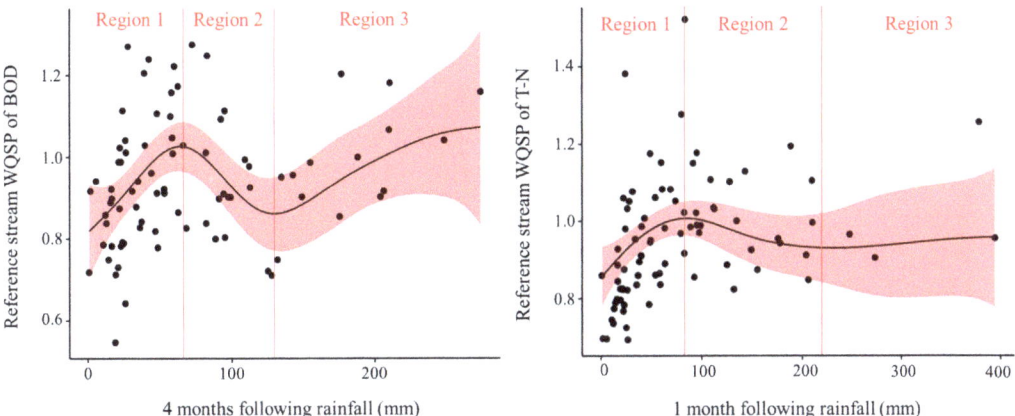

Figure 7. Smooth functions and classification regions by smooth function patterns. Smooth functions of the relationship between the WQSP of reference streams for BOD and four months following rainfall (**left**); smooth functions of the relationship between the WQSP of reference streams for T-N and one month following rainfall (**right**). The dot represents the system performance value, and the red area represents the confidence interval.

4.2. Robustness Comparison of Reference and Impaired Streams

To compare the robustness and rapidity of the reference and impaired streams, the relationships between rainfall and WQSP, variability, and recovery period were examined, and it was determined that the impaired streams had lower resilience. The difference in resilience between the reference and impaired streams concerning rainfall was considered to be due to the permeability and runoff of the watershed, according to its land cover. To identify the difference in the land cover between the reference and impaired streams, a t-test was conducted on the land cover proportions within a 1 km buffer of the reference and impaired streams (Table 5). The reference streams had the highest proportion of forest area, and the impaired streams had the most urban area. As a result of the t-test, a difference in land cover was found for urban, forest, and grassy areas for the reference and impaired streams.

This difference in land cover affects the permeability of the soil and the rainfall–runoff and runoff rate [29–32]. Urban areas are highly impermeable and respond more quickly to rainfall due to the low penetration of rainfall, resulting in massive amounts of runoff into streams [33–35]. By modeling watershed imperviousness, runoff, and peak discharge, Huang et al. [36] and Braud et al. [37] confirmed that the higher the impermeability is, the faster the peak discharge is reached and the greater the amount of runoff. In addition, it was confirmed that a large amount of runoff from rainfall dilutes the T-N, T-P, BOD, and COD concentrations of streams that maintain high levels during the dry season [26]. Therefore, the impaired streams have a high land cover in the urban area, so there is a large outflow at once during rainfall, and this outflow water quickly dilutes the stream concentration, which is judged to have a large and steep increase in the WQSP of the impaired streams.

Table 5. *t*-test results for land cover of reference and impaired streams.

Classification	Average		Standard Deviation		*t*-Value
	Reference	Impaired	Reference	Impaired	
Urban area	8.14	49.37	8.22	22.12	7.306 **
Agricultural area	19.02	20.10	10.95	20.78	0.195
Forest area	55.30	11.85	19.08	11.92	−8.219 **
Grassy area	3.25	6.93	3.43	3.53	3.278 **
Wetland	3.65	1.16	4.86	1.70	−2.013
Bare soil	4.67	3.43	3.79	3.16	−1.091
Water	5.97	7.16	4.83	9.50	0.510

** $p < 0.01$.

However, in the dry season after rainfall, the water quality is polluted due to nonpoint pollution continuously flowing out from the urban and agricultural areas around the streams, and the WQSP of the impaired streams, which increased during rainfall, drops sharply and is judged to have a high fluctuation range [33,38]. Additionally, impaired streams have a short time lag for rainfall because they are near highly impermeable urban areas, and it takes very little time for runoff to reach these streams [31,32]. On the other hand, in the case of the reference streams, the water permeability and penetration rate are high due to the forest cover that is dominant in the area, so the peak flow is alleviated such that the increase in WQSP for rainfall is relatively low and the time lag is longer than that for the impaired streams [39,40]. Additionally, since the proportions of urban and agricultural areas are small, the inflow of nonpoint pollution in the dry season is less than that of impaired streams, so the fluctuation range of WQSP is not large [41,42].

5. Conclusions

To compare the sensitivity and stability of resilience according to the aquatic ecological condition of the stream ecosystem in the Han River watershed, this study identified the rapidity and robustness of the BOD, T-N, and T-P WQSP of reference and impaired streams. The rapidity and robustness of the reference and impaired streams were derived from the time lag for rainfall and the rainfall period, showing a significant relationship between WQSP variability and the recovery period. The findings of this study suggested that the reference streams were less sensitive to rainfall than the impaired streams and provided a more stable ecosystem and, thus, had better resilience. Our research has elucidated that reference streams exhibit markedly lower sensitivity to rainfall variability when compared to their impaired counterparts, resulting in enhanced ecosystem stability and resilience. The diminished resilience of impaired streams calls for strategic management interventions to mitigate their heightened vulnerability to precipitation and reinforce their structural and ecological integrity. Particularly, impaired streams exhibit heightened fluctuations in resilience attributed to urban runoff, displaying swift responses to rainfall; this necessitates measures to mitigate rapid runoff and nonpoint pollution inflow into streams. Proactive strategies, including the establishment of robust waterside vegetation, the creation of small-scale wetlands, and the integration of retention ponds, are recommended to bolster the resilience of these streams. The findings of this study furnish foundational insights for the formulation of comprehensive management plans. However, there is a limitation in that flow data and soil permeability data that can confirm the dilution effect by rainfall are lacking, so further analysis of changes in WQSP considering the flow rate and geological effect is required.

Author Contributions: Conceptualization: Y.P., J.L. and S.-W.L. Data curation and software: Y.P. Formal analysis and writing—original draft: Y.P. Writing—review and editing: Y.P., J.L., S.-R.P. and S.-W.L. All authors have read and agreed to the published version of the manuscript.

Funding: This study received no external funding.

Data Availability Statement: Data will be made available on request.

Acknowledgments: This study was supported by Konkuk University in 2022.

Conflicts of Interest: The authors declare no conflict of interest.

References

1. An, Z.; Wu, G.; Li, J.; Sun, Y.; Liu, Y.; Zhou, W.; Cai, Y.; Duan, A.; Li, L.; Mao, J. Global monsoon dynamics and climate change. *Annu. Rev. Earth Planet. Sci.* **2015**, *43*, 29–77. [CrossRef]
2. Lee, J.J.; Kim, Y.J. Analysis of flow duration characteristics due to environmental change in Korea river basin. *J. Korean Soc. Hazard. Mitig.* **2011**, *11*, 67–75. [CrossRef]
3. Kim, J.S.; Park, S.Y.; Hong, H.P.; Chen, J.; Choi, S.J.; Kim, T.W.; Lee, J.H. Drought risk assessment for future climate projections in the Nakdong River Basin, Korea. *Int. J. Climatol.* **2020**, *40*, 4528–4540. [CrossRef]
4. Matthews, W.J.; Marsh-Matthews, E. Effects of drought on fish across axes of space, time and ecological complexity. *Freshw. Biol.* **2003**, *48*, 1232–1253. [CrossRef]
5. Wilkinson, C.L.; Yeo, D.C.; Tan, H.H.; Hadi Fikri, A.; Ewers, R.M. Resilience of tropical, freshwater fish (*Nematabramis everetti*) populations to severe drought over a land-use gradient in Borneo. *Environ. Res. Lett.* **2019**, *14*, 045008. [CrossRef]
6. Ching, Y.C.; Lee, Y.H.; Toriman, M.E.; Abdullah, M.; Yatim, B.B. Effect of the big flood events on the water quality of the Muar River, Malaysia. *Sustain. Water Resour. Manag.* **2015**, *1*, 97–110. [CrossRef]
7. Son, M.W. Influences of An Extreme Flood on Habitual Environment of Aquatic Ecosystem of Urban Stream. *J. Korean Assoc. Reg. Geogr.* **2008**, *14*, 105–113.
8. Ministry of Environment (MOE). *Aquatic Ecosystem Reference Stream Suitability Assurance and Utilization Research*; Korean Literature; Ministry of Environment: Incheon, Republic of Korea, 2016.
9. Folke, C. Resilience: The emergence of a perspective for social-ecological systems analyses. *Glob. Environ. Chang.* **2006**, *16*, 253–267. [CrossRef]
10. Holling, C.S. Resilience and stability of ecological systems. *Annu. Rev. Ecol. Evol. Syst.* **1973**, *4*, 1–23. [CrossRef]
11. Timmerman, P. Vulnerability, Resilience, and the Collapse of Society. In *Environmental Monograph 1*; Institute for Environmental Studies, Toronto University: Toronto, ON, Canada, 1981.
12. Korea Research Institute for Human Settlements (KRIHS). *Research on Regional Resilience Improvement Coping with Flooding Disaster by Climate Change Effect (I) -Development and Application of Resilience Assessment Methods-*; Korean Literature; Korea Research Institute for Human Settlements: Seojong, Republic of Korea, 2016.
13. Korea Land and Housing Corporation (LH). *A Research on Urban Resilience for Urbam Regeneration*; Korean Literature; Land and Housing Institute: Jinju, Republic of Korea, 2015.
14. Kang, S.J.; Jo, S.H.; Hong, S.Y. *A Policy Implication for Community Resilience from Natural Disasters*; Gyeonggi Research Institute: Suwon, Republic of Korea, 2013.
15. Tran, H.T.; Balchanos, M.; Domerçant, J.C.; Mavris, D.N. A framework for the quantitative assessment of performance-based system resilience. *Reliab. Eng. Syst. Saf.* **2017**, *158*, 73–84. [CrossRef]
16. Vasistha, P.; Ganguly, R. Water quality assessment of natural lakes and its importance: An overview. *Mater. Today Proc.* **2020**, *32*, 544–552. [CrossRef]
17. Wang, F.; Wang, Y.; Zhang, K.; Hu, M.; Weng, Q.; Zhang, H. Spatial heterogeneity modeling of water quality based on random forest regression and model interpretation. *Environ. Res.* **2021**, *202*, 111660. [CrossRef] [PubMed]
18. Ministry of Environment (MOE); National Institute of Environmental Research (NIER). *Waterwide Aquatic Ecological Monitoring Program (V)*; Korean Literature; Ministry of Environment and National Institute of Environmental Research: Incheon, Republic of Korea, 2012.
19. National Institute of Environment Research (NIER). *A Study on Reference Stream Criteria Enhancing the Applicability of River Ecosystem Health Assessment*; Korean Literature; National Institute of Environmental Research: Incheon, Republic of Korea, 2015.
20. Hashimoto, T.; Stedinger, J.R.; Loucks, D.P. Reliability, resiliency, and vulnerability criteria for water resource system performance evaluation. *Water Resour. Res.* **1982**, *18*, 14–20. [CrossRef]
21. Simonovic, S.P.; Peck, A. Dynamic resilience to climate change caused natural disasters in coastal megacities quantification framework. *Int. J. Environ. Clim. Chang.* **2013**, *3*, 378–401. [CrossRef] [PubMed]
22. Lee, J.H. *Stream Engineering*; Korean Literature; Goomiseokwan: Seoul, Republic of Korea, 2018.

23. Choi, Y.; Kim, Y.; Kim, S.; Kim, M. Management of Ponding Depth and Discharge Filtration from Paddy Fields for Controlling Nonpoint Source Pollution. *J. Korean Soc. Agric. Eng.* **2015**, *57*, 125–130.
24. Park, S.M.; Shin, Y.K. The Impact of Monsoon Rainfall on the Water Quality in the Upstream Watershed of Southern Han River *Korean J. Ecol. Environ.* **2011**, *44*, 373–384.
25. Song, I.-H.; Kang, M.-S.; Hwang, S.-H.; Song, J.-H. Characteristics and EMCs of NPS pollutants runoff from a forest-paddy composite watershed. *J. Korean Soc. Agric. Eng.* **2012**, *54*, 9–17. [CrossRef]
26. Kang, S.; An, K. Spatio-temporal variation analysis of physico-chemical water quality in the Yeongsan-river watershed. *J. Ecol. Environ.* **2006**, *39*, 73–84.
27. Won, C.; Choi, Y.; Seo, J.; Kim, K.; Shin, M.; Choi, J. Determination of EMC and unit loading of rainfall runoff from forestry-crops field. *J. Korean Soc. Water Environ.* **2009**, *25*, 615–623. Available online: https://www.koreascience.or.kr/article/JAKO200910103491724.page (accessed on 13 November 2023).
28. Lee, M.; Young-shin, L. A study on runoff characteristics of nonpoint pollutant with rainfall intensity—A case of fowls manure. *J. Wetl. Res.* **2009**, *11*, 91–97.
29. Liu, L.; You, X. Water quality assessment and contribution rates of main pollution sources in Baiyangdian Lake, northern China *Environ. Impact Assess. Rev.* **2023**, *98*, 106965. [CrossRef]
30. Sun, D.; Wang, X.; Yu, M.; Ouyang, Z.; Liu, G. Dynamic evolution and decoupling analysis of agricultural nonpoint source pollution in Taihu Lake Basin during the urbanization process. *Environ. Impact Assess. Rev.* **2023**, *100*, 107048. [CrossRef]
31. Ten Veldhuis, J.A.E.; Skovgard Olsen, A. Hydrological response times in lowland urban catchments characterised by looped drainage systems. In Proceedings of the 9th International Workshop on Precipitation in Urban Areas: Urban Challenges in Rainfall Analysis, UrbanRain 2012, St Moritz, Switzerland, 6–9 December 2012.
32. Walsh, C.J.; Roy, A.H.; Feminella, J.W.; Cottingham, P.D.; Groffman, P.M.; Morgan, R.P. The urban stream syndrome: Current knowledge and the search for a cure. *J. N. Am. Benthol. Soc.* **2005**, *24*, 706–723. [CrossRef]
33. Miller, J.D.; Kim, H.; Kjeldsen, T.R.; Packman, J.; Grebby, S.; Dearden, R. Assessing the impact of urbanization on storm runoff in a peri-urban catchment using historical change in impervious cover. *J. Hydrol.* **2014**, *515*, 59–70. [CrossRef]
34. McGrane, S.J. Impacts of urbanisation on hydrological and water quality dynamics, and urban water management: A review. *Hydrol. Sci. J.* **2016**, *61*, 2295–2311. [CrossRef]
35. Orta-Ortiz, M.S.; Geneletti, D. What variables matter when designing nature-based solutions for stormwater management? A review of impacts on ecosystem services. *Environ. Impact Assess. Rev.* **2022**, *95*, 106802. [CrossRef]
36. Huang, H.; Cheng, S.; Wen, J.; Lee, J. Effect of growing watershed imperviousness on hydrograph parameters and peak discharge *Hydrol. Process.* **2008**, *22*, 2075–2085. [CrossRef]
37. Braud, I.; Breil, P.; Thollet, F.; Lagouy, M.; Branger, F.; Jacqueminet, C.; Kermadi, S.; Michel, K. Evidence of the impact of urbanization on the hydrological regime of a medium-sized periurban catchment in France. *J. Hydrol.* **2013**, *485*, 5–23. [CrossRef]
38. Lee, S.-H.; Bae, S.-K. Long-term trend of groundwater recharge according to urbanization. *J. Environ. Sci.* **2010**, *19*, 779–785 [CrossRef]
39. Burges, S.J.; Wigmosta, M.S.; Meena, J.M. Hydrological effects of land-use change in a zero-order catchment. *J. Hydrol. Eng.* **1998**, *3*, 86–97. [CrossRef]
40. Hümann, M.; Schüler, G.; Müller, C.; Schneider, R.; Johst, M.; Caspari, T. Identification of runoff processes—The impact of different forest types and soil properties on runoff formation and floods. *J. Hydrol.* **2011**, *409*, 637–649. [CrossRef]
41. Gang, D.H.; Na, J.H.; Park, I.H.; Kim, J.H.; Gwon, O.S.; Lee, S.J. Ecosystem service valuation of forests using sewage reservoir calculation equation. In Proceedings of the Korean Institute of Landscape Architecture Conference, Mokpo, Korea, 12–14 October 2016
42. Neary, D.G.; Ice, G.G.; Jackson, C.R. Linkages between forest soils and water quality and quantity. *For. Ecol. Manag.* **2009**, *258*, 2269–2281. [CrossRef]

Disclaimer/Publisher's Note: The statements, opinions and data contained in all publications are solely those of the individual author(s) and contributor(s) and not of MDPI and/or the editor(s). MDPI and/or the editor(s) disclaim responsibility for any injury to people or property resulting from any ideas, methods, instructions or products referred to in the content.

Article

Delineation of Urban Development Boundary and Carbon Emission Effects in Xuzhou City, China

Haitao Ji [1], Xiaoshun Li [1,2], Yiwei Geng [1], Xin Chen [3,4,*], Yuexiang Wang [1], Jumei Cheng [1] and Zhuang Chen [1]

1. School of Public Policy and Management, China University of Mining and Technology, Xuzhou 221116, China; jihaitao@cumt.edu.cn (H.J.); lxshun@cumt.edu.cn (X.L.); gengywei@cumt.edu.cn (Y.G.); wyx@hytc.edu.cn (Y.W.); ts21090050a31ld@cumt.edu.cn (J.C.); tb17160024b0@cumt.edu.cn (Z.C.)
2. Research Center for Transition Development and Rural Revitalization of Resource-Based Cities in China, China University of Mining and Technology, Xuzhou 221116, China
3. Ministry of Education Key Laboratory for Earth System Modeling, Department of Earth System Science, Tsinghua University, Beijing 100084, China
4. Institute of Loess Plateau, Shanxi University, Taiyuan 030006, China
* Correspondence: xinchenthu@mail.tsinghua.edu.cn; Tel.: +86-0516-8359-1322

Abstract: Urban development boundary (UDB) has always served as a crucial aspect of urban sprawl research. The objective of this paper is to investigate boundary delineation and carbon emission effects. Firstly, we examined the patterns of land use changes. Additionally, this paper utilized the FLUS model and land use carbon emission calculation model to delineate UDB and calculate carbon emission effects within UDB under typical scenarios. The research results are as follows: (1) Xuzhou city witnessed a significant increase in both forest land and construction land from 2010 to 2020. (2) Under the CPS, the area of cultivated land increased by 217.05 km^2 compared to the NDS. The UDB area under the NDS and CPS was 971.50 km^2 and 968.99 km^2, respectively. (3) Compared to the NDS, the CPS led to a net carbon emission increase of 4759.93 t within the UDB. Therefore, we should enhance the carbon sequestration and emission reduction capacity of the agricultural system. This study is beneficial for expanding the depth of research on the UDB and guiding the low-carbon urban development. We sincerely encourage readers to download this paper to improve this paper.

Keywords: urban development boundary; carbon emission effects; Xuzhou city; FLUS model

1. Introduction

Since the Industrial Revolution, urbanization has been accelerating globally [1]. Indeed, urban land expansion was occurring at a much faster rate than population growth [2], with China being particularly pronounced in this aspect. Since the implementation of economic reform and opening up, the urbanization rate in China has significantly increased by 3.6 times from 1978 to 2022. The construction area has also expanded from 0.7 × 10^4 km^2 in 1981 to 6.2 × 10^4 km^2 in 2021. Unplanned urban expansion has brought significant social and environmental challenges on a global scale [3]. Many cities are experiencing uncontrolled land development and excessive sprawl of urban land, resulting in a severe imbalance in urban development patterns. This has led to various issues such as biodiversity loss, deterioration of natural environment, increased air pollution, soil degradation, and a reduction in cultivated land [4–6]. Therefore, there is growing recognition for the need to control the uncontrolled and rapid expansion of cities and guide their orderly expansion, aiming for sustainable urban spatial development. As a result, the urban development boundary (UDB) has become a crucial measure for regulating urban sprawl and achieving optimized urban spatial planning. At the same time, it is particularly noteworthy that rapid and chaotic urban expansion caused an increase in greenhouse gas emission, which is impacting climate change [7] and accelerating global warming [8], posing a great threat to both the natural world and humanity [9]. A great deal of global

carbon dioxide emissions can be attributed to urban areas, and this proportion might rise further with ongoing urbanization [10]. At the same time, China has already surpassed other countries to become the world's largest annual emitter of carbon dioxide [11]. In order to address this urgent situation and achieve the coordinated pattern between urban expansion and low-carbon development, China has announced its commitment to strive for dual carbon goal [12]. This solemn commitment imposes new requirements on the urban sprawl pattern and management. Therefore, integrating the UDB with carbon emission effects and systematically exploring the carbon emission effects after the delineation of the UDB is crucial. This is an important approach to reconcile the conflicting goals of urban sprawl and low-carbon development, and to address the crucial issue and prerequisite of regulating the orderly and low-carbon sprawl of urban space.

Through reading a large amount of literature, we found that many existing studies primarily concentrate on defining the concept of UDB and delineating it through simulation. With regard to the conceptual definition of the UDB, the concept of urban development boundary (UDB) originated from urban growth boundary (UGB) and has been widely recognized as one of the most useful tools globally to manage and restrict urban sprawl [13]. It was derived from the "Garden City" theory proposed by European scholar Howard in the 19th century. The research on the conceptual definition of UDB can generally be divided into two primary areas. On the one hand, from the physical boundary perspective, it is considered that the UDB is seen as limits or restrictions that determine where urban settlements should not exceed [14]. It can also refer to the boundaries where there is a clear distinction in construction density between inside and outside the boundary [15]. On the other hand, from the practical function perspective, it is considered that the UDB is effective for controlling the urban sprawl and protecting non-urban land beyond the city's territory [16,17], which is conducive to the coordination of the conflicts between land use plans [18]. Building on these studies, this paper considered UDB as regional limit that guides and regulates urban development within a specific time period, particularly for urban development and construction.

With regard to the simulation delineation of UDB, the first delineation of UDB occurred in Salem, USA, in 1976, as a means to address conflicts between urban and rural land management [19]. In China, the content regarding the delineation of UDB in central urban areas first appeared in the "Urban Planning Compilation Methods" implemented on 1 April 2006 [20]. Upon reviewing the existing literature, we found that the methods and criteria for delineating urban development boundary have not been fully clarified and standardized, which can be mainly divided into two primary areas. On the one hand, from the ecological security perspective, methods such as ecological suitability assessment, resource and environmental carrying capacity assessment, or landscape safety pattern establishment [21,22] are employed to preliminarily determine ecologically fragile areas (such as restricted or prohibited development zones), which in turn influence the scope of urban sprawl. On the other hand, from the inherent urban sprawl perspective, models such as the ANN model [23], CA model [24], CLUE-S model [25], SLEUTH model [26], and the FLUS model [27] are utilized to simulate urban expansion in terms of construction land spatial growth.

From the above literature review, we concluded that scholars dedicated significant research efforts to conceptualizing and simulating the delineation of the UDB [28,29]. However, the impact and effects after the boundary delineation have received limited attention in the existing literature [30], particularly regarding carbon emission effects. Therefore, considering the needs of low-carbon development in the current context and the existing research shortcomings, this study aims to carry out the research on the carbon emission effect after the UDB delineation, which is conducive to deepening the research content of UDB and regulating the orderly and low-carbon expansion of urban space.

In view of this, this paper focused on Xuzhou City and examined the land use changes from 2010 to 2020. Simultaneously, by setting the NDS and the CPS, we utilized the FLUS model to delineate the UDB under different scenarios. Furthermore, we calculated the carbon emission effects within UDB using the land use carbon emission calculation model. This study is beneficial for enriching the research on UDB, providing strategic recommendations for constraining urban expansion in Xuzhou City and contributing to the achievement of low-carbon development and the "dual carbon" goal.

2. Materials and Methods

2.1. Study Area

Xuzhou is located in Jiangsu Province, China, and is situated in the Huang Huai Plain. It is divided into five municipal districts, namely Tongshan, Jiawang, Quanshan, Gulou, and Yunlong (Figure 1). The city boasts diverse landforms, expansive plains, and abundant natural, cultural, and tourist resources. As a major regional center and a resource-exhausted city [31], Xuzhou is the central city of the Huaihai Economic Zone. It acts as a national comprehensive transportation hub, connecting the railroad network in all directions. Xuzhou is a significant old industrial base, energy supply base, and power transmission and distribution base in China. With strong momentum in urbanization, the city is experiencing rapid development. From 2000 to 2020, the built-up area of Xuzhou City experienced a significant expansion, increasing from 71.7 km^2 to 289.6 km^2, experiencing a fourfold increase. This growth propelled Xuzhou City to become the fourth largest city in Jiangsu Province, following Nanjing, Suzhou, and Wuxi.

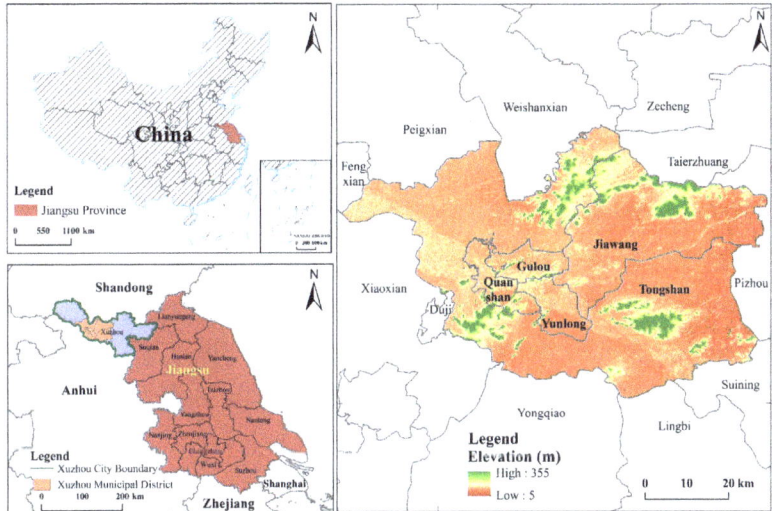

Figure 1. Location map of the study area.

2.2. Data and Preprocessing

Considering the physical conditions of the study area and accessibility of the research date, this paper mainly utilized data from four parts and the details are shown in Table 1. The land use data were reclassified according to the Current Land Use Classification (GB/T 21010-2017) into seven types, namely cultivated land (L1), garden land (L2), forest land (L3), grass land (L4), construction land (L5), water area (L6), and other land (L7). All the driving factor data underwent normalization processing (Figure 2). The projection coordinate system used was WGS_1984_UTM_Zone_50N. The resolution of all grid data was uniformly set at 30 × 30 m, with consistent row and column numbers. It is worth noting that the land use data were provided by the China Land Surveying and Planning Institute, which is an official government department in China. The terrain factor data were sourced from Geospatial Data Cloud, a big data platform created by the Chinese Academy of Sciences. Therefore, the accuracy of the land use data and terrain factor data can be guaranteed.

Table 1. Information on data sources.

Data Attribute	Data Name	Data Source
Land use data	Land use data of Xuzhou in 2010, 2015, and 2020	China Land Surveying and Planning Institute
Terrain factor data	DEM Aspect Slope	Geospatial Data Cloud (https://www.gscloud.cn/, accessed on 18 April 2022)
Transportation accessibility factor data	To town To airport To highway To railway station To city To waterway To water To railway To main road	Open Street Map (https://www.openstreetmap.org/, accessed on 17 April 2022)
Socio-economic factor data	1 km × 1 km grid level GDP 1 km × 1 km grid level population	Resource and Environment Science and Data Center (https://www.resdc.cn/, accessed on 19 April 2022)

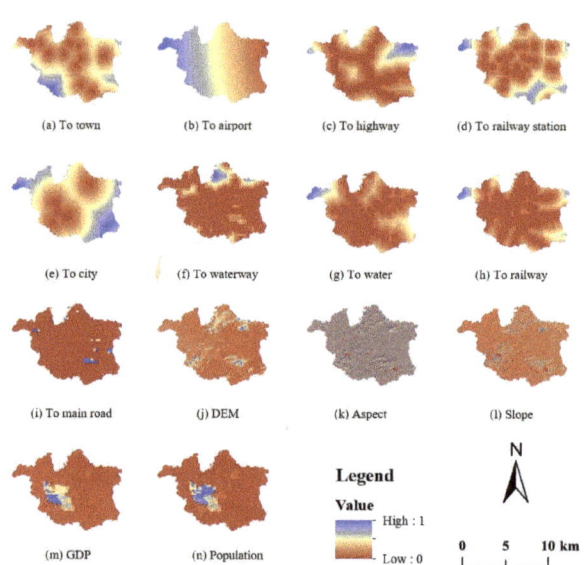

Figure 2. Land use driving factors.

2.3. Methodology

This paper utilized methods such as the land use transfer matrix, land use dynamic degree, FLUS model, and the land use carbon emission calculation model. We analyzed the land use changes, delineated the UDB, and explored the carbon emission effects within the UDB under different scenarios. To present the research methodology and approach in more clear manner, this paper incorporated a flowchart illustrating the framework of the research methodology (Figure 3).

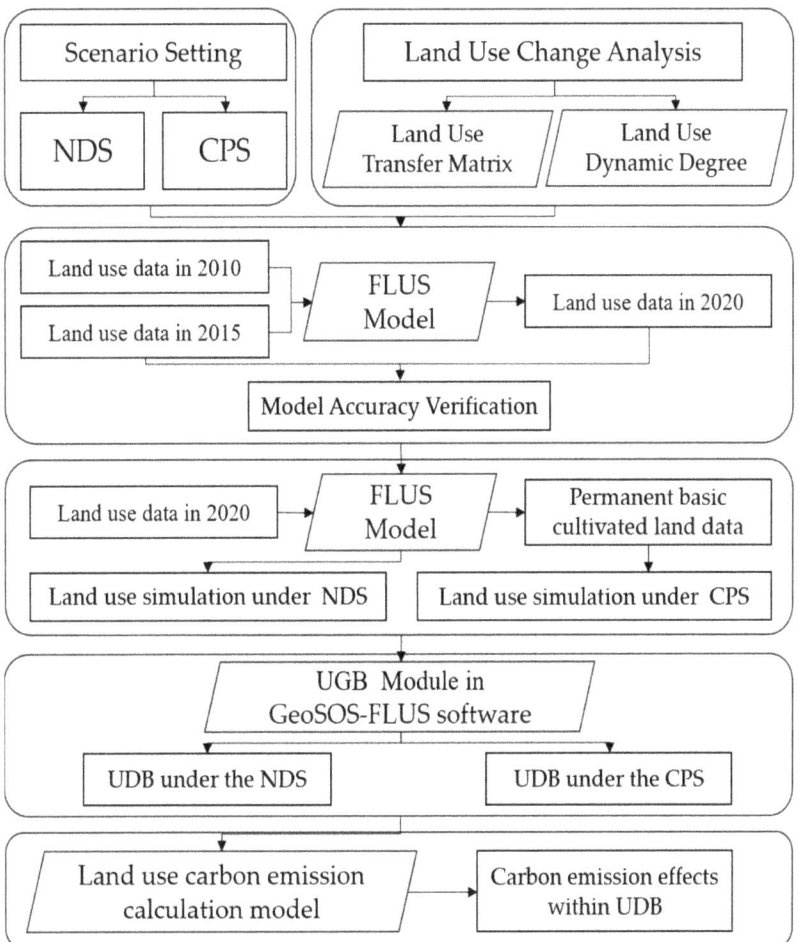

Figure 3. The flowchart of the methodology.

2.3.1. Scenario Setting

(1) Scenario 1: Natural Development Scenario (NDS). The basis for setting the NDS is that it does not consider artificial planning control, protected areas, or limitations on the conversion between different types. It solely focuses on the development of cities within the framework of the existing natural conditions.

(2) Scenario 2: Cultivated Land Protection Scenario (CPS). The basis for setting the CPS was conducted by introducing permanent basic cultivated land data as a restricting conversion factor into the model for the simulation, and strictly prohibited the situation that permanent basic cultivated land converts to other types.

2.3.2. Methodology for Land Use Change Analysis

(1) Land Use Transfer Matrix

The land use transfer matrix can describe the conversion between different types during a specific period [32], and has been widely used in many countries such as China [33],

Japan [34], Rwanda [35], and Turkey [36]. It can indicate the transfer and changes during the study period [37]. The calculation formula is as follows:

$$S_{ij} = \begin{bmatrix} S_{11} & S_{12} & S_{13} & \cdots & S_{1n} \\ S_{21} & S_{22} & S_{23} & \cdots & S_{2n} \\ S_{31} & S_{32} & S_{33} & \cdots & S_{3n} \\ \vdots & \vdots & \vdots & \vdots & \vdots \\ S_{n1} & S_{n2} & S_{n3} & \cdots & S_{nn} \end{bmatrix} \quad (1)$$

where S_{ij} represents the convert area of land use type i to type j; n represents the number of land use types.

(2) Land Use Dynamic Degree

Land use dynamic degree can reflect the speed and magnitude of changes in various types over a certain period of time [38,39], and has been widely used in many countries such as China [40], Greece [41], and India [42]. This includes both comprehensive and single land use dynamic degrees. The former refers to the speed of change in all types over a certain study period, while the latter refers to the changes in a specific type. The calculation formula is as follows:

$$K = (\sum_{i=1}^{n} \Delta U_{i-j}/2 \sum_{i}^{n} U_{it1}) \times (t_2 - t_1)^{-1} \times 100\% \quad (2)$$

$$K_i = ((U_{it2} - U_{it1})/U_{it1}) \times (t_2 - t_1)^{-1} \times 100\% \quad (3)$$

where K represents the comprehensive land use dynamic degree; K_i represents the single land use dynamic degree of the i-th land use type; $t_2 - t_1$ represent the study period; ΔU_{i-j} represents the changed area.

2.3.3. FLUS Model

The FLUS model, developed by Liu et al. in 2017 [43], is a simulation model for predicting future land use changes. In 2018, Liang et al. further expanded the model by adding the UGB (urban growth boundary) module [44], which enables effective land use simulation and UDB delineation. Due to the advantages of higher accuracy, faster processing speed, and ease of use, the FLUS model has gained widespread use in many countries such as China [45], Myanmar [46], Turkey [47], and Colombia [48]. The FLUS model consists of three components: (1) Probability-of-occurrence estimation using artificial neural network; (2) Cellular automata based on self-adaptive inertia and competition mechanism; (3) Boundary delineation based on morphological erosion and dilation. For detailed formulas, please see the references [43,44].

2.3.4. Calculation of Land Use Carbon Emission

Land use carbon emission can be divided into direct carbon emission and indirect carbon emission [49], and has been widely used in many countries such as China [50] and India [51]. Direct carbon emission from land use refers to those generated by human activities on the land. The calculation formula is as follows:

$$E = \sum T_i \times f_i \quad (4)$$

where E represents the direct carbon emission from land use; i represents the land use type; T_i represents the area; f_i represents the carbon emission coefficient. Due to the inherent characteristics of L1, L2, L3, L4, L6, and L7, their carbon emission and absorption rates change minimally over a certain period. Therefore, for the study period, the carbon emission coefficient for L1, L2, L3, L4, L6, and L7 is assumed to be constant. Based on previous research findings [52,53], the coefficients for L1, L2, L3, L4, L6, and L7 are shown in Table 2.

Table 2. Carbon emission coefficient for land use.

Land Use Type	Carbon Emission Coefficient	Unit
L1	42.2	t C/km^2
L2	−73	t C/km^2
L3	−57.8	t C/km^2
L4	−2.1	t C/km^2
L6	−25.2	t C/km^2
L7	−0.5	t C/km^2

Indirect carbon emission from land use primarily refers to the carbon emission associated with construction land. These emissions are estimated indirectly by considering the consumption of different energy types [54]. The calculation formula is as follows:

$$CE = \sum_{i=1}^{n} A_i \times B_i \times C_i \tag{5}$$

where CE represents the carbon emissions from construction land; i, n, A_i, B_i, C_i represent the energy source type, the number of energy source types, the consumption, the carbon emission coefficient, and the coal equivalent coefficient, respectively. The coal equivalent coefficients are sourced from the "China Energy Statistical Yearbook 2013", which lists the coal equivalent coefficients for various energy source types. Due to the study area being the municipal district of Xuzhou City, it is difficult to obtain comprehensive and accurate data on energy consumption. Therefore, this paper indirectly estimated the carbon emission from construction land in the municipal district of Xuzhou City by using the ratio of GDP between the municipal district and the entire city of Xuzhou [55]. Based on the result from construction land in Xuzhou City from 2005 to 2020, the GM(1,1) model was used for grey prediction, combined with the projected construction land area in Xuzhou City in 2035, yielding a coefficient value of approximately 15,468.33 t C/km^2 for the construction land in Xuzhou City in 2035.

3. Results

3.1. Analysis of the Status of Land Use Change

3.1.1. Changes in Land Use Area

Firstly, we reclassified the original land use data of Xuzhou City for the periods of 2010, 2015, and 2020 into seven types (Figure 4). From Figure 4, we found that the land use types in Xuzhou are predominantly composed of L1, L5, and L6. In the year 2020, these three types collectively constituted around 84.58% of the total area. On the other hand, L3, L2, L7, and L4 were relatively small in area, comprising only 15.42%. The land use area in Xuzhou experienced different degrees of change, leading to a transformation in the whole structure. The areas of L1, L7, L6, L4, and L2 decreased by 131.40 km^2, 26.66 km^2, 23.52 km^2, 10.41 km^2, and 0.69 km^2, respectively. However, L3 and L5 increased by 142.31 km^2 and 50.37 km^2, respectively. Among them, L3 increased most prominently with a growth rate of 127.39% over the 10-year period. On the other hand, there was a dramatic decrease in the area of L7, with a decrease of 32.55%. The area of garden land slightly decreased, but overall, the change was the least significant compared to other types.

3.1.2. Changes in Land Use Type Transition

From Table 3, we found that from 2010 to 2020, Xuzhou City experienced significant land use transitions, especially with frequent conversions among L1, L3, and L5. In terms of land area transferred out, the types with the largest to smallest transfer amounts were as follows: L1 > L5 > L6 > L2 > L7 > L4 > L3. Cultivated land was the dominant type, accounting for 40.63% of the transferred-out area. On the other hand, in terms of land area transferred in, the types with the largest to smallest transfer amounts were as follows:

L1 > L5 > L3 > L6 > L > L7 > L4. Construction land was the main recipient, accounting for 25.65% of the transferred-in area.

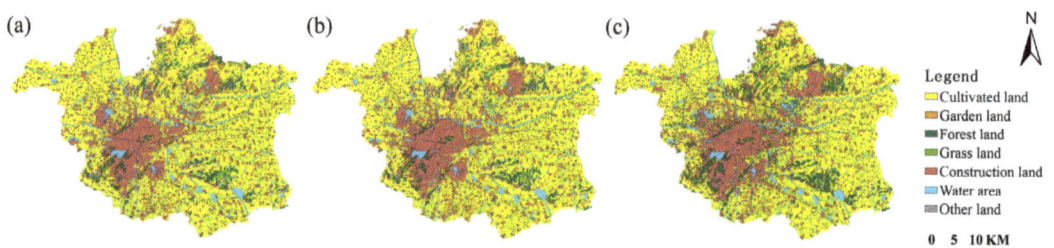

Figure 4. The land use types in Xuzhou City in 2010, 2015, and 2020. (a) Land use types in Xuzhou City in 2010; (b) Land use types in Xuzhou City in 2015; (c) Land use types in Xuzhou City in 2020.

Table 3. Xuzhou City Land Use Transfer Matrix from 2010 to 2020 (km^2).

	L1	L2	L3	L4	L5	L6	L7	Total
L1	1148.95	50.12	65.50	3.62	144.96	76.64	24.64	1514.43
L2	40.40	27.95	10.39	0.98	13.38	7.50	2.70	103.29
L3	6.21	1.71	89.53	2.89	8.19	1.36	2.06	111.96
L4	12.88	2.96	18.04	8.02	7.29	2.10	1.40	52.68
L5	77.50	10.01	30.77	17.50	609.60	36.11	5.02	786.51
L6	79.64	6.31	11.68	3.28	44.29	161.62	3.00	309.82
L7	15.82	3.78	26.65	5.86	12.65	1.76	13.83	80.35
Total	1381.40	102.84	252.56	42.14	840.35	287.10	52.65	2959.03

In general, the land use transitions were primarily focused on L1, L5, and L3. Therefore, the analysis focused on these three types. Regarding L1, from 2010 to 2020, 24.13% of L1 had converted to other land types, with the majority of it being converted to L5. From L1 to L5, the transferred-out area reached 144.96 km^2. Regarding L5, a total of 230.75 km^2 of land from other land use types had converted to L5. This indicated a high level of urban expansion in Xuzhou City. Regarding L3, from 2010 to 2020, a total of 163.03 km^2 of land from other types had converted to L3. This is strongly related to the implementation of Xuzhou City's policy of promoting afforestation and artificial forestation.

3.1.3. Changes in Land Use Dynamic Degree

The land use dynamic degree is an indicator that reflects the speed of land use change Based on the Equations (2) and (3), we obtained the land use dynamic degree in Xuzhou (Table 4).

Table 4. Xuzhou City Land Use Dynamic Degree from 2010 to 2020 (%).

Period	Single Land Use Dynamic Degree							Comprehensive Land Use Dynamic Degree
	L1	L2	L3	L4	L5	L6	L7	
2010–2015	−0.30	−0.42	−0.36	−0.02	0.88	−0.58	0.34	0.24
2015–2020	−1.48	0.30	26.30	−3.88	0.38	−0.89	−6.74	1.09
2010–2020	−0.88	−0.07	12.74	−1.94	0.64	−0.72	−3.26	0.65

From the perspective of comprehensive land use dynamic degree, the speed of land use change in Xuzhou was constantly changing. From 2010 to 2015, the degree in Xuzhou was 0.24%, indicating a relatively slow rate. However, from 2015 to 2020, the degree in Xuzhou city increased to 1.09%. Compared to the period of 2010 to 2015, the rate of change in land use types exhibited an increasing trend, suggesting that socioeconomic activities had a greater influence on the land pattern during the period of 2015 to 2020. From the perspective of single land use dynamic degree, the various types of land use exhibit dynamic fluctuations in trends during the study period. Among them, forest land changed the most, with a dynamic degree of 12.74%, while the smallest change was in garden land, with an intense degree of −0.07%. Forest land and construction land demonstrated an upward trend, while other types decreased. Among them, other land experienced the most significant decline, showing a considerable decrease with a dynamic degree of −3.26%.

3.2. Delineation of UDB in Xuzhou City under Multiple Scenarios

3.2.1. Model Accuracy Verification

Firstly, we simulated the spatial distribution of different land use types in 2015 (Figure 5b) based on the date of 2010. Subsequently, compared to the actual data of 2015 (Figure 5a), it can be observed that the simulated results generally correspond to the distribution of different types in the actual data. Afterwards, we selected 10% of the total number of simulated results' pixels for accuracy verification. By calculating accuracy verification metrics, we obtained that the Kappa value is 0.7618 and the overall accuracy is 0.8427. This result suggested that FLUS model exhibited a high accuracy level in simulating urban land use changes. Therefore, we can utilize the model to simulate changes in various types in Xuzhou city, and it serves as an important foundational model for this study.

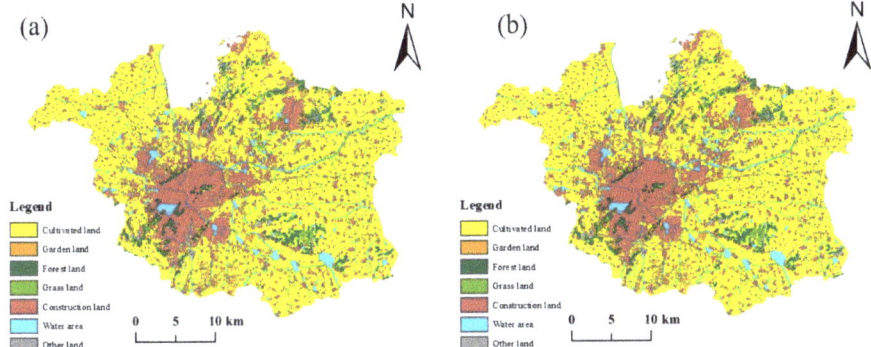

Figure 5. Comparison of real and simulated land use in Xuzhou City in 2015. (**a**) Real land use types in Xuzhou City in 2015; (**b**) Simulated land use types in Xuzhou City in 2015.

3.2.2. Future Land Use Simulation

Firstly, we estimated the probability-of-occurrence in 2020 based on the real data in 2020 of Xuzhou. Secondly, based on the land use data of 2015 and 2020 of Xuzhou City, we used the built-in Markov chain module in the GeoSOS-FLUS software v. 2.4.1. for determining the pixel count of each type in Xuzhou City in 2035. Finally, the FLUS model was applied to simulate the future land use in Xuzhou in 2035 under different scenarios (Figure 6).

It can be observed that L1 and L3 show significant changes under different scenarios. In the NDS, compared to the current area of Xuzhou City in 2020, there is a decreasing trend in L1 and L7. Specifically, the L1 area has decreased most, by 197.49 km². Moreover, the L3, L5, L4, and L2 show an increasing trend. Among them, L3 shows a particularly evident upward trend, with an increase in area of 191.70 km², representing a growth rate of 75.47%. The water area (L6) remained relatively stable. In the CPS, compared to the current area of Xuzhou City in 2020, there is a decreasing trend in L6, L2, L7, and L4. The reduction in water area is particularly significant, with a decrease of 30.08 km², representing a reduction rate of 9.98%. Moreover, the L1, L3, and L5 show an increasing trend. Specifically, the L1 area has increased by 19.56 km². It is worth emphasizing that, compared to the simulation results of the NDS in 2035, there is an additional increase of 217.05 km² in the

L1 area. This indicates that the CPS can effectively achieve the expected goal of cultivated land protection.

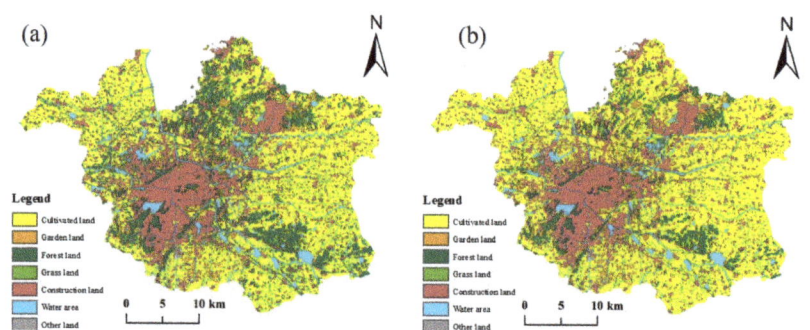

Figure 6. Simulation of land use in Xuzhou City in 2035. (**a**) Land use simulation under NDS; (**b**) Land use simulation under CPS.

3.2.3. Delineation of Urban Development Boundary (UDB)

The UDB is delineated based on the future land development status of the city. In this paper, we utilized the UGB module in the GeoSOS-FLUS software to delineate the UDB under different scenarios in Xuzhou City, using morphological erosion and dilation operations in the opening–closing mathematical operation method.

Through comprehensive comparison of the results obtained using $3 \times 3, 5 \times 5, 7 \times 7$, and 9×9 structuring elements, it was found that using the 7×7 structuring element for delineating the UDB results in a smoother and more continuous boundary. The boundary line can better integrate with the urban area contours. Therefore, a window size of 7×7 was chosen for boundary delineation. Additionally, considering the actual situation, we removed scattered patches (less than or equal to 0.4 km^2) and filled in small patches within large patches. The delineation results are shown in Figure 7.

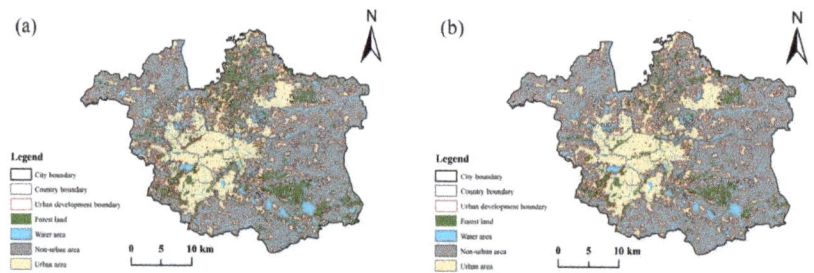

Figure 7. UDB under different scenarios in 2035. (**a**) UDB under the NDS; (**b**) UDB under the CPS.

From Figure 7, it can be observed that the UDB of Xuzhou in 2035 demonstrates a strong concentration and integrity. The main distribution is in Tongshan District and Jiawang District, while the UDB areas covered by the Yunlong District, Quanshan District, and Gulou District were quite similar. Under the NDS, the UDB area of Xuzhou City was 971.50 km^2, accounting for 32.82% of the total area of Xuzhou. Under the CPS, the UDB area of Xuzhou City was 968.99 km^2, accounting for 32.74% of the total area of Xuzhou City. The UDB area under the CPS was reduced by 2.51 km^2 compared to the NDS. This indicated that the CPS can better control urban land sprawl, resulting in a lower proportion of boundary area to the total area.

3.3. Carbon Emission Effects within UDB under Different Scenarios

3.3.1. Aggregate Analysis of Carbon Emission Effects

According to the results of UDB delineation in Xuzhou City in 2035, we extracted the spatial distribution and area of different types within the UDB under the NDS and the CPS. These data were then incorporated into the calculation model to obtain the carbon emission effects within the UDB of Xuzhou City in 2035 (Table 5).

Table 5. Carbon Emission Effects within UDB under Different Scenarios (t).

Scenario	Carbon Emission of Different Land Use Types within UDB							Net Carbon Emission	Carbon Source	Carbon Sink
	L1	L2	L3	L4	L5	L6	L7			
NDS	4641.80	−1477.59	−4708.43	−35.79	10,473,420.62	−1430.61	−4.45	10,470,405.55	10,478,062.42	−7656.88
CPS	5702.13	−1387.19	−3608.42	−34.63	10,475,773.36	−1275.73	−4.04	10,475,165.47	10,481,475.48	−6310.01

Regarding the net carbon emission, under the NDS, the net carbon emission within the urban development boundary of Xuzhou City in 2035 was estimated to be 10,470,405.55 t. Under the CPS, the result was 10,475,165.47 t. The CPS showed an increase of 4759.93 t compared to the NDS. This is mainly because the CPS can effectively protect cultivated land, while cultivated land exhibits carbon source characteristics, leading to higher net carbon emission compared to the NDS. Regarding the perspective of carbon source, the carbon source under the NDS and the CPS were estimated to be 10,478,062.42 t and 10,481,475.48 t, respectively. L1 and L5 are both carbon sources, with L5 being the main contributor, accounting for 99.95% of the total carbon source. Regarding the carbon sink, the carbon sink under the NDS and the CPS were estimated to be 7656.88 t and 6310.01 t, respectively. The L2, L3, L4, L6, and L7 act as carbon sinks. Among them, L3 was the main source of carbon sink. Under the NDS, L3 accounted for 61.49% of the total carbon sink, while under the CPS, L3 accounted for 57.19% of the total carbon sink.

3.3.2. Analysis of Regional Differences in Carbon Emission Effects

In this paper, we utilized ArcGIS software v.10.8.2 to partite the UDB of Xuzhou city under the NDS and CPS in 2035. Following that, we calculated the carbon emission effects in each municipal district within the UDB of Xuzhou city. In terms of net carbon emission, the municipal districts within the urban development boundary showed regional differentiation with Tongshan > Jiawang > Gulou > Quanshan > Yunlong District under both the NDS and the CPS. Among them, Tongshan District had the highest net carbon emission, reaching 4,602,517.47 t under the NDS, accounting for 43.96% of the total net carbon emission, and, reaching 4,560,670.41 t under the CPS, accounting for approximately 43.54% of the total net carbon emission. In terms of per-unit net carbon emission intensity, the municipal districts within the urban development boundary of Xuzhou city exhibited regional differentiation with Gulou > Quanshan > Yunlong > Jiawang > Tongshan District under both the NDS and the CPS. Among them, Gulou District had the highest per-unit net carbon emission intensity, reaching 12,819.09 tC/km^2 under the NDS and 12,466.28 tC/km^2 under the CPS. This is mainly because Gulou District serves as the main and central urban area of Xuzhou city, with active socioeconomic activities and a dense population, resulting in the highest carbon emission intensity.

In terms of carbon source, the municipal districts still exhibited the same regional differentiation as the total net carbon emissions, with Tongshan > Jiawang > Gulou > Quanshan > Yunlong District, and Tongshan District was the largest carbon source district. In terms of carbon sink, under the NDS, the municipal districts within the urban development boundary of Xuzhou city exhibited regional differentiation with Tongshan > Jiawang > Yunlong > Quanshan > Gulou District. Under the CPS, the municipal districts exhibited regional differentiation with Tongshan > Jiawang > Gulou > Yunlong > Quanshan District, where Yunlong District's rank had changed significantly. Under the NDS, the carbon sink in Yunlong District reached 572.18 t, accounting for approximately 7.47% of the total carbon sink. Under the CPS, it reached 489.91 t, accounting for 7.76% of the total carbon sink. This is mainly because Yunlong District has various natural attractions such as Yunlong Mountain and Dalong Lake, which have strong carbon sequestration capability and high carbon sink value. It is worth noting that analyzing regional Differences facilitates the development of more targeted measures.

4. Discussion

4.1. Significance and Innovation

The delineation and management of UDB serve as effective tools to guide city development and restrict unplanned urban sprawl [56], contributing to sustainable urban development and more rational spatial planning [27]. Since the initiation of land spatial planning in China [57], UDB has been an integral part of the work. Increasingly, scholars have conducted in-depth research on this topic [58,59]. Existing studies have explored the definition of urban development boundary and have investigated how to utilize higher-precision models or more comprehensive data to achieve more accurate boundary delineation. However, some of the existing studies only considered the delineation of flexible boundary under the state of natural urban development [17], without taking into account major national strategies such as cultivated land protection. Based on this, this paper set up CPS, and substituted permanent basic cultivated land as the data restricting the transformation into the model for simulation [60], which realized in the combination of rigid boundary and flexible boundary. Based on the study results, compared to the NDS, it can be observed that the area of the cultivated land increased by 217.05 km^2 and the area of UDB decreased by 2.51 km^2 under the CPS. The CPS demonstrated better control over urban land sprawl while fully achieving the objective of cultivated land protection. The findings are consistent with the previous study [61].

Moreover, we found that there has been limited exploration of the effects of UGB delineation in existing studies [30]. Only a limited number of studies had examined the impact of UDB on land prices and housing prices [62,63]. Therefore, the innovative contribution of this paper lies in combining the urban development boundary with carbon emission effects, in line with China's recent "dual carbon" goal. This paper utilized the land use carbon emission calculation model to estimate carbon emissions within the boundary, with the intention of achieving both rational and orderly urban expansion while evaluating carbon emissions levels. This evaluation enabled the adoption of targeted low-carbon development measures, thereby reconciling conflicts between urban expansion and low-carbon development. This paper seeks to guide a low-carbon and orderly expansion of urban space, ultimately achieving sustainable urban development. This represents the significance and innovation of this study. Indeed, due to the carbon source attribute of cultivated land as a whole [50], the CPS led to an increase of 4759.93 t carbon emission compared to the NDS. Therefore, in subsequent policy recommendations, it is crucial to focus on strengthening the efficient use of cultivated land, enhancing the carbon sequestration and emission reduction capacity of the agricultural land system [64] and supporting low-carbon development in cities.

4.2. Policy Proposals for the Future Development of Cities

Based on the results of this paper, the following policy proposals are proposed to realize the future low-carbon sustainable development of cities:

(1) During the process of delineating and managing UDB, it is important to integrate regional circumstances and base the delineation on accuracy. Furthermore, efforts should be made to strengthen the construction of management and control mechanisms. We should enhance the professional competence of planning and management personnel to ensure that the urban development boundary effectively fulfills their intended purpose and achieves rational control over the urban sprawl.

(2) It is important to innovate the methods and approaches for land spatial planning from the perspective of low carbon [65]. This involves the rational planning of future land

use layout and methods, as well as precise delineation of UDB. By doing so, the delineation of low-carbon urban development boundary becomes more targeted, enabling low-carbon and sustainable urban development.

(3) We should develop strategies to enhance carbon sequestration and emission reduction based on the carbon source and sink attributes of different types. On one hand, efforts should be made to further promote efficient and intensive use of cultivated land and construction land [66], strengthen low-carbon agricultural technology innovation, promote industrial upgrading and transformation, and reduce regional carbon emission. On the other hand, we need increase the proportion of forest land and other carbon sink types, and enhance the region's carbon sequestration and sink capacity.

4.3. Problem Statement and Future Work

Although this paper explored the carbon emission effects of urban development boundary delineation under different scenarios in Xuzhou City in 2035, there are still some problems that need to be stated and some future works that need to be conducted. Firstly, this paper only investigated the carbon emission effects within urban development boundary from a top-down perspective, without approaching it from a bottom-up perspective to study the delineation of urban development boundary under low-carbon constraints [65]. This limitation of the study contributes to the lack of more targeted policies and should be considered as the future research direction. Secondly, due to limitations in data availability, only the NDS and CPS were considered without fully considering other scenarios such as ecological conservation scenarios [19,22]. This limitation of the study contributes to the lack of more comprehensive policies and needs to be further explored in future research. Thirdly, the accuracy of future land simulation can be improved by considering additional influencing factors or using more precise data to achieve more accurate boundary delineation.

5. Conclusions

This paper took Xuzhou City as an example and adopted various methods to analyze the land use changes in Xuzhou City from 2010 to 2020, to delineate the UDB in Xuzhou City in 2035, and to explore the carbon emission effects within the UDB under different scenarios. The main conclusions are as follows:

(1) From 2010 to 2020, there was an increasing trend in forest land and construction land. Forest land showed the greatest changes from 2010 to 2020. It can be seen that Xuzhou's urban sprawl is dramatic. Additionally, due to the implementation of afforestation policies, the area of forest land experienced substantial growth in Xuzhou City.

(2) Compared to the NDS, the CPS leaded to an increase of 217.05 km^2 in cultivated land area. This indicated that the CPS has effectively achieved the intended goal of cultivated land protection. Moreover, the delineation of UDB in Xuzhou City in 2035 showed that compared to the NDS, the CPS can better control urban sprawl.

(3) The analysis of total carbon emissions showed that CPS produced more carbon emission than NDS due to the fact that cultivated land as a whole behaves as a carbon source. Therefore, in addition to implementing urban development boundary delineation, it is crucial to strengthen the efficient use of cultivated land, enhance the carbon sequestration and emission reduction capacity of the cultivated land system, and contribute to the low-carbon development of the city.

In conclusion, the main contribution of this paper lies in exploring the carbon emission effects after the delineation of UDB, bridging the gap in existing research. It provides valuable insights for the rational regulation of low-carbon urban development and lays a foundation for achieving the 'dual carbon' goal successfully.

Author Contributions: Conceptualization, H.J. and X.L.; methodology, H.J., Y.G. and Y.W.; software, H.J. and Y.W.; validation, H.J., J.C. and X.C.; formal analysis, H.J. and X.C.; investigation, H.J. and Z.C.; resources, H.J., X.L. and Z.C.; data curation, H.J.; writing—original draft preparation, H.J.; writing—review and editing, H.J. and Y.G.; visualization, X.L.; supervision, X.L.; project administration, X.L.

and J.C.; funding acquisition, X.L., X.C., Y.W. and J.C. All authors have read and agreed to the published version of the manuscript.

Funding: This research was funded by National Natural Science Foundation Program (Grant No. 71874192; No. 42201279), Fundamental Research Funds for the Central Universities (Grant No. 2020ZDPY0219; No. 2022ZDPYSK08), Major Project of philosophy and Social Science Research in Colleges and Universities of Jiangsu Province (Grant No. 23WTA-001), Huai'an Natural Science Research Program Joint Special (HABL202201) and Graduate Research and Innovation Projects of Jiangsu Province (Grant No. KYCX23_2628).

Informed Consent Statement: Informed consent was obtained from all subjects involved in the study.

Data Availability Statement: The data used in this study can be found in the relevant publications and have already been cited in the text for illustration. The data presented in this study are available on request from the corresponding author.

Conflicts of Interest: The authors declare no conflict of interest.

References

1. Seto, K.C.; Sánchez-Rodríguez, R.; Fragkias, M. The New Geography of Contemporary Urbanization and the Environment. *Annu. Rev. Environ. Resour.* **2010**, *35*, 167–194. [CrossRef]
2. Angel, S. Urban Forms and Future Cities: A Commentary. *Urban Plan.* **2017**, *2*, 1–5. [CrossRef]
3. Liu, Y.; Gao, Y.; Liu, L.; Song, C.; Ai, D. Nature-based solutions for urban expansion: Integrating ecosystem services into the delineation of growth boundaries. *Habitat Int.* **2022**, *124*, 102575. [CrossRef]
4. Liu, X.; Ou, J.; Li, X.; Ai, B. Combining system dynamics and hybrid particle swarm optimization for land use allocation. *Ecol. Model.* **2013**, *257*, 11–24. [CrossRef]
5. Peng, J.; Zhao, S.; Dong, J.; Liu, Y.; Meersmans, J.; Li, H.; Wu, J. Applying ant colony algorithm to identify ecological security patterns in megacities. *Environ. Model. Softw.* **2019**, *117*, 214–222. [CrossRef]
6. Ren, Q.; He, C.; Huang, Q.; Zhang, D.; Shi, P.; Lu, W. Impacts of global urban expansion on natural habitats undermine the 2050 vision for biodiversity. *Resour. Conserv. Recycl.* **2023**, *190*, 106834. [CrossRef]
7. Chen, X.; Yu, L.; Du, Z.; Liu, Z.; Qi, Y.; Liu, T.; Gong, P. Toward sustainable land use in China: A perspective on China's national land surveys. *Land Use Policy* **2022**, *123*, 106428. [CrossRef]
8. Luo, M.; Lau, N.C. Urban Expansion and Drying Climate in an Urban Agglomeration of East China. *Geophys. Res. Lett.* **2019**, *46*, 6868–6877. [CrossRef]
9. Sun, W.; Huang, C. How does urbanization affect carbon emission efficiency? Evidence from China. *J. Clean. Prod.* **2020**, *272*, 122828. [CrossRef]
10. Guo, S.; Zhang, Y.; Qian, X.; Ming, Z.; Nie, R. Urbanization and CO_2 emissions in resource-exhausted cities: Evidence from Xuzhou city, China. *Nat. Hazards* **2019**, *99*, 807–826. [CrossRef]
11. Cai, B.; Liu, H.; Zhang, X.; Pan, H.; Zhao, M.; Zheng, T.; Nie, J.; Du, M.; Dhakal, S. High-resolution accounting of urban emissions in China. *Appl. Energy* **2022**, *325*, 119896. [CrossRef]
12. Wu, H.; Fang, S.; Zhang, C.; Hu, S.; Nan, D.; Yang, Y. Exploring the impact of urban form on urban land use efficiency under low-carbon emission constraints: A case study in China's Yellow River Basin. *J. Environ. Manag.* **2022**, *311*, 114866. [CrossRef] [PubMed]
13. Tan, R.; Liu, P.; Zhou, K.; He, Q. Evaluating the effectiveness of development-limiting boundary control policy: Spatial difference-in-difference analysis. *Land Use Policy* **2022**, *120*, 106229. [CrossRef]
14. Gennaio, M.-P.; Hersperger, A.M.; Bürgi, M. Containing urban sprawl—Evaluating effectiveness of urban growth boundaries set by the Swiss Land Use Plan. *Land Use Policy* **2009**, *26*, 224–232. [CrossRef]
15. Tayyebi, A.; Perry, P.C.; Tayyebi, A.H. Predicting the expansion of an urban boundary using spatial logistic regression and hybrid raster–vector routines with remote sensing and GIS. *Int. J. Geogr. Inf. Sci.* **2013**, *28*, 639–659. [CrossRef]
16. Hepinstall-Cymerman, J.; Coe, S.; Hutyra, L.R. Urban growth patterns and growth management boundaries in the Central Puget Sound, Washington, 1986–2007. *Urban Ecosyst.* **2013**, *16*, 109–129. [CrossRef]
17. Tayyebi, A.; Pijanowski, B.C.; Tayyebi, A.H. An urban growth boundary model using neural networks, GIS and radial parameterization: An application to Tehran, Iran. *Landsc. Urban Plan.* **2011**, *100*, 35–44. [CrossRef]
18. Wang, W.; Jiao, L.; Zhang, W.; Jia, Q.; Su, F.; Xu, G.; Ma, S. Delineating urban growth boundaries under multi-objective and constraints. *Sustain. Cities Soc.* **2020**, *61*, 102279. [CrossRef]
19. Han, N.; Hu, K.; Yu, M.; Jia, P.; Zhang, Y. Incorporating Ecological Constraints into the Simulations of Tropical Urban Growth Boundaries: A Case Study of Sanya City on Hainan Island, China. *Appl. Sci.* **2022**, *12*, 6409. [CrossRef]
20. Wang, S.-N.; Liu, C.-G.; Sun, W. Simulation research of urban development boundary based on ecological constraints: A case study of Nanjing. *J. Nat. Resour.* **2021**, *36*, 2913–2925. [CrossRef]
21. Ouyang, X.; Xu, J.; Li, J.; Wei, X.; Li, Y. Land space optimization of urban-agriculture-ecological functions in the Changsha-Zhuzhou-Xiangtan Urban Agglomeration, China. *Land Use Policy* **2022**, *117*, 106112. [CrossRef]

2. Yang, X.; Bai, Y.; Che, L.; Qiao, F.; Xie, L. Incorporating ecological constraints into urban growth boundaries: A case study of ecologically fragile areas in the Upper Yellow River. *Ecol. Indic.* **2021**, *124*, 107436. [CrossRef]
3. Bakshi, A.; Esraz-Ul-Zannat, M. Application of urban growth boundary delineation based on a neural network approach and landscape metrics for Khulna City, Bangladesh. *Heliyon* **2023**, *9*, e16272. [CrossRef] [PubMed]
4. Chen, Y.; Li, X.; Liu, X.; Huang, H.; Ma, S. Simulating urban growth boundaries using a patch-based cellular automaton with economic and ecological constraints. *Int. J. Geogr. Inf. Sci.* **2018**, *33*, 55–80. [CrossRef]
5. Huang, D.; Huang, J.; Liu, T. Delimiting urban growth boundaries using the CLUE-S model with village administrative boundaries. *Land Use Policy* **2019**, *82*, 422–435. [CrossRef]
6. Liu, J.; Zhang, G.; Zhuang, Z.; Cheng, Q.; Gao, Y.; Chen, T.; Huang, Q.; Xu, L.; Chen, D. A new perspective for urban development boundary delineation based on SLEUTH-InVEST model. *Habitat Int.* **2017**, *70*, 13–23. [CrossRef]
7. Liu, X.; Wei, M.; Li, Z.; Zeng, J. Multi-scenario simulation of urban growth boundaries with an ESP-FLUS model: A case study of the Min Delta region, China. *Ecol. Indic.* **2022**, *135*, 108538. [CrossRef]
8. Xu, H.; Song, Y.; Tian, Y. Simulation of land-use pattern evolution in hilly mountainous areas of North China: A case study in Jincheng. *Land Use Policy* **2022**, *112*, 105826. [CrossRef]
9. Zhang, D.; Liu, X.; Lin, Z.; Zhang, X.; Zhang, H. The delineation of urban growth boundaries in complex ecological environment areas by using cellular automata and a dual-environmental evaluation. *J. Clean. Prod.* **2020**, *256*, 120361. [CrossRef]
10. Tan, R.; Liu, Y.; Liu, Y.; He, Q. A literature review of urban growth boundary: Theory, modeling, and effectiveness evaluation. *Prog. Geogr.* **2020**, *39*, 327–338. [CrossRef]
11. Liang, X.; Ji, X.; Guo, N.; Meng, L. Assessment of urban heat islands for land use based on urban planning: A case study in the main urban area of Xuzhou City, China. *Environ. Earth Sci.* **2021**, *80*, 308. [CrossRef]
12. Chen, L.; Zhou, B.; Man, W.; Liu, M. Landsat-Based Monitoring of the Heat Effects of Urbanization Directions and Types in Hangzhou City from 2000 to 2020. *Remote Sens.* **2021**, *13*, 4268. [CrossRef]
13. Chen, X.; Yu, L.; Cao, Y.; Xu, Y.; Zhao, Z.; Zhuang, Y.; Liu, X.; Du, Z.; Liu, T.; Yang, B.; et al. Habitat quality dynamics in China's first group of national parks in recent four decades: Evidence from land use and land cover changes. *J. Environ. Manag.* **2023**, *325*, 116505. [CrossRef]
14. Guan, D.; Gao, W.; Watari, K.; Fukahori, H. Land use change of Kitakyushu based on landscape ecology and Markov model. *J. Geogr. Sci.* **2008**, *18*, 455–468. [CrossRef]
15. Li, C.; Yang, M.; Li, Z.; Wang, B. How Will Rwandan Land Use/Land Cover Change under High Population Pressure and Changing Climate? *Appl. Sci.* **2021**, *11*, 5376. [CrossRef]
16. Reis, S. Analyzing Land Use/Land Cover Changes Using Remote Sensing and GIS in Rize, North-East Turkey. *Sensors* **2008**, *8*, 6188–6202. [CrossRef]
17. Cao, J.; Li, T. Analysis of spatiotemporal changes in cultural heritage protected cities and their influencing factors: Evidence from China. *Ecol. Indic.* **2023**, *151*, 110327. [CrossRef]
18. Xiao, X.; Huang, X.; Jiang, L.; Jin, C. Empirical study on comparative analysis of dynamic degree differences of land use based on the optimization model. *Geocarto Int.* **2022**, *37*, 9847–9864. [CrossRef]
19. Zhao, Y.; Li, R.; Wu, M. Correlation Studies between Land Cover Change and Baidu Index: A Case Study of Hubei Province. *ISPRS Int. J. Geo-Inf.* **2020**, *9*, 232. [CrossRef]
20. Chen, W.; Chi, G.; Li, J. The spatial association of ecosystem services with land use and land cover change at the county level in China, 1995–2015. *Sci. Total Environ.* **2019**, *669*, 459–470. [CrossRef]
21. Sklavou, P.; Karatassiou, M.; Parissi, Z.; Galidaki, G.; Ragkos, A.; Sidiropoulou, A. The Role of Transhumance on Land Use/Cover Changes in Mountain Vermio, Northern Greece: A GIS Based Approach. *Not. Bot. Horti Agrobot. Cluj-Napoca* **2017**, *45*, 589–596. [CrossRef]
22. Yadav, V.; Ghosh, S.K. Assessment and prediction of urban growth for a mega-city using CA-Markov model. *Geocarto Int.* **2019**, *36*, 1960–1992. [CrossRef]
23. Liu, X.; Liang, X.; Li, X.; Xu, X.; Ou, J.; Chen, Y.; Li, S.; Wang, S.; Pei, F. A future land use simulation model (FLUS) for simulating multiple land use scenarios by coupling human and natural effects. *Landsc. Urban Plan.* **2017**, *168*, 94–116. [CrossRef]
24. Liang, X.; Liu, X.; Li, X.; Chen, Y.; Tian, H.; Yao, Y. Delineating multi-scenario urban growth boundaries with a CA-based FLUS model and morphological method. *Landsc. Urban Plan.* **2018**, *177*, 47–63. [CrossRef]
25. Liao, W.; Liu, X.; Xu, X.; Chen, G.; Liang, X.; Zhang, H.; Li, X. Projections of land use changes under the plant functional type classification in different SSP-RCP scenarios in China. *Sci. Bull.* **2020**, *65*, 1935–1947. [CrossRef] [PubMed]
26. Jin, Y.; Li, A.; Bian, J.; Nan, X.; Lei, G. Modeling the Impact of Investment and National Planning Policies on Future Land Use Development: A Case Study for Myanmar. *ISPRS Int. J. Geo-Inf.* **2023**, *12*, 22. [CrossRef]
27. Şenik, B.; Kaya, H.S. Landscape sensitivity-based scenario analysis using flus model: A case of Asarsuyu watershed. *Landsc. Ecol. Eng.* **2021**, *18*, 139–156. [CrossRef]
28. Cuellar, Y.; Perez, L. Assessing the accuracy of sensitivity analysis: An application for a cellular automata model of Bogotá's urban wetland changes. *Geocarto Int.* **2023**, *38*, 2186491. [CrossRef]
29. Zhao, L.; Yang, C.-H.; Zhao, Y.-C.; Wang, Q.; Zhang, Q.-P. Spatial Correlations of Land Use Carbon Emissions in Shandong Peninsula Urban Agglomeration: A Perspective from City Level Using Remote Sensing Data. *Remote Sens.* **2023**, *15*, 1488. [CrossRef]

50. Zhang, P.; He, J.; Hong, X.; Zhang, W.; Qin, C.; Pang, B.; Li, Y.; Liu, Y. Carbon sources/sinks analysis of land use changes in China based on data envelopment analysis. *J. Clean. Prod.* **2018**, *204*, 702–711. [CrossRef]
51. Ghosh, S.; Dinda, S.; Chatterjee, N.D.; Dutta, S.; Bera, D. Spatial-explicit carbon emission-sequestration balance estimation and evaluation of emission susceptible zones in an Eastern Himalayan city using Pressure-Sensitivity-Resilience framework: An approach towards achieving low carbon cities. *J. Clean. Prod.* **2022**, *336*, 130417. [CrossRef]
52. Rong, T.; Zhang, P.; Zhu, H.; Jiang, L.; Li, Y.; Liu, Z. Spatial correlation evolution and prediction scenario of land use carbon emissions in China. *Ecol. Inform.* **2022**, *71*, 101802. [CrossRef]
53. Sun, H.; Liang, H.; Chang, X.; Cui, Q.; Tao, Y. Land Use Patterns on Carbon Emission and Spatial Association in China. *Econ. Geogr.* **2015**, *35*, 154–162. [CrossRef]
54. Zhao, C.; Liu, Y.; Yan, Z. Effects of land-use change on carbon emission and its driving factors in Shaanxi Province from 2000 to 2020. *Environ. Sci. Pollut. Res. Int.* **2023**, *30*, 68313–68326. [CrossRef] [PubMed]
55. Zhang, C.-Y.; Zhao, L.; Zhang, H.; Chen, M.-N.; Fang, R.-Y.; Yao, Y.; Zhang, Q.-P.; Wang, Q. Spatial-temporal characteristics of carbon emissions from land use change in Yellow River Delta region, China. *Ecol. Indic.* **2022**, *136*, 108623. [CrossRef]
56. Schuster Olbrich, J.P.; Vich, G.; Miralles-Guasch, C.; Fuentes, L. Urban sprawl containment by the urban growth boundary: The case of the Regulatory Plan of the Metropolitan Region of Santiago of Chile. *J. Land Use Sci.* **2022**, *17*, 324–338. [CrossRef]
57. Yue, W.; Hou, B.; Ye, G.; Wang, Z. China's land-sea coordination practice in territorial spatial planning. *Ocean Coast. Manag.* **2023**, *237*, 106545. [CrossRef]
58. He, F.; Yang, J.; Zhang, Y.; Sun, D.; Wang, L.; Xiao, X.; Xia, J. Offshore Island Connection Line: A new perspective of coastal urban development boundary simulation and multi-scenario prediction. *GIScience Remote Sens.* **2022**, *59*, 801–821. [CrossRef]
59. Ma, S.; Zhao, Y.; Tan, X. Exploring Smart Growth Boundaries of Urban Agglomeration with Land Use Spatial Optimization: A Case Study of Changsha-Zhuzhou-Xiangtan City Group, China. *Chin. Geogr. Sci.* **2020**, *30*, 665–676. [CrossRef]
60. Ouyang, X.; He, Q.; Zhu, X. Simulation of Impacts of Urban Agglomeration Land Use Change on Ecosystem Services Value under Multi-Scenarios: Case Study in Changsha-Zhuzhou-Xiangtan Urban Agglomeration. *Econ. Geogr.* **2020**, *40*, 93–102. [CrossRef]
61. Lai, Z.; Chen, C.; Chen, J.; Wu, Z.; Wang, F.; Li, S. Multi-Scenario Simulation of Land-Use Change and Delineation of Urban Growth Boundaries in County Area: A Case Study of Xinxing County, Guangdong Province. *Land* **2022**, *11*, 1598. [CrossRef]
62. Mathur, S. Impact of an urban growth boundary across the entire house price spectrum: The two-stage quantile spatial regression approach. *Land Use Policy* **2019**, *80*, 88–94. [CrossRef]
63. Ball, M.; Cigdem, M.; Taylor, E.; Wood, G. Urban growth boundaries and their impact on land prices. *Environ. Plan. A Econ. Space* **2014**, *46*, 3010–3026. [CrossRef]
64. Zhang, L.; Lin, X.; Xiao, Y.; Lin, Z. Spatial and structural characteristics of the ecological network of carbon metabolism of cultivated land based on land use and cover change: A case study of Nanchang, China. *Environ. Sci. Pollut. Res. Int.* **2023**, *30*, 30514–30529. [CrossRef] [PubMed]
65. Li, J.; Guldmann, J.-M.; Gong, J.; Su, H. Urban growth boundaries optimization under low-carbon development: Combining multi-objective programming and patch cellular automata models. *J. Environ. Manag.* **2023**, *340*, 117934. [CrossRef]
66. Wu, H.; Qiu, Y.; Yin, L.; Liu, S.; Zhao, D.; Zhang, M. Effects of China's land-intensive use on carbon emission reduction: A new perspective of industrial structure upgrading. *Front. Environ. Sci.* **2022**, *10*, 1073565. [CrossRef]

Disclaimer/Publisher's Note: The statements, opinions and data contained in all publications are solely those of the individual author(s) and contributor(s) and not of MDPI and/or the editor(s). MDPI and/or the editor(s) disclaim responsibility for any injury to people or property resulting from any ideas, methods, instructions or products referred to in the content.

Article

New Insights into Urbanization Based on Global Mapping and Analysis of Human Settlements in the Rural–Urban Continuum

Xiyu Li [1], Le Yu [1,2,3,*] and Xin Chen [1,4]

1. Ministry of Education Key Laboratory for Earth System Modeling, Department of Earth System Science, Tsinghua University, Beijing 100084, China; lixiyu21@mails.tsinghua.edu.cn (X.L.); xinchenthu@mail.tsinghua.edu.cn (X.C.)
2. Ministry of Education Ecological Field Station for East Asian Migratory Birds, Beijing 100084, China
3. Tsinghua University (Department of Earth System Science)-Xi'an Institute of Surveying and Mapping Joint Research Center for Next-Generation Smart Mapping, Beijing 100084, China
4. Institute of Loess Plateau, Shanxi University, Taiyuan 030006, China
* Correspondence: leyu@tsinghua.edu.cn

Abstract: The clear boundary between urban and rural areas is gradually disappearing, and urban and rural areas are two poles of a gradient with many continuous human settlements in between, which is a concept known as the rural–urban continuum. Little is known about the distribution and change trajectories of the various types in the rural–urban continuum across the globe. Therefore, using global land-cover data (FROM-GLC Plus) and global population data (Worldpop) based on the decision-making tree method, this study proposed a method and classification system for global rural–urban continuum mapping and produced the mapping results on a global scale in the Google Earth Engine platform. With the expansion of built-up areas and the increase in population, the global human settlements follow the pattern that develops from wildland to villages (isolated—sparse—dense), and then to towns (sparse—dense), and finally to urban areas (edge—center). From a regional perspective, there are some obvious differences: Africa is dominated by sparse villages; Asia has the highest proportion of densely clustered towns; the proportion of dense villages in Europe is high. Rural–urban continuum mapping and analysis provide a database and new insights into urbanization and differences between urban and rural areas around the world.

Keywords: rural–urban continuum; land system; urban expansion; regional differences; change trajectories

1. Introduction

A wide variety of human settlements have been created as the result of the great human transformation of natural landscapes [1,2]. Urban areas are the most intense transformation of natural landscapes among all the human settlements, and a common urban–rural dichotomy refers to human settlements outside urban areas as rural areas [3,4]. Increasing urbanization has led to the formation of complex geographies [5]. Many regional studies have found the limitations of a simple urban–rural dichotomy to reveal the real situation of human settlements [6–8]. In addition, the criteria and thresholds for dividing urban and rural areas vary widely in each country and region [9]. The clear boundary between urban and rural areas is gradually disappearing; urban and rural areas are two poles of a gradient with many continuous transitional landscapes in between [4].

This concept, which emphasizes the continuous transitional landscapes between urban and rural areas, is called the rural–urban continuum. Since the early 20th century, the theory of a rural–urban continuum has been continuously enriched, developed, and matured in practice [3,10–14]. In 1970, the United States Department of Agriculture began to develop the Rural–Urban Continuum Code (RUCC) to classify counties [15]. Similarly, landscape features of the rural–urban continuum have been noted in several regions of the globe, including Europe [6,8], China [16,17], and India [7]. In Europe, much of physical territory is regarded as areas between urban and rural, and there have been proposed definitions and mapping criteria for these peri-urban areas [6,8]. In addition, the traditional Chinese settlement system has been seen as a town–village–field continuum [16]. There are studies that have defined the settlement system in China and mapped its change trajectory between 1990 and 2010 [17]. However, the criteria and thresholds for dividing urban and rural areas vary widely in each country and region. Empirical analysis and conceptual comparisons show that towns and cities are likely to be classified as rural areas in Africa and Asia, and as urban areas in other parts of the world [18]. Therefore, it is necessary to construct a unified classification system on a global scale to define and map the various types of areas in the rural–urban continuum. Globally, the Global Human Settlement Layer (GHSL) produced by the Joint Research Centre (JRC) of the European Commission has provided mapping results at 5-year intervals beginning in 1975 and predicts changes over the next decade [19].

The concept of the rural–urban continuum and its extensive practice provide a new perspective for the study of urban areas and the urbanization process. In the classification results of remote sensing images, the data basis for a large number of analyses, the area mainly covered by impervious pixels is assigned to the built-up area, while the other areas are regarded as non-built-up areas [4]. There have been many studies that have estimated the distribution of urban areas around the world and the populations that live in them. It is estimated that about 2% of the terrestrial land can be characterized as urban [20], and these urban areas contain more than half of the global population, which is expected to increase in the near future [21]. In addition, there is also an overall understanding of urban growth rates and regional differences. Gong et al. mapped global artificial impervious areas (GAIAs) and analyzed the trend of urban expansion in various countries and regions around the world [22]. Liu et al. analyzed global urban change from 1985 to 2015 and found that global urban extent has expanded at an unprecedented rate that is notably faster than that of population growth [23]. However, based on the perspective of the urban–rural continuum, urban areas are no longer just a single type, but a state of continuous development. From the perspective of the rural–urban continuum, the relationship between urban land expansion and population growth also needs to be viewed more carefully. It is also meaningful to understand the distribution, size, population, and historical trends of various types of human settlements from the rural–urban continuum perspective across the globe.

Furthermore, in the system of the urban–rural continuum, urbanization is no longer just the transformation of non-built-up area to built-up area, but the complex process of rural–urban development, which should be studied in depth. There are studies that compare urban patterns and urbanization in different regions [24–26]. Schmitt et al. found that Germany's metropolitan areas are characterized by numerous small, equal-sized clusters, while the United States contains one large and dominant center [24]. An uneven distribution of urbanization was observed at different economic levels manifested by varying rates of population growth and expansion of built-up areas [25]. Tian et al. found that the dominant characteristics of urbanization in Asia and Africa are increased population density and built-up patch density, while urbanization in Europe and North America took a rather steady pace, combined with widespread greening [26]. It is said that human settlement changes follow multiple different trajectories in different locations, in small and incremental steps toward urban areas [17]. However, the specific characteristics of the development process have not yet provided conclusions on a global scale and still need to be systematically and comprehensively analyzed from the rural–urban continuum.

To understand the developing patterns and rules of human settlements, different types of human settlements must be defined and mapped in a continuous trajectory. Additionally, the study of human settlements has great implications, such as the disturbance of human activities in nature [27,28], equality of access to services [3], and the effectiveness of resource utilization [29,30]). Therefore, this study proposed a classification system for global rural–urban continuum mapping and produced global maps in the Google Earth Engine platform using global land-cover data (FROM-GLC Plus) and global population data (Worldpop) based on the decision-making tree method. We further analyzed the current state of the rural–urban continuum (distribution, area, unit number, population, etc.) as well as historical trends over the past two decades and tried to understand the developing patterns and rules of human settlements and their geographic differentiation.

2. Materials and Methods

2.1. A Review of the Human Settlement Classification Systems

Unlike a clearly defined and widely acknowledged classification system in land-cover mapping such as dividing land categories into cropland, impervious, forest, and so on, how to generate a continuous and integrated classification system and accurately capture the differences among all types of human settlements still remains a problem. Population is a commonly used indicator when separating urban and rural areas [31]. GHSL combines population size and population density thresholds to capture the full settlement hierarchy (including urban center, dense urban cluster, semi-dense urban cluster, suburban grid cell, rural cluster, low-density rural grid cell, very low-density grid cell) [32]. Population as a classification indicator for human settlements does not apply to areas with high population density but to low levels of urbanization, such as in India [33]. Built-up areas are another commonly chosen indicator, which are regarded as the most direct modification of the natural surface by human activities. Based on built-up land, cluster density, and cluster size, Li et al. divided China's land surface into large cities, urban landscapes, suburban landscapes, densely clustered towns, sparsely clustered towns, dense villages, sparse villages, isolated villages, and deep rural areas [17]. However, this classification system cannot avoid errors in areas where the indicators of built-up area and actual resident population are inconsistent, such as ghost cities [34] and hollow villages [35]. Therefore, population, built-up areas, and other indicators should be combined to provide an integrated classification framework.

The classification system and methodology in the above studies are of great reference value [17,32], but these classification systems only distinguish the various types of urban to rural landscapes and cannot reflect the gradual transition of rural to wilderness. Ellis et al. mapped the anthropogenic biomes based on population density and land use in a long time series (from 10,000 BC to 2015 AD) [1]. In this classification system, non-built-up land was first classified into cropland, rangeland, woodland, etc., and then further classified them into residential, populated, and remote according to the level of population density. Moreover, time continuity, spatial scope, spatial resolution, and the resolution of the classification system greatly affect the use of data products. Data products from Li et al. (2019) focused on China and were mapped in ten-year intervals [17]. The anthropogenic biomes product has global spatial coverage and historical continuity; it still has the problem of coarse spatial resolution (10 km) whose use is limited, especially in the last 20 years. Although GHSL covers the global space scope and provides data series at five-year intervals, it ignores the detailed transitional change from rural to wild (Table 1). Mu et al. developed a global annual human footprint dataset from 2000 to 2018 with a resolution of 1 km, dividing the land surface into the wilderness (Human Footprint < 1), the intact areas (Human Footprint < 4), and the highly modified region (Human Footprint $\geq$ 4) [36]. Although continuous values are assigned to the land grids representing the intensity of human activity, they are roughly grouped into three categories, limiting further use of the dataset. Tian et al. used population, built-up land, and greenness to classify global urbanization types into greenness loss, steady urbanization, population decrease, and so on [26]. Although the change types of human settlement have been interpreted in various aspects, a clear mapping result to explain the current state of human settlement is still lacking.

Table 1. Comparison of datasets in human settlement mapping studies.

Dataset	Classification System	Time Period	Spatial Coverage	Spatial Resolution
Li et al. (2019) [17]	Rural–urban continuum	1990, 2000, 2010	China	2 km
GHSL	Rural–urban continuum	1975–2030	World	1 km
Anthropogenic Biomes	Urban, rural–wild continuum	10,000 BCE—2015 CE	World	10 km
Mu et al. (2022) [36]	Unified Values from 1–10, but 3 classes	2000–2018	World	1 km
Tian et al. (2022) [26]	Urbanization pattern types	1975, 1990, 2015	World	5 km

Therefore, we adopted a classification system that integrates features of previous studies. As for the rural–urban continuum, we classified land as urban centers, urban landscapes, densely clustered towns, sparsely clustered towns, dense villages, sparse villages, and isolated villages. As for the cropland, rangeland, and woodland, we further classified them into residential, populated, and remote according to the level of population density. Other land types include wildland, water, and ice snow.

2.2. Data and Processing

Data used in human settlements mapping include the following three types (Table 2):

Table 2. Datasets used in human settlement mapping and analysis.

Dataset Type	Name	Spatial Resolution	Duration	Usage
Land cover	FROM-GLC-Plus	30 m	2000–2020	Mapping
Population	Worldpop	100 m	2000–2020	Mapping
Elevation	SRTM	30 m	/	Mapping
Settlements samples	World Cities Database	/	/	Analysis
GDP	Gridded Global GDP dataset	5 arc-min	1990–2015	Analysis
Administrative boundaries	GeoBoundaries	/	/	Analysis

Global annual 30 m land-cover dataset. The land-cover dataset used in this study is FROM-GLC-Plus, which describes the dynamics of global land-cover change from 2000 to 2020 and divides the global land cover into the following 10 categories: cropland, forest, grassland, shrubland, wetland, water, tundra, impervious surface (built-up), bare land, snow and ice, with high spatial and temporal resolution [37].

Global population dataset. This study used Worldpop, which decomposed census-based count results into 100 m resolution grid cells using machine learning methods and the relationship between population density and a series of geospatial layers [38].

Global digital elevation data from the Shuttle Radar Topography Mission (SRTM), provided by NASA JPL laboratory, is at 1 arc second resolution (approximately 30 m) [39].

There are an additional three datasets helping us to analyze the final mapping result:

World Cities Database. This is a simple, accurate, and up-to-date database of all the populated places in the world (about 4.3 million). It is built from the ground up using authoritative sources such as the NGIA, US Geological Survey, US Census Bureau, and NASA (Online resources: https://simplemaps.com/data/world-cities, accessed on 13 August 2023).

Gridded Global GDP dataset. This provides annual gridded datasets for GDP per capita (PPP) and total GDP (PPP) for the whole world at 5 arc-min resolution for the 25-year period of 1990–2015 [40]. To provide a consistent product over time and space, the sub-national data were only used indirectly, scaling the reported national value and thus remaining representative of the official statistics.

Global Database of Political Administrative Boundaries (GeoBoundaries) is an online, open license resource of boundaries for every country in the world. All boundary types were ingested and include the following, with admin level varying from 0–4 (online resources: https://www.geoboundaries.org/, accessed on 13 August 2023) [41]. Here, we used political administrative boundaries with admin level varying from 0–2.

All the data need to be at the same resolution. Therefore, the global annual 30 m land-cover dataset was resampled at 1 km scale to calculate the proportion of each land-cover type on a 1 km grid, the global population 100 m dataset was resampled at 1 km scale to calculate the total population on a 1 km grid, and global slope data were calculated based on SRTM digital elevation data using terrain analysis algorithms in the Google Earth Engine [42].

2.3. Method

In order to better reflect the gradient difference in human activities' intensity among the various types of human settlements, we calculated indicators based on population and land-cover datasets, including built-up density, population density, population size, etc. These indicators have proven to be effective in mapping the gradient difference between urban and rural areas in previous studies [1,17,19,26,43].

2.3.1. Index Calculation Based on Population and Built-Up Area

(1) ABDI (Adjusted built-up density index)

Based on the global slope data and land-cover dataset, suitable construction land was defined as all areas that are not water or permanent snow and that have a slope that does not exceed 25% [17]. The 25% threshold is based on previous studies in China, which indicate that areas with a slope greater than 25% are not suitable for construction. The proportion of the suitable construction area was calculated at the scale of 1 km. ABDI was defined as the proportion of impervious surface (built-up area) divided by the proportion of suitable construction area at the 1 km scale. Li et al. defined the four intervals divided by ABDI at the three thresholds of 0.35, 0.2, and 0.05 as the density differences of built-up areas in urban areas, suburban/towns, villages and deep rural areas [17].

(2) Spatial connectivity indicators

First, the parts larger than 1500, 300, and 20 of the global 1 km population grid were selected and exported separately to generate three different grid layers, which are respectively used as the partition thresholds of the 1 km population grid for urban areas, towns, and villages [32]. Spatial connectivity analysis was further used to calculate the number of connected grids in a cluster. More specifically, functions in the Google Earth Engine (GEE) were used to generate an image where each pixel contains the number of 4-connected or 8-connected neighbors (including itself). The image of the connected grid count was used to distinguish whether the total population reached the above standard. An urban grid cluster was defined as a cluster with connected grids of more than 1500 people and a total population of more than 50,000 people, which means this cluster must contain more than 34 grids. Similarly, a town grid cluster was defined as a cluster with connected grids of more than 300 people and a total population of more than 5000 people (more than 17 grids). Village grid clusters were defined as clusters with connected grids of more than 20 people and a total population of more than 100 people (more than 5 grids).

2.3.2. Construction of Classification Decision Tree

In this part, different types of human settlements were defined one by one by overlapping layers, basically from urban to rural and then to wild. Layers were stacked according to the order in Figure 1. Later defined types were defined to an extent out of the earlier defined type and would not cause overlap or conflict. The determination of threshold value referred to the previous work, and the specific situation was explained in Section 2.3.1.

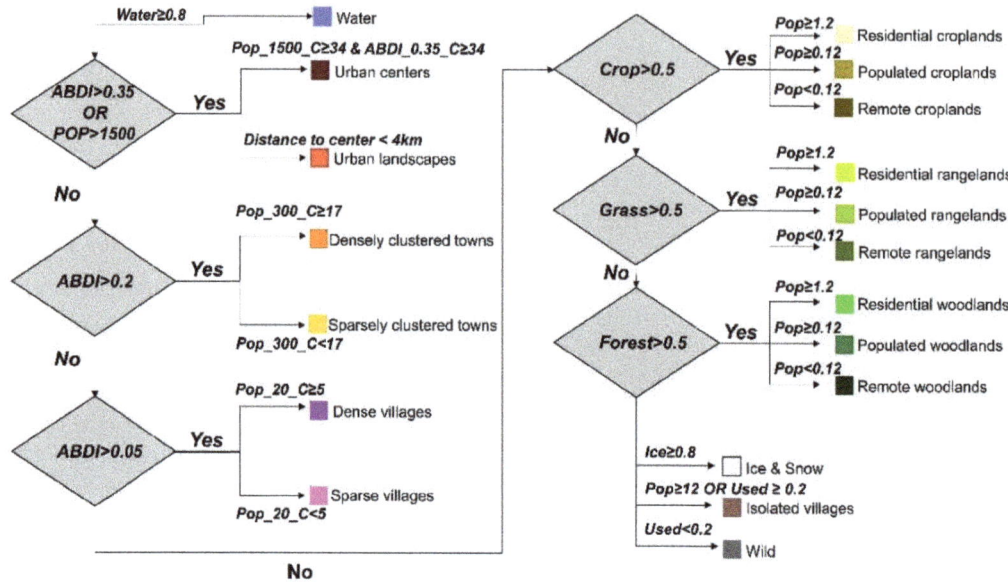

Figure 1. Human settlement type mapping workflow.

First, areas with a water proportion greater than 0.8 were defined as water. And then grids with an ABDI greater than 0.35 or a population of more than 1500 were defined as urban areas. The urban center was defined as a cluster of more than 34 connected grids whose ABDIs were greater than 0.35 and population more than 1500. That is, the urban center had a total population of more than 50,000 people. The remaining urban areas within 4 km from the urban center were defined as urban landscapes (Figure 1).

Areas with ABDIs greater than 0.2 were defined as town areas; densely clustered towns were defined as clusters of more than 17 connected grids whose populations were more than 300 within town areas, and other town areas were defined as sparsely clustered towns. Areas with ABDIs greater than 0.05 were defined as rural areas, dense villages were defined as clusters of more than 5 connected grids whose populations were more than 20 within rural areas, and other rural areas were defined as sparse villages. The areas with proportions of cropland, grassland, and forest exceeding 0.5 in the 1 km grid were defined as cropland, rangeland, and woodland, respectively, and further divided into three levels according to the population. Taking cropland as an example, cropland with a population of more than 1.2 was defined as residential cropland, while cropland with a population of more than 0.12 was defined as populated cropland. The other cropland was defined as remote cropland. Finally, the type "Ice and snow" was defined as areas with a proportion of ice and snow higher than 0.8; isolated villages were defined as areas with a population of more than 12 or used land proportion higher than 0.2. Used land was defined as impervious surface or cropland. Wild areas were finally defined as used land proportion lower than 0.2. Thresholds used in the natural land system refer to the anthropogenic biomes mapping [1].

3. Results

3.1. Global Human Settlement Mapping Results in the Rural–Urban Continuum

A global map of human settlements in 2020 is shown in Figure 2a. Combining population data to show the gradient distribution of human activity intensity, this map shows more detailed information than land-cover maps, which can help us better understand the relationship between human activity and land use, urbanization, and its regional differences (Figure 2a).

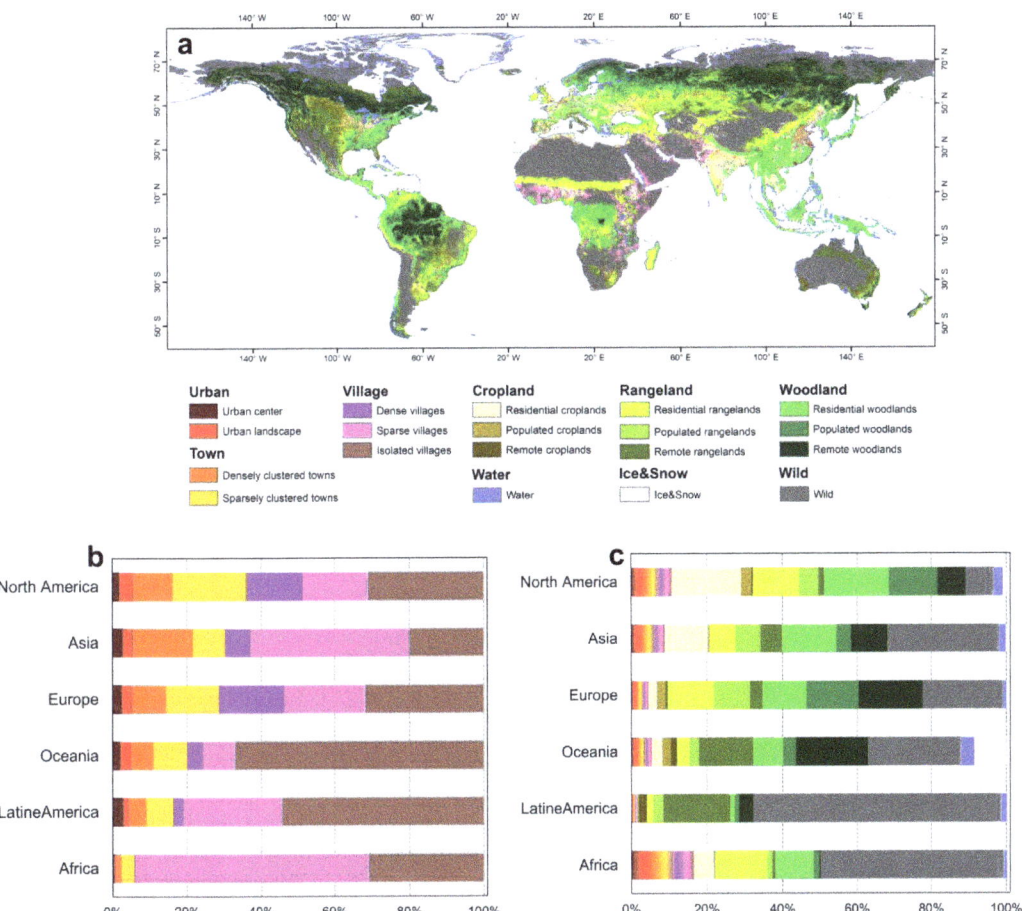

Figure 2. Distribution of global human settlements in 2020. (**a**) Mapping results of human settlements (2020). (**b**) Percentage of land area in 2020 occupied by each human settlement type of the rural–urban continuum in each continent. (**c**) Percentage of land area in 2020 occupied by each human settlement type of global land in each continent.

The distribution of human settlements varied greatly among continents. The only thing in common is that the proportion of the rural–urban continuum is very small, but the proportion of different types in the rural–urban continuum is also very different, which reflects the different levels of urbanization across continents. In 2020, North America, Asia, and Europe respectively had the highest proportions of urban areas and towns (nearly 30%), but in Oceania and Latin America, this ratio was less than 20%. About 10% of Africa is urban areas and towns, and over 60% of the rural–urban continuum is dominated by sparse villages (Figure 2b).

Every continent has its most prevalent human settlements. Africa is dominated by sparse villages. Asia has the highest proportion of densely clustered towns. The proportion of dense villages in Europe is extremely high. Oceania and Latin America have a high proportion of isolated villages. And the rural–urban continuum of North America is relatively evenly proportioned (Figure 2b).

The higher the level of urbanization, the higher the degree of development of the natural land system. In continents with high levels of urbanization, such as North America, Asia, and Europe, there are higher proportions of residential cropland, rangeland, and woodland. In contrast, the proportion of uninhabited lands is higher in Oceania, Latin America, and Africa. Remote rangeland and woodland are the main parts, rather than residential areas. Africa has a very high proportion of sparse villages, resulting in a high proportion of its rural–urban continuum. Actually, most of the sparse villages in Africa are developed and utilized on bare land, whose development and utilization levels are only comparable to that of residential cropland and rangeland in other continents (Figure 2c).

3.2. Global Human Settlement Change Trajectory in the Rural–Urban Continuum

Not only are there regional differences in the distribution status of various human settlements, but their change rates over the past two decades have also varied considerably. Urban areas in Africa and Asia have been developed exceeding or near the rate of 100%, while almost none of the human settlement types have been developed more than 50% in Oceania and Latin America (Figure 3a).

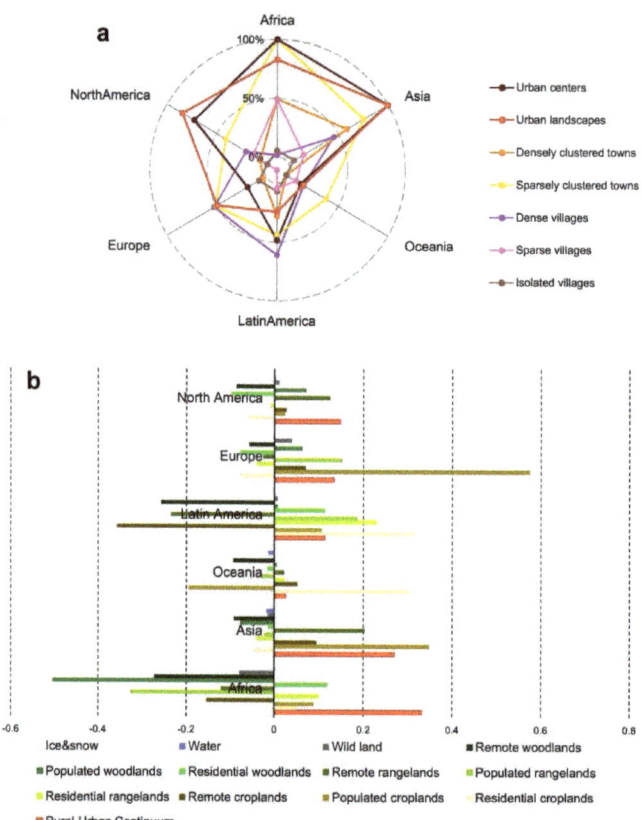

Figure 3. Change rate of land area in human settlements by each continent from 2000 to 2020. (a) Each continent's change rate for human settlements in the rural–wild continuum from 2000 to 2020. (b) Each continent's change rate for human settlements in global land from 2000 to 2020.

In general, urban areas have the highest growth rates in most regions except Europe and Latin America. More dense villages and sparsely clustered towns have been developed in Europe for the last two decades (Figure 3a), which indicates that the process of urbanization in Europe no longer needs to focus on the urban center but enriches the diverse peri-urban types in the rural–urban continuum. Dense villages are also the top growth type in Latin America, and they may take longer than two decades to develop into urban areas. The growth rate of sparsely clustered towns is the highest except for urban areas, which is consistent in each continent (Figure 3a). Sparse villages and isolated villages have expanded the least land area because this may only be the first station of long urbanization process over a period of 20 years. Further evidence will be found in an analysis of differences in the world continents' human settlement change trajectory, which will provide an overall and detailed explanation for regional differences.

With the expansion of built-up areas and the increase in population, global human settlements are generally developing along a wild-rural–urban trajectory. In all of the changed global land, only about 35% became less affected by human activity (labeled as purple), which was mainly distributed in the north of the Eurasian continent (Figure 4a,b).

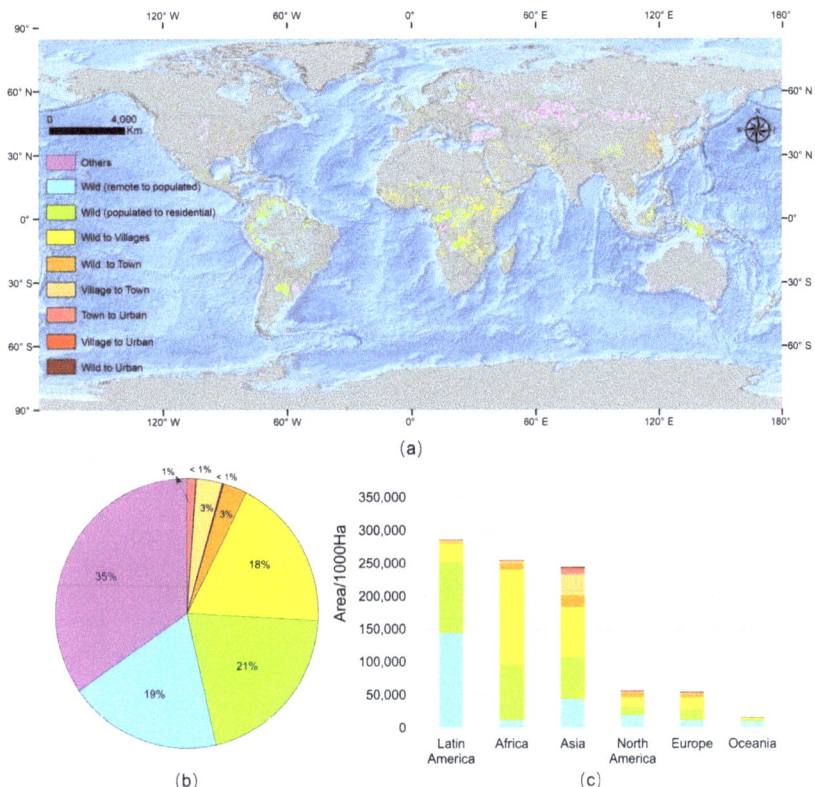

Figure 4. Distribution of human settlement change types (2000–2020). (**a**) Mapping results of human settlement change types (2000–2020). Wild refers to the types of human settlements outside of the rural–urban continuum (urban, town, and village). (**b**) Percentage of land area occupied by each human settlement change type. (**c**) Land area of each human settlement change type in each continent (except other change type, which refers to backward flow against wild-rural–urban direction).

Globally, there were respectively 19%, 21%, and 18% of the changed global land completing the developing trajectory of wild (remote to populated), wild (populated to residential), wild to villages. Urbanization becomes more difficult and slows down when it comes to the developing stage of towns. Only less than 8% of the changed land developed into towns or urban areas. The origin of the town was about half cropland (mainly residential cropland) and half villages while urban areas almost completely came from towns (Figure 4b).

As vast and slowly developing continents, Latin America, Africa, and Asia were the top three continents with the largest change areas of land along the forward wild-rural–urban trajectory. Most of Africa's wild area developed into villages over a 20-years interval while Latin America's wildland recently became more intensely affected by human activities. Asia had the largest area of towns that developed from 2000 to 2020, especially towns developed from villages (Figure 4c).

From the perspective of the 10-year interval, there was a certain pattern of transformation among the types of human settlements. In general, the global human settlements followed the pattern that developed from wildland to villages (isolated—sparse—dense), and then to towns (sparse—dense), and finally to urban areas (edge—center) (Figure 5a,b).

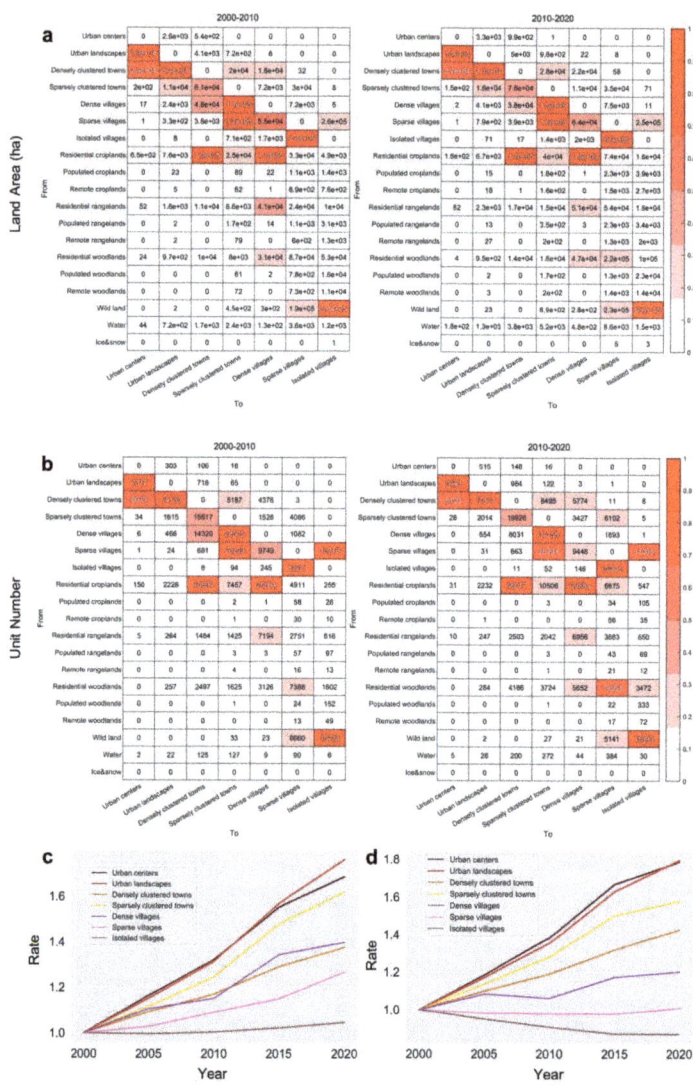

Figure 5. Change between human settlement types. (a) Land-area change matrixes of human settlements (ha, 2000–2010, 2010–2020). (b) Unit number change matrixes of human settlements (2000–2010, 2010–2020). The color of each cell represents the ratio of the value of the cell divided by the maximum value of the column in which the cell resides. (c) Land-area change rate of rural–urban settlements (2000–2020). (d) Unit number change rate of rural–urban settlements (2000–2020).

More specifically, urban centers developed from the urban landscape and densely clustered towns. Densely clustered towns and dense villages originated from residential cropland mainly, while sparsely clustered towns used to be mainly dense villages and sparse villages. The future of isolated villages is sparse villages.

Urbanization was not only reflected in the expansion of the land area in the rural–urban continuum, but also in the phenomenon that the number of cities and towns increased while the number of villages decreased (Figure 5c,d). Over the past 20 years, all types of the rural–urban continuum have expanded in land area, but with different changes in quantity: the number of urban areas has increased by 80%, the number of towns has increased by between 40% and 60%, and the number of villages has increased slowly (20%), remained constant, or even decreased. These increased rural–urban continuum types were all converted from the natural land system in less than 20 years (Figure 5c,d).

3.3. Continent-Scale Urbanization Characteristics

From a regional perspective, human settlements mainly followed the pattern that developed from wildland to villages (isolated—sparse—dense), then to towns (sparse—dense), and finally to urban areas (edge—center), but there were some obvious differences in the structural features of human settlement development progress: Africa is dominated by sparse villages; Asia has the highest proportion of dense clustered towns; the proportion of dense villages in Europe is extremely high; Oceania and Latin America have a high proportion of isolated villages; and the rural–urban continuum of North America is relatively evenly proportioned (Figure 6).

Not all types of human settlement in each continent followed established development patterns. Generally, sparse villages and not isolated villages were formed first in Europe and Oceania. Sparse villages and dense villages were formed in the second phase of Asia's human settlement development progress parallelly not progressively. Only human settlements in three continents (North America, Latin America, Africa) followed established patterns, although there were still some great differences in their development progress. In Africa, the process of human settlement development is stuck at the stage of sparse villages. Very few from this stage continue to develop into dense villages or towns, let alone urban areas. This difficult situation is alleviated in Latin America, whereas in North America there is no such problem, where dense villages and sparsely clustered towns are the main stages in the development process (Figure 6).

There were also some differences in the source of the rural–urban continuum. Wildland was the main source of the rural–urban continuum in Africa; both cropland and wildland contributed to the formation of the rural–urban continuum in Asia; and as for North America, the greatest contribution came from woodland. Unlike in other continents, grassland was also a major contributor in Europe and Oceania (Figure 6).

Not all urbanization processes end in becoming an urban center. From the detailed mapping result, the patterns of urbanization in different regions were captured more intuitively. In Europe, a lot of dense villages emerged outside of urban areas or in transition zones between urban areas and towns. Densely clustered towns are the most common human settlements in Asia; they are usually developed from residential cropland and continue to expand the land area further rather than becoming urban areas with a greater density of built-up area and population. In Africa, sparse villages are the most easily developed human settlements type from natural land systems, but their urbanization process probably ends for this status (Figure 7).

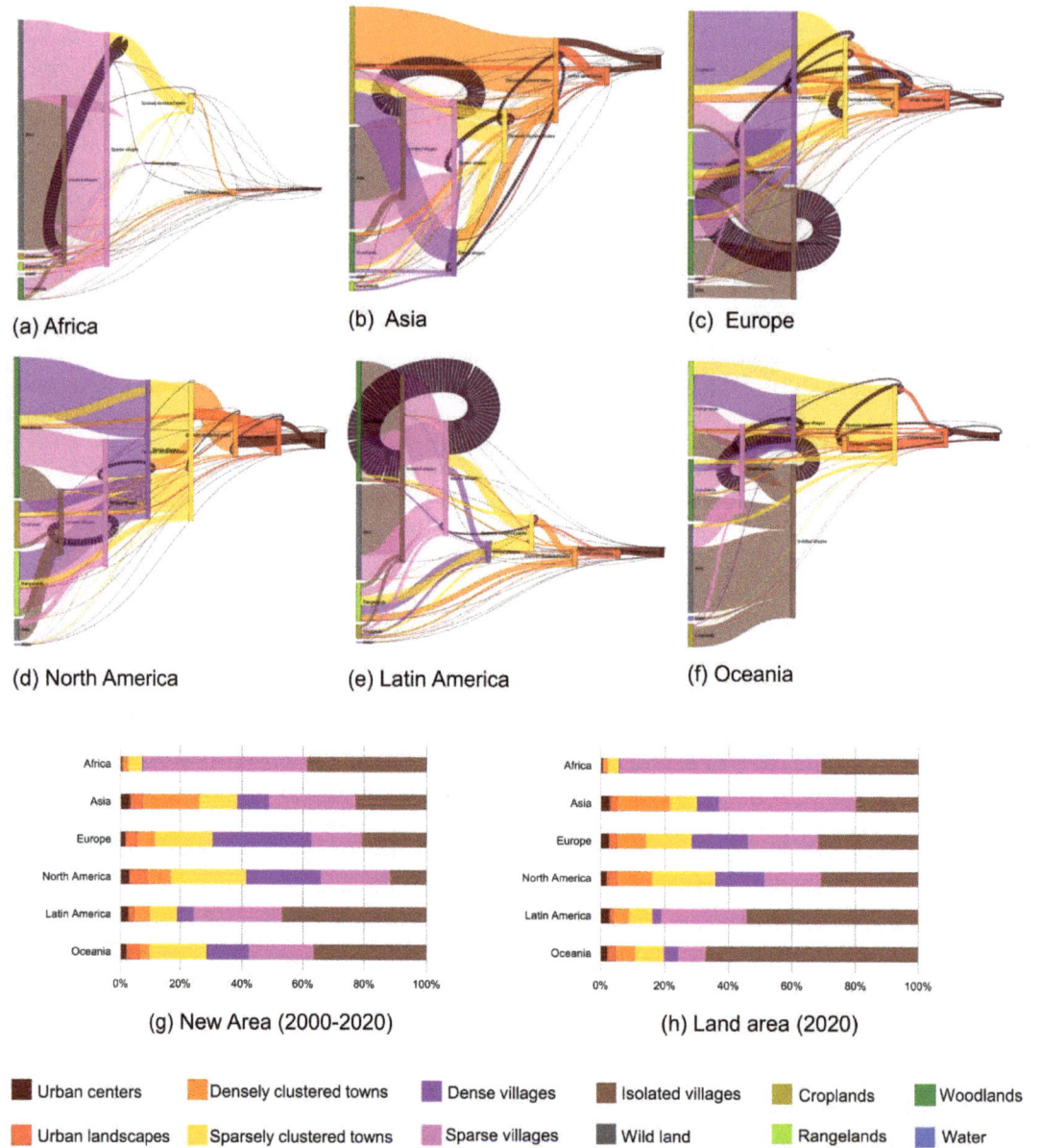

Figure 6. Sankey maps of human settlement development progress (2000–2020, (**a**): Africa, (**b**): Asia, (**c**): Europe, (**d**): North America, (**e**): Latin America, (**f**): Oceania) and percentages of new areas ((**g**), during 2000–2020) and percentages of areas in 2020 (**h**) occupied by each human settlement type in each continent.

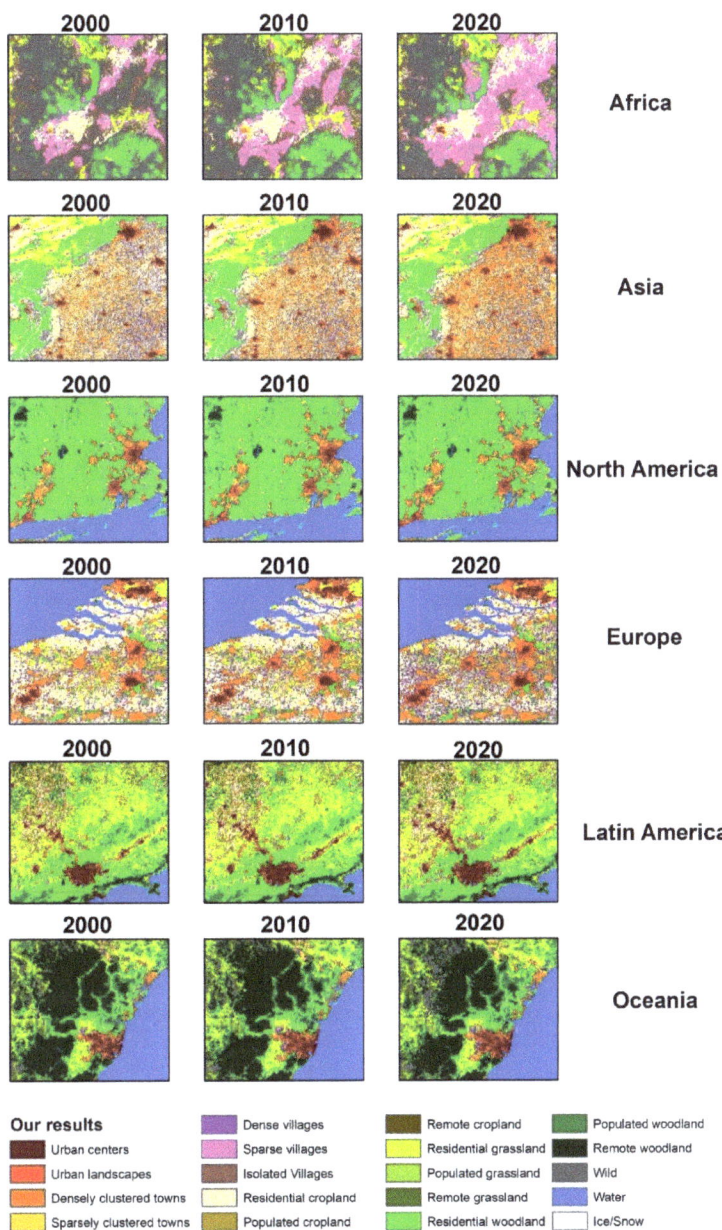

Figure 7. Detailed mapping result of human settlements change trajectory.

3.4. Land and Population in the Rural–Urban Continuum

Human activity reshapes the land surface unevenly. On the whole, the degree of reshaping increases step by step from wild, rural to urban, specifically manifested as the increase in population and land expansion (Figure 8). The rural–urban continuum is the most active part of all the global land, which only covers less than 7% of the global land area but its area has increased by 20% in the past two decades, while other land types have shown weak increases (<5%) or decreases. In addition to urban areas (increasing nearly

70%), the largest increase in the rural–urban continuum was also observed in sparsely clustered towns (~60%) (Figure 4c).

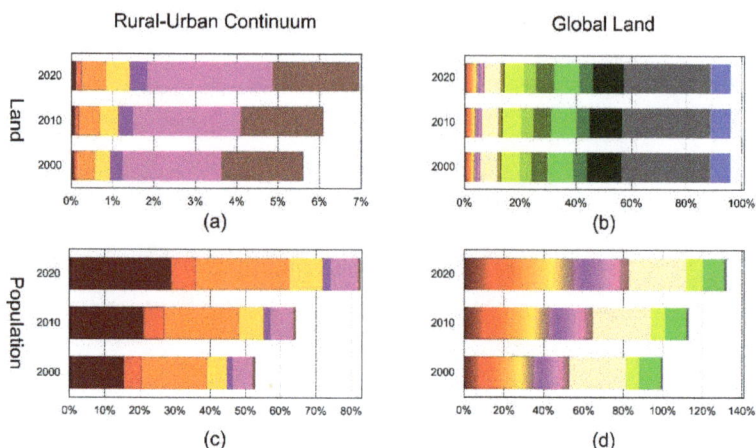

Figure 8. Land occupied by and population distributed in each type of human settlement. (**a**) Percentage of area occupied by each human settlement type in the rural–urban continuum. (**b**) Percentage of area occupied by each land type globally. (**c**) Percentage of population distributed in each human settlement type of the rural–urban continuum. (**d**) Percentage of population distributed in each land type globally. Percentage of population is calculated based on world total population in 2000.

With the land-area expansion of the rural–urban continuum, more and more people live in the rural–urban continuum, especially in urban and peri-urban areas. In 2020, more than 60% of the world's population lived on less than 7% of the land, and 86% of this population lived mainly in urban areas and towns (approximately 20% of the land area) (Figure 8a,c). In urban areas and towns, the growth rate of the land tends to increase every decade, and the growth rate of the population (30%) is even greater than that of the land (25%).

3.5. Economic Growth in Each Human Settlement

In our study, types in the rural–urban continuum are defined and divided using built-up density and population density. The type of rural–urban continuum reflects the combination of built-up density and population density at different levels, whose geographical differences are obvious. Economic factors are also important indicators that distinguish between urban and rural areas. It is worth exploring the role of various human settlement types in economic production, especially its differences in geography. Analysis of economic growth may also provide new insights into the different trajectories of urbanization in each continent.

GDP output per unit area of different human settlement is different. Generally, its value increases step by step along the rural–urban continuum, which is consistent across different continents. Urban centers have the highest production efficiency, while in second place are urban landscapes and densely clustered towns; they differ little in Africa, North America, and Oceania (Figure 9a). In rural areas, it is more difficult to convert land into economic value, which is consistent with our understanding of the intensity of talent and land needed for economic production. Although the relative difference in GDP among types in the rural–urban continuum per unit of land is consistent in all continents, the absolute amount of GDP per unit of land still differs a lot in all continents, which is related to the layout of the global industrial chain and international trade.

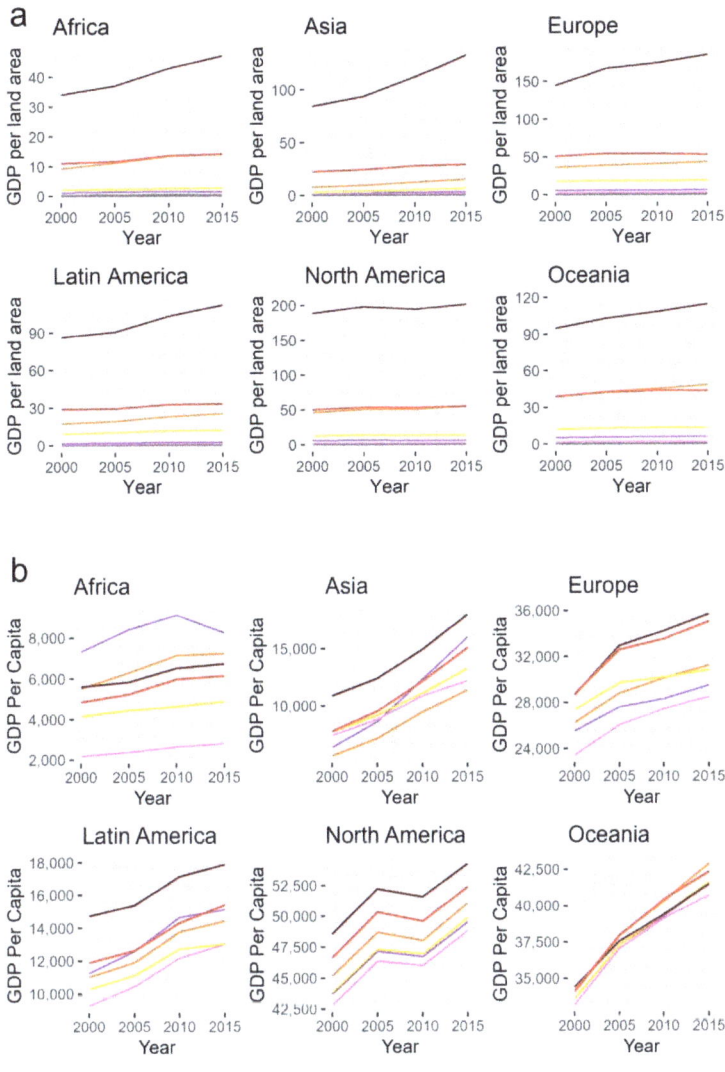

Figure 9. GDP growth in each human settlement from 2000 to 2015. (**a**) GDP per land area (USD/square meter) in each human settlement from 2000 to 2015. (**b**) GDP per capita (USD per capita) in each human settlement from 2000 to 2015. Legend is the same as Figure 5c,d.

Unlike GDP per land area, the GDP per capita of different human settlements is more complex across world continents. Consistent with the evenly proportioned area of human settlements in its development process (Figure 6), the GDP per capita in North America also increases step by step along the rural–urban continuum. In Asia, Africa, and Latin America, which are mostly developing countries, the GDP per capita in dense villages is quite high. In Asia and Latin America, the GDP per capita in dense villages is near that of urban landscapes while the GDP per capita of dense villages is the highest in Africa. This phenomenon was not found in any other continent (Figure 9b). To some extent, this shows the importance of rural revitalization for developing countries. As for the inequality, in Oceania, the GDP per capita differs little in the rural–urban continuum, while Africa has the widest gap in GDP per capita in the rural–urban continuum; the GDP per capita in dense villages is nearly four times higher than in sparse villages (Figure 9b).

4. Discussion

4.1. Implications

The urbanization process is taking place in different and distinctive ways around the world. In our research, this difference in the urbanization process was viewed as the difference in the human settlement change trajectory, which was represented by the change in population and built-up area.

This provides new insights when looking at changes in human settlements from the perspective of the rural–urban continuum. There is only a small percentage of new land area expanded in urban areas, even in North America, the most urbanized region, at less than 10%. The vast majority of new built-up areas occur in rural landscapes like towns and villages. This conclusion was confirmed in other regional scale studies. Most of the built-up land in European countries is dominated by rural landscape [44], most of the built-up area expansion in South America occurs in small cities and around rural areas, not only concentrated around major cities [45]. And in China, the new built-up area is mostly concentrated in rural landscapes rather than big cities [17]. A rural revival is needed to counter urbanization around the globe [46]. As a key initial or intermediate stage in the process of urbanization, towns and villages need additional attention in the future for better coordination and balance between built-up areas, population, and economic growth.

Our study also provides insights into the differences in urbanization paths in different regions of the globe. Sun et al. found that large cities in the low-income and lower-middle-income countries had the highest urban population growth, and built-up area expansion in the upper-middle-income countries was more than three times that of the high-income countries [25]. Tian et al. found that the increases in population density and built-up patch density were the main characteristics of Asia and Africa [26]. Our results also indicated that the growth rate of urban areas in regions dominated by developing countries such as Asia and Africa was much higher than that of North America, Europe, and Oceania (Figure 3a). Tian et al. also found that urbanization in Europe and North America took a rather steady pace, combined with widespread greening [26]. This is consistent with our findings that the urbanization process in Europe over the past two decades has been dominated by the development of dense villages, indicating slower urbanization in Europe. A general trend of population decline in Europe since 2000 has been observed in recently published literature [47]. And the rural–urban continuum of North America is relatively evenly proportioned in the urbanization process (Figure 6).

There could be multiple drivers leading to differences in the urbanization path across the globe. Original land cover varies widely across continents, with more than half of human settlements in Asia and Europe originating in agricultural land systems, while the vast majority of human settlements in Africa have been developed from wild bare land (Figure 6). While there is no direct evidence of causal inference, it is likely that the original land system is related to the most prevalent human settlement types in the urbanization process. Economic development and urbanization complement each other. The population share living in metropolitan areas above 1 million is roughly four times higher in high-income (47%) than in low-income countries (12%). While urbanization does not necessarily lead to economic development, economic development does not happen without urbanization [9]. Aggregation of population and land (especially built-up area) can generate scale effects and boost economic development, thus obtaining the feedback of capital to support and promote the construction of real estate and infrastructure. As a result, economically developed countries can often enter a virtuous cycle of urbanization, as the example of North America shows (Figure 6). In addition, policy factors are also a very important part. Over last two decades, the Chinese government has developed many macro strategies in support of regional sustainable and coordinated development, including the China Western Development Plan (1999), Northeast Area Revitalization Plan (2004), and Rise of Central China Plan (2016) [30]. The trajectory of urbanization is deeply influenced by the management measures issued by the government. Policy in China has truly promoted the development of rural areas, which could be seen in the characteristic landscape of densely clustered towns distributed on the cropland plain (Figure 8).

The detailed mapping results of human settlements in this study can be used as analyzable raster images to meet the diversified needs of land-use data input in the future. A clear definition of the rural–urban continuum and pixel-level global maps provide a database for analyzing inequality in social welfare and infrastructure, including services [3], education, income [48], health care [49], and green space [25,26].

4.2. Scale Effects of Human Settlement Development Progress

In this section, we identified the development stages with the largest proportion of land-area growth at three geographic scales. We found that with the geographic scale becoming increasingly deeper, the unified characteristics of the continents were further diluted, and the different characteristics of the local areas appeared (Figure 10).

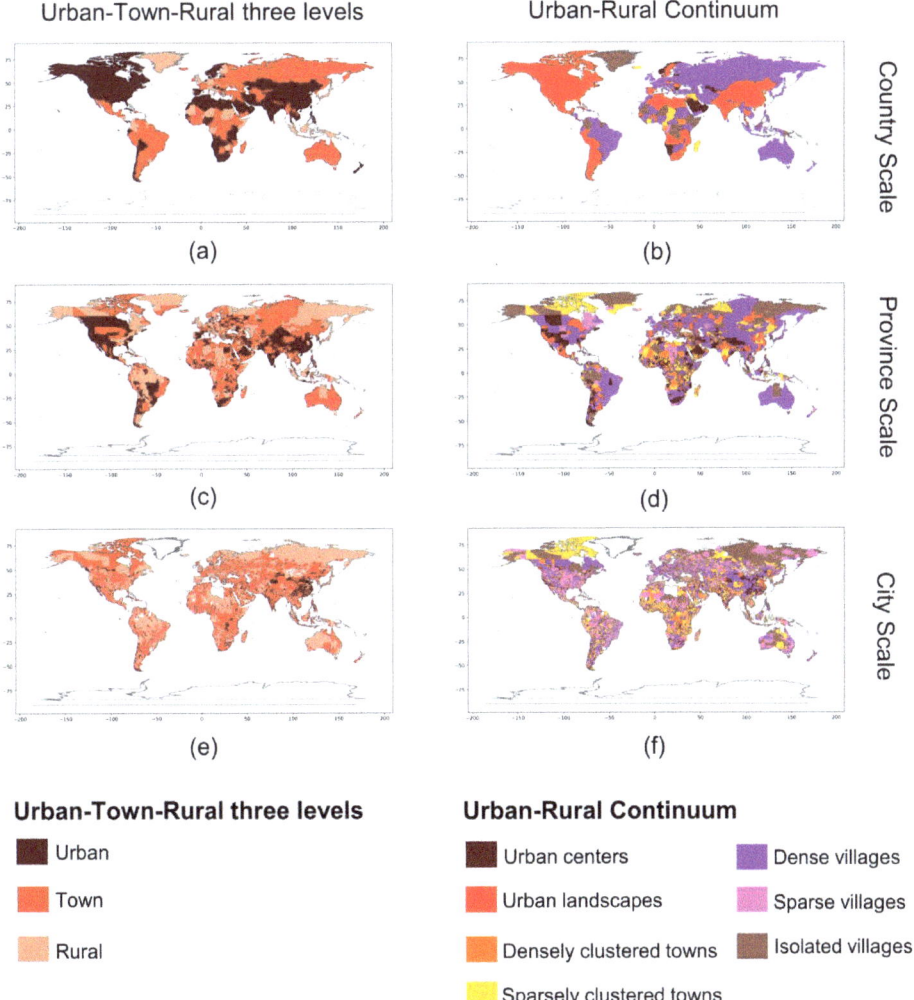

Figure 10. The largest percentage of increasing human settlement identified in the progress of urbanization at country scale (**a,b**), at province scale (**c,d**), at city scale (**e,f**) in the urban-town-rural three levels (**a,c,e**) and in the urban–rural continuum (**b,d,f**).

At country scale, countries with the largest increases in urban area are spread across all continents. Contrary to the observed characteristics of the continents, there are also some African countries with the largest proportion of urban area growth, but Africa is the continent with the largest differences among countries, for there are also some countries stuck in the stage of town and rural development (Figure 10a,b).

It is worth noting that in such a multiscale analysis, geographical units grouped into the same category according to the development process of human settlements may have completely different levels of economic development. For example, in the United States and some third world countries, there is the highest growth rate in urban areas. However, from the overall analysis on the continental scale in Section 3.3, it can be seen that the original urban area of North America is large, and all types of human settlements develop in a balanced and coordinated way, while countries in Africa that originally had a very low level of urbanization have just entered the period of rapid development in the past two decades.

By the same token, the reasons behind some geographical units that are also classified as rural are quite different. For example, in some areas of Europe, urbanization is almost complete, so they can only focus on rural development at this stage. However, some countries in Asia, Africa, and Latin America are still in the early stage of urbanization, trying to revitalize the countryside.

4.3. Uncertainties and Limitations

The criteria and thresholds for the division of urban and rural areas have always been difficult to determine and are different across countries worldwide. Therefore, it is very meaningful to propose a classification system of the global rural–urban continuum and produce the mapping result, so that we can compare the differences in the urbanization process on a global scale. Inevitably, some local specificities have been overlooked in this global effort. So, we compared the spatial distribution of our data products with other similar human settlement mapping products (Figure 11). Although different classification systems are not completely the same, it can be seen from the results that the data products in this study have a strong consistency with other data products in spatial distribution, which can reflect the continuous changes in urban expansion in a long time series. In addition, it also provides finer details of the urbanization progress in the rural–urban continuum.

The differences in human settlement types in the rural–urban continuum are mainly represented by the difference in population density and built-up area density. Based on the World Cities Database, which includes almost all populated places in the world (about 4.3 million), we validated the mapping results by two factors (population density and built-up area density). Results showed that the classification system adopted in this study can effectively reflect the difference in the continuous levels between population density and built-up area density in the rural–urban continuum (Figure 12).

However, there are still some deficiencies in terms of data and methods. In terms of data, the population data used in this study decomposed census-based count results into grid cells using machine learning methods, in which a certain amount of the population was decomposed into cropland, grassland, and woodland, which did not completely correspond to the built-up area data layer, although this information provided us with an understanding of the level of human activities in natural land systems. Similarly, in landcover data, there is also the problem of poor identification of built-up areas in natural land systems. To solve this problem, except for the main mapping process, we post-processed the grid whose population density met the corresponding conditions while the built-up density did not. Since the research was carried out on a scale of 1 km, there may be bias in the numerical estimation of the land area due to the coarse resolution.

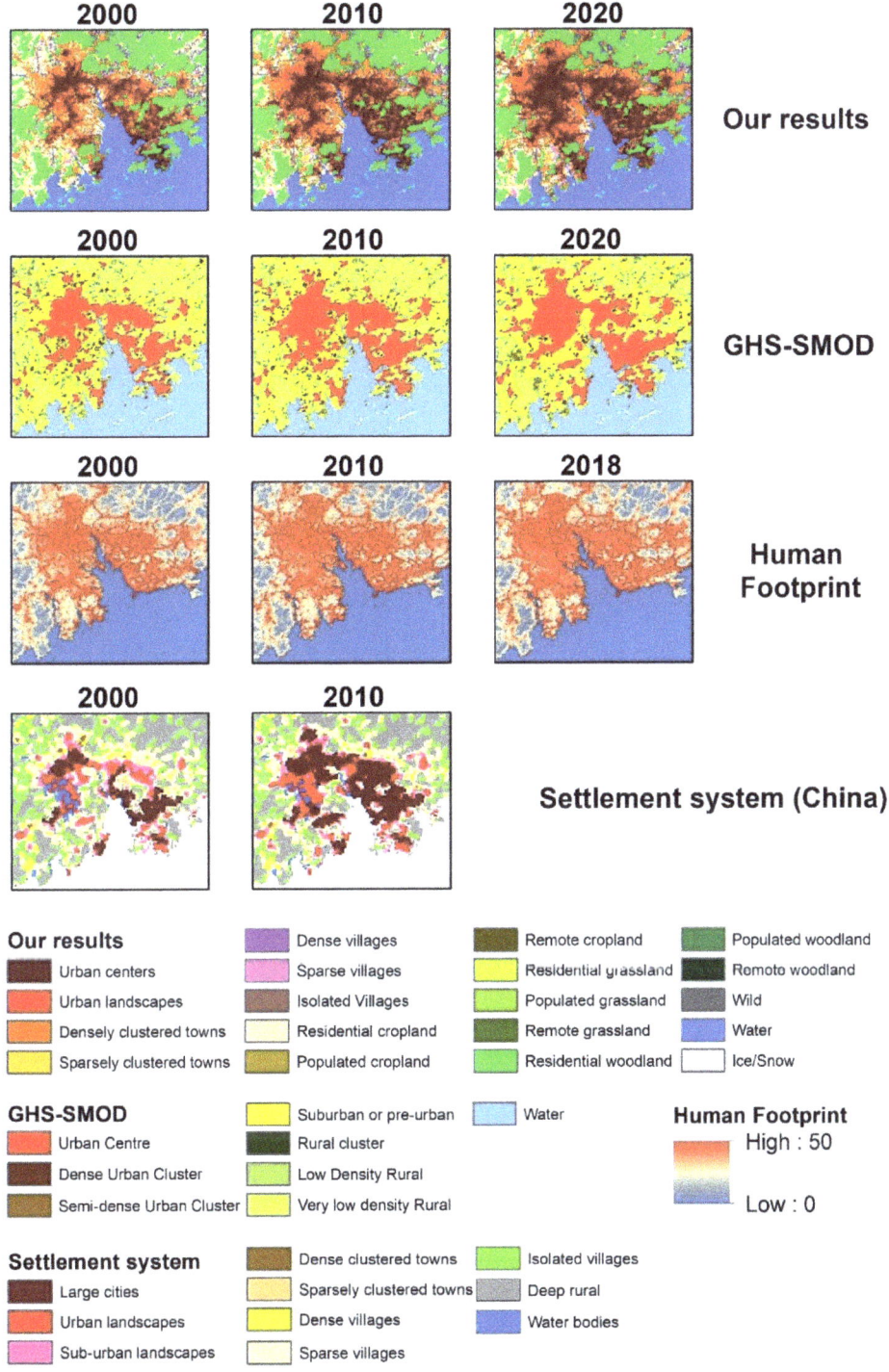

Figure 11. Comparing our dataset with other related datasets.

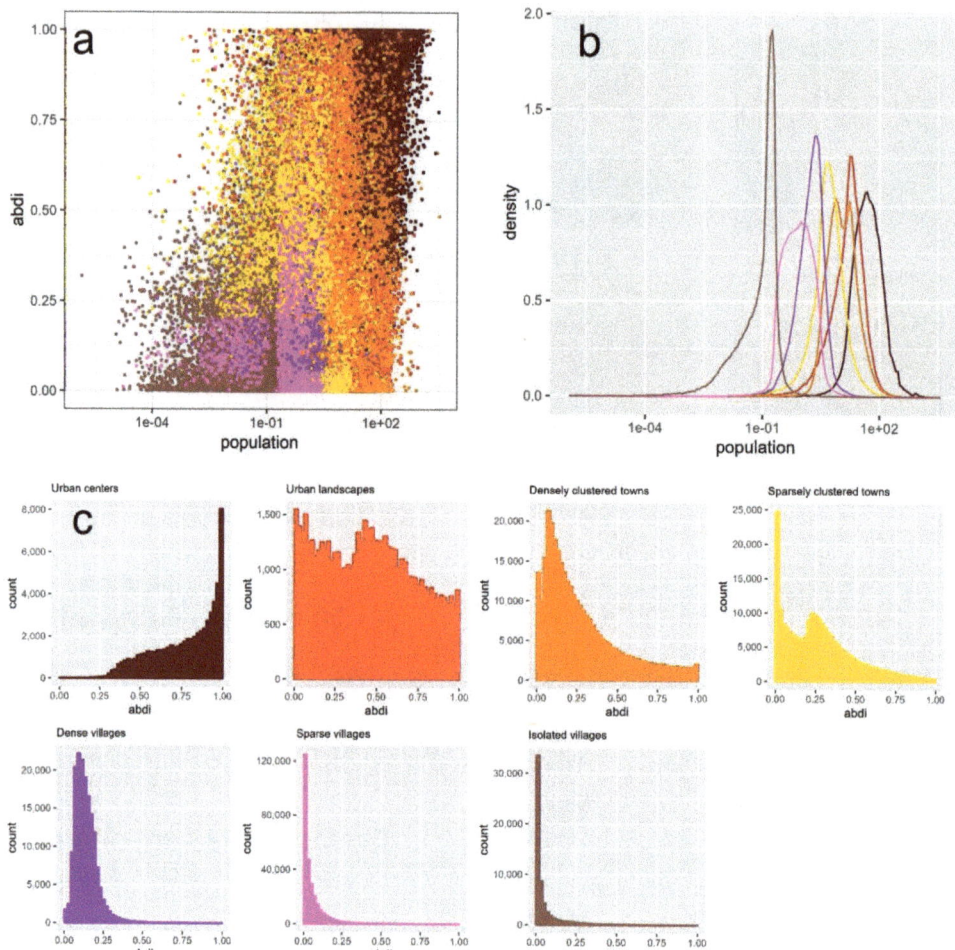

Figure 12. Population density and built-up density (ABDI, Adjusted Built-up Density Index) of human settlements samples in the rural–urban continuum. (**a**) Scatter plot of ABDI against the population density. (**b**) Distribution of population density in the rural–urban continuum. (**c**) Sample counts of different ABDI levels in the rural–urban continuum.

As for the method, corresponding to the threshold problem, the formulation of the classification system also ignores the particularity of the local landscape. This study only considered the two factors (population and built-up area) in the process of mapping the rural–urban continuum. In Section 4.3, it was also mentioned that the economic factor is also the main difference between urban and rural areas. In addition, the urban–rural differences in human welfare such as in income, education, and health are also worth exploring. The simple decision tree classification system in this study can be applied in the Google Earth Engine to efficiently produce mapping products. However, the complex differences between urban and rural areas around the world still need to be deeply and carefully excavated.

5. Conclusions

From the perspective of the rural–urban continuum, urban and rural areas are two poles of a gradient with many continuous human settlements in between. Using global land-cover data (FROM-GLC Plus) and global population data (Worldpop) based on the decision tree method, this study proposed a method and classification system for global rural–urban continuum mapping and produced the mapping results on a global scale in the Google Earth Engine platform.

Our results indicated that global human settlements follow the pattern that develops from wildland to villages (isolated—sparse—dense), then to towns (sparse—dense), and finally to urban areas (edge—center). Over the past 20 years, all types of cities, towns, and villages have expanded in land area but with different trends in quantity; contrasted with the dramatic growth of urban areas (80%) and towns (40–60%), the number of villages has increased slowly (20%) or even decreased. From a regional perspective, North America, Asia, and Europe respectively had the highest proportions of urban areas and towns (nearly 30%) in the rural–urban continuum, but in Oceania and Latin America, this ratio was less than 20%. Urban areas and towns in Africa comprise only about 10%. There are also some obvious differences in the structural features of urbanization progress: Africa is dominated by sparse villages; Asia has the highest proportion of densely clustered towns; the proportion of dense villages in Europe is extremely high; Oceania and Latin America have a high proportion of isolated villages; and the rural–urban continuum of North America is relatively evenly proportioned. Rural–urban continuum mapping and analysis provide a database and new insights into urbanization and the differences between urban and rural areas around the world.

Author Contributions: X.L. and L.Y. designed the experiments and X.L. carried them out. X.L. prepared the manuscript with contributions from L.Y. and X.C. All authors have read and agreed to the published version of the manuscript.

Funding: This work was supported by the National Key R&D Program of China (grant number: 2022YFE0209400) and Tsinghua University Initiative Scientific Research Program (grant number: 20223080017).

Data Availability Statement: The Worldpop dataset is available at https://www.worldpop.org (accessed on 13 August 2023). and SRTM DEM data are available at https://doi.org/10.1029/2005RG0 00183 (accessed on 13 August 2023). The two datasets used in this study are also available at the GEE platform (ee.ImageCollection ("WorldPop/GP/100m/pop"); ee.Image ("USGS/SRTMGL1_003")). The Gridded Global GDP dataset is available at https://doi.org/10.5061/dryad.dk1j0 (accessed on 13 August 2023). The GeoBoundaries dataset is available at https://www.geoboundaries.org/ (accessed on 13 August 2023).

Acknowledgments: The authors wish to thank all the anonymous reviewers for their constructive comments.

Conflicts of Interest: The authors declare no conflict of interest.

References

1. Ellis, E.C.; Beusen, A.H.W.; Goldewijk, K.K. Anthropogenic Biomes: 10,000 BCE to 2015 CE. *Land* **2020**, *9*, 129. [CrossRef]
2. Ellis Erle, C.; Kaplan Jed, O.; Fuller Dorian, Q.; Vavrus, S.; Klein Goldewijk, K.; Verburg Peter, H. Used planet: A global history. *Proc. Natl. Acad. Sci. USA* **2013**, *110*, 7978–7985. [CrossRef] [PubMed]
3. Cattaneo, A.; Nelson, A.; McMenomy, T. Global mapping of urban–rural catchment areas reveals unequal access to services. *Proc. Natl. Acad. Sci. USA* **2021**, *118*, e2011990118. [CrossRef]
4. van Vliet, J.; Birch-Thomsen, T.; Gallardo, M.; Hemerijckx, L.-M.; Hersperger, A.M.; Li, M.; Tumwesigye, S.; Twongyirwe, R.; van Rompaey, A. Bridging the rural-urban dichotomy in land use science. *J. Land Use Sci.* **2020**, *15*, 585–591. [CrossRef]
5. Lang, R.E.; Dhavale, D. *Beyond Megalopolis: Exploring America's New "Megapolitan" Geography*; Brookings Mountain West: Las Vegas, NV, USA, 2005; pp. 1–33.
6. Alexander Wandl, D.I.; Nadin, V.; Zonneveld, W.; Rooij, R. Beyond urban–rural classifications: Characterising and mapping territories-in-between across Europe. *Landsc. Urban Plan.* **2014**, *130*, 50–63. [CrossRef]
7. Cyriac, S.; Firoz, C.M. Dichotomous classification and implications in spatial planning: A case of the Rural-Urban Continuum settlements of Kerala, India. *Land Use Policy* **2022**, *114*, 105992. [CrossRef]
8. Shaw, B.J.; van Vliet, J.; Verburg, P.H. The peri-urbanization of Europe: A systematic review of a multifaceted process. *Landsc. Urban Plan.* **2020**, *196*, 103733. [CrossRef]

9. OECD & European Commission. Cities in the World: A New Perspective on Urbanisation. 2020. Available online: https://www.oecd.org/publications/cities-in-the-world-d0efcbda-en.htm (accessed on 13 August 2023).
10. Benet, F. Sociology Uncertain: The Ideology of the Rural-Urban Continuum. *Comp. Stud. Soc. Hist.* **1963**, *6*, 1–23. [CrossRef]
11. Dewey, R. The Rural-Urban Continuum: Real but Relatively Unimportant. *Am. J. Sociol.* **1960**, *66*, 60–66. [CrossRef]
12. Pahl, R.E. THE RURAL-URBAN CONTINUUM1. *Sociol. Rural.* **1966**, *6*, 299–329. [CrossRef]
13. Sorokin, P.A.; Zimmerman, C.C. *Principles of Rural-Urban Sociology*; Henry Holt: New York, NY, USA, 1929.
14. Wirth, L. Urbanism as a Way of Life. *Am. J. Sociol.* **1938**, *44*, 1–24. [CrossRef]
15. Hines, F.K.; Brown, D.L.; Zimmer, J.M. *Social and Economic Characteristics of the Population in Metro and Nonmetro Counties, 1970*; Economic Research Service, US Department of Agriculture: Washington, DC, USA, 1975.
16. Yang, X.; Wang, Q.; Zhou, Q. Regional habitat units in the context of urban-rural China: Concept, mechanism and features. *Habitat Int.* **2022**, *128*, 102668. [CrossRef]
17. Li, M.; van Vliet, J.; Ke, X.; Verburg, P.H. Mapping settlement systems in China and their change trajectories between 1990 and 2010. *Habitat Int.* **2019**, *94*, 102069. [CrossRef]
18. Dijkstra, L.; Florczyk, A.J.; Freire, S.; Kemper, T.; Melchiorri, M.; Pesaresi, M.; Schiavina, M. Applying the Degree of Urbanisation to the globe: A new harmonised definition reveals a different picture of global urbanisation. *J. Urban Econ.* **2021**, *125*, 103312. [CrossRef]
19. Schiavina, M.; Melchiorri, M.; Pesaresi, M.; Politis, P.; Freire, S.; Maffenini, L.; Florio, P.; Ehrlich, D.; Goch, K.; Tommasi, P. *GHSL Data Package 2022*; Publications Office of the European Union: Luxembourg, 2022.
20. van Vliet, J.; Eitelberg, D.A.; Verburg, P.H. A global analysis of land take in cropland areas and production displacement from urbanization. *Glob. Environ. Change* **2017**, *43*, 107–115. [CrossRef]
21. UN Habitat. The Strategic Plan 2020–2023. 2019. Available online: https://unhabitat.org/the-strategic-plan-2020-2023 (accessed on 13 August 2023).
22. Gong, P.; Li, X.; Wang, J.; Bai, Y.; Chen, B.; Hu, T.; Liu, X.; Xu, B.; Yang, J.; Zhang, W.; et al. Annual maps of global artificial impervious area (GAIA) between 1985 and 2018. *Remote Sens. Environ.* **2020**, *236*, 111510. [CrossRef]
23. Liu, X.; Huang, Y.; Xu, X.; Li, X.; Li, X.; Ciais, P.; Lin, P.; Gong, K.; Ziegler, A.D.; Chen, A.; et al. High-spatiotemporal-resolution mapping of global urban change from 1985 to 2015. *Nat. Sustain.* **2020**, *3*, 564–570. [CrossRef]
24. Schmitt, A.; Uth, P.; Standfuß, I.; Heider, B.; Siedentop, S.; Taubenböck, H. Quantitative assessment and comparison of urban patterns in Germany and the United States. *Comput. Environ. Urban Syst.* **2023**, *100*, 101920. [CrossRef]
25. Sun, L.; Chen, J.; Li, Q.; Huang, D. Dramatic uneven urbanization of large cities throughout the world in recent decades. *Nat. Commun.* **2020**, *11*, 5366. [CrossRef]
26. Tian, Y.; Tsendbazar, N.-E.; van Leeuwen, E.; Fensholt, R.; Herold, M. A global analysis of multifaceted urbanization patterns using Earth Observation data from 1975 to 2015. *Landsc. Urban Plan.* **2022**, *219*, 104316. [CrossRef]
27. Di Marco, M.; Venter, O.; Possingham, H.P.; Watson, J.E.M. Changes in human footprint drive changes in species extinction risk. *Nat. Commun.* **2018**, *9*, 4621. [CrossRef] [PubMed]
28. Pecl Gretta, T.; Araújo Miguel, B.; Bell Johann, D.; Blanchard, J.; Bonebrake Timothy, C.; Chen, I.C.; Clark Timothy, D.; Colwell Robert, K.; Danielsen, F.; Evengård, B.; et al. Biodiversity redistribution under climate change: Impacts on ecosystems and human well-being. *Science* **2017**, *355*, eaai9214. [CrossRef]
29. Cumming, G.S.; Buerkert, A.; Hoffmann, E.M.; Schlecht, E.; von Cramon-Taubadel, S.; Tscharntke, T. Implications of agricultural transitions and urbanization for ecosystem services. *Nature* **2014**, *515*, 50–57. [CrossRef] [PubMed]
30. Jiang, H.; Sun, Z.; Guo, H.; Weng, Q.; Du, W.; Xing, Q.; Cai, G. An assessment of urbanization sustainability in China between 1990 and 2015 using land use efficiency indicators. *npj Urban Sustain.* **2021**, *1*, 34. [CrossRef]
31. Li, X.; Gong, P. Urban growth models: Progress and perspective. *Sci. Bull.* **2016**, *61*, 1637–1650. [CrossRef]
32. European, C.; Joint Research, C.; Freire, S.; Corbane, C.; Zanchetta, L.; Schiavina, M.; Politis, P.; Kemper, T.; Ehrlich, D.; Pesaresi, M.; et al. *GHSL Data Package 2019: Public Release GHS P2019*; Publications Office: Luxembourg, 2019.
33. Balk, D.; Montgomery, M.R.; Engin, H.; Lin, N.; Major, E.; Jones, B. Urbanization in India: Population and Urban Classification Grids for 2011. *Data* **2019**, *4*, 35. [CrossRef] [PubMed]
34. Jin, X.; Long, Y.; Sun, W.; Lu, Y.; Yang, X.; Tang, J. Evaluating cities' vitality and identifying ghost cities in China with emerging geographical data. *Cities* **2017**, *63*, 98–109. [CrossRef]
35. Sun, H.; Liu, Y.; Xu, K. Hollow villages and rural restructuring in major rural regions of China: A case study of Yucheng City, Shandong Province. *Chin. Geogr. Sci.* **2011**, *21*, 354–363. [CrossRef]
36. Mu, H.; Li, X.; Wen, Y.; Huang, J.; Du, P.; Su, W.; Miao, S.; Geng, M. A global record of annual terrestrial Human Footprint dataset from 2000 to 2018. *Sci. Data* **2022**, *9*, 176. [CrossRef]
37. Yu, L.; Du, Z.; Dong, R.; Zheng, J.; Tu, Y.; Chen, X.; Hao, P.; Zhong, B.; Peng, D.; Zhao, J.; et al. FROM-GLC Plus: Toward near real-time and multi-resolution land cover mapping. *GIScience Remote Sens.* **2022**, *59*, 1026–1047. [CrossRef]
38. Tatem, A.J. WorldPop, open data for spatial demography. *Sci. Data* **2017**, *4*, 170004. [CrossRef] [PubMed]
39. Farr, T.G.; Rosen, P.A.; Caro, E.; Crippen, R.; Duren, R.; Hensley, S.; Kobrick, M.; Paller, M.; Rodriguez, E.; Roth, L.; et al. The Shuttle Radar Topography Mission. *Rev. Geophys.* **2007**, *45*. [CrossRef]
40. Kummu, M.; Taka, M.; Guillaume, J.H.A. Gridded global datasets for Gross Domestic Product and Human Development Index over 1990–2015. *Sci. Data* **2018**, *5*, 180004. [CrossRef] [PubMed]

1. Runfola, D.; Anderson, A.; Baier, H.; Crittenden, M.; Dowker, E.; Fuhrig, S.; Goodman, S.; Grimsley, G.; Layko, R.; Melville, G.; et al. geoBoundaries: A global database of political administrative boundaries. *PLoS ONE* **2020**, *15*, e0231866. [CrossRef] [PubMed]
2. Safanelli, J.L.; Poppiel, R.R.; Ruiz, L.F.; Bonfatti, B.R.; Mello, F.A.; Rizzo, R.; Demattê, J.A.M. Terrain Analysis in Google Earth Engine: A Method Adapted for High-Performance Global-Scale Analysis. *ISPRS Int. J. Geo-Inf.* **2020**, *9*, 400. [CrossRef]
3. Cao, Y.; Carver, S.; Yang, R. Mapping wilderness in China: Comparing and integrating Boolean and WLC approaches. *Landsc. Urban Plan.* **2019**, *192*, 103636. [CrossRef]
4. van Vliet, J.; Verburg, P.H.; Grădinaru, S.R.; Hersperger, A.M. Beyond the urban-rural dichotomy: Towards a more nuanced analysis of changes in built-up land. *Comput. Environ. Urban Syst.* **2019**, *74*, 41–49. [CrossRef]
5. Andrade-Núñez, M.J.; Aide, T.M. Built-up expansion between 2001 and 2011 in South America continues well beyond the cities. *Environ. Res. Lett.* **2018**, *13*, 084006. [CrossRef]
6. Liu, Y.; Li, Y. Revitalize the world's countryside. *Nature* **2017**, *548*, 275–277. [CrossRef]
7. Newsham, N.; Rowe, F. Understanding trajectories of population decline across rural and urban Europe: A sequence analysis. *Popul. Space Place* **2023**, *29*, e2630. [CrossRef]
8. Thiede, B.C.; Butler, J.L.W.; Brown, D.L.; Jensen, L. Income Inequality across the Rural-Urban Continuum in the United States, 1970–2016. *Rural. Sociol.* **2020**, *85*, 899–937. [CrossRef]
9. Sibley, L.M.; Weiner, J.P. An evaluation of access to health care services along the rural-urban continuum in Canada. *BMC Health Serv. Res.* **2011**, *11*, 20. [CrossRef]

Disclaimer/Publisher's Note: The statements, opinions and data contained in all publications are solely those of the individual author(s) and contributor(s) and not of MDPI and/or the editor(s). MDPI and/or the editor(s) disclaim responsibility for any injury to people or property resulting from any ideas, methods, instructions or products referred to in the content.

 *land*

Article

Analysis of the Contribution of Land Registration to Sustainable Land Management in East Gojjam Zone, Ethiopia

Abebaw Andarge Gedefaw

Institute of Land Administration, Debre Markos University (DMU), Debre Markos 269, Ethiopia; abebaw.andarge@dmu.edu.et

Abstract: Land registration programs on a large scale aimed at strengthening the land rights of farm households in Ethiopia have been executed in different degrees across different regions since 1998. This study investigates the contribution of land registration on the perceived tenure security of farmers, farmer confidence, women and marginalized groups, and sustainable land-management practice after receiving a land holding certificate in the dryland areas of East Gojjam Zone, Ethiopia. Face-to-face interviews were conducted with 385 households selected by using stratified random sampling techniques. Furthermore, focus group discussions and key informants are primary data sources. According to an investigation of qualitative and quantitative data, 163 households have a mean of 0.40 ha of agricultural land on steep slope areas, and approximately 26% of households are afraid of land redistribution and farm loss in the next five years. Moreover, 22% of households fear the government taking their farm plot at any time. Respondents, on the other hand, believe that land registration has reduced the landlessness of women, the disabled, and the poorest of the poor while increasing the landlessness of youths. After land registration, household participation in land-management practices increased by 15%. Despite this, the difference in the mean of major crop yields per household is insignificant, except for wheat, which decreased significantly at the $p < 0.1$ level. The study determined household head age, household size, land management training and advice, livestock holdings, and the mean distance from farm to settlement as influential factors for increasing construction of water-harvesting systems. Land registration, in general, enhances land tenure security, land-management practice, and land rights of women and marginalized groups of societies, but did not improve crop productivity. The findings should persuade policymakers to address potential sources of insecurity, such as future land redistribution issues.

Keywords: land registration; sustainable land management; land tenure security; water-harvesting system; dryland areas; East Gojjam Zone; Ethiopia

Citation: Gedefaw, A.A. Analysis of the Contribution of Land Registration to Sustainable Land Management in East Gojjam Zone, Ethiopia. *Land* 2023, *12*, 1157. https://doi.org/10.3390/land12061157

Academic Editor: Le Yu

Received: 10 April 2023
Revised: 29 May 2023
Accepted: 29 May 2023
Published: 31 May 2023

Copyright: © 2023 by the author. Licensee MDPI, Basel, Switzerland. This article is an open access article distributed under the terms and conditions of the Creative Commons Attribution (CC BY) license (https://creativecommons.org/licenses/by/4.0/).

1. Introduction

1.1. Background of the Study

Ethiopia is currently in the process of economic transformation with the goal of becoming a lower-middle income economy by 2025. Agriculture is arguably the most important focus of this process, as developing the agricultural sector is one of the best ways to stimulate rapid, inclusive economic growth. However, this development would have proven impossible if not for land registration. The International Federation of Surveyors (FIG) defines land registration as the official recording of legally recognized interests in land [1]. Land registration is important in understanding the impact of human societies on natural systems, which also has psychological implications [2]. Land degradation entails soil erosion [3,4], desertification [5], pollution [6] and inappropriate land-management practices [7], among others. Land degradation is also caused by human intervention in natural ecosystems [8,9]. Environmental and socioeconomic issues such as high population pressure, land degradation, unsustainable farming practices, and land tenure insecurity impede Ethiopia's agricultural development [10–14]. Across Africa, land tenure insecurity limits agricultural production and livelihood improvements [15,16]. Government efforts to achieve their development goals can be hampered by tenure insecurity, which is seen to affect agricultural productivity [17–19].

Land is an essential component of household socioeconomic capital, especially in Africa, where agriculture supports most households. More importantly, secure access to land is critical. Long-term investments in sustainable livelihoods by rural households are required for sustainable agricultural development [20]. Most African communities rely on land for survival, and land resources are the cornerstone of achieving many of the UN Sustainable Development Goals (SDGs) [21]. Moreover, securing land rights has been identified as an important strategy for achieving SDGs [22]. The 2030 United Nations Development Goals, specifically Goal 1 (poverty), Goal 2 (hunger, food security, nutrition, and sustainable agriculture), Goal 5 (addressing gender equality and the empowerment of women and girls), and Goal 15 (issue on life on land), emphasize the importance of access to and control over land, as well as sustainable land management and associated resources. As a result, a modern land administration system, including formal land registration, titling, and certification, has been viewed as a prerequisite for ensuring property rights and agricultural development [23–25]. Land tenure should be properly administered for positive societal changes by establishing formal land titling procedures [26]. It is argued that tenure security has a positive impact on land investment by improving holding rights and providing a sense of stability, which encourages farmers to make sustainable land investments and increase yield [27]. The need to divert private resources to protect property rights is decreased by improved tenure security [28]. The main finding of empirical research is that land tenure security improves land-related investment [29–32] by strengthening land claims and enhancing farmers' credit access [33,34] and agricultural productivity [35,36]. Titling, on the other hand, can enhance intensification and other unsustainable land practices by fueling land contestation, particularly in legally pluralistic contexts [37–40], and reinventing local common-pool resource problems that communities may or may not be willing to address [41,42].

Contrarily, tenure insecurity is a major barrier to the adoption of sustainable land management, contributing to increased environmental degradation across Sub-Saharan Africa, including Ethiopia [43–47]. It has long been recognized that unclear and insecure property rights can discourage farmers from making land-improvement investments due to the uncertainty and future expropriation risk by the government [48–50]. Furthermore, theoretical and empirical studies suggest that a lack of secure access to land is frequently seen as a significant factor in food insecurity, limited livelihood opportunities, and, consequently, poverty [20,51].

Thus, calls for land titling are widespread and have been going on for a long time in Africa, despite the fact that early land reform programs were frequently unsuccessful [52,53]. However, the growing need in Africa for the formalization of land rights and a well-regulated land management system is highlighted by the increasing pressure on farmland brought on by population growth and foreign investor demand for large-scale agricultural land [54].

Due to the importance of land as a source of livelihood and political power in Ethiopia, the land tenure system has been at the forefront of policy debates for generations [55,56]. In the decades prior to and during the imperial era, land was concentrated in the hands of absentee landlords, and arbitrary evictions posed a serious threat to tenant farmers [56,57]. After overthrowing the imperial regime of Haile Selassie through a military coup (1974), the socialist Derg regime implemented radical reforms that altered the agrarian structure and access to land, transferring land ownership to the state [56,58].

Following the fall of the Derg regime in 1991, the current government began to liberalize the economy. However, the reform package largely "overlooked" the land issue mainly land reform, and the legacy of the Derg continued to define key elements of current land policy [55,56]. Land rights are still held by the state. On the other hand, the current administration has made several changes. First, responsibility for land issues was devolved to regions. Second, the frequency of land redistribution (where the aim was to redistribute land according to the needs (family size) of households and to provide land for young married couples, women, marginalized groups, and youth) has been reduced, but it is not entirely off the agenda. Third, while land rentals are officially permitted, some regions still impose restrictions on the terms of rental contracts. Overall, the state continues to be a source of tenure insecurity. The government remains critical of privatizing land holdings, retaining a discourse of social equity and protection of land concentration in the hands of the few. However, some have argued that the government uses land rather as a "carrot and stick" to achieve political goals [55,56].

In the past, Ethiopia experienced frequent land redistribution, which led to land fragmentation, underutilization of land, and tenure insecurity [29,59–63]. Furthermore, land redistribution was primarily carried out in the years immediately following the 1975 governmental change, but additional land redistributions have occurred since then (constitutionally this requires a significant majority to demand a land redistribution to take place) [64]. As a result of these legal changes, and significant land holding shifts, smallholders did not perceive that they had a high degree of land tenure security—the land redistribution after all was only usufruct rights, not ownership rights. This tenure system was largely continued with the entrance of the new government in 1991, which made only minor changes to the ability to rent land on a short-term basis. In 1997, the Amhara National Regional State made significant land redistributions. Following this, there was much debate in Ethiopia about the consequences of this redistribution. Farmers have been discouraged from making improvements to their land due to the perception that land redistribution undermines tenure security [65]. Therefore, it is thought that this fragmentation and reallocation of land holdings will negatively affect land management activities [65]. Ethiopia's government is currently focusing on landscape restoration and sustainable land management.

In response to the negative effects of tenure insecurity on sustainable land-management practices, the Ethiopian government executed a large-scale land registration program in 1998. Ethiopia has one of Africa's most extensive, rapid, and low-cost land registration reforms, and has been cited as a model for land certification in Africa [66]. Across the four regions (i.e., Amhara, Oromia, Tigray and (SNNP) Southern Nations, Nationalities, and Peoples), some 15 million parcels of the total 50 million parcels had been registered and certificates distributed to landholders. From these, about 25% of the parcels are solely owned by women and 55% jointly held by husbands and wives. Only 20% of the total parcels registered were under name of male landholders alone [67]. Previous research on the effects of land registration in Ethiopia has focused primarily on the Tigray region. According to these studies, land registration is associated with higher levels of land-related investment and productivity [68], improved welfare [58], increased land rental market participation [58] and reduced border conflicts [69]. Similar to this, the Amhara region of Ethiopia has also documented the positive and significant effects of land registration on household perceived tenure security, investment, and land market participation [29]. In addition to this, [57,66] used data from four major regions of Ethiopia and discovered that land registration has a positive effect on land management.

According to one of the preambles of the Amhara Region Rural Land Administration and Land Use Proclamation No. 133/2006, the establishment of land ownership enhances landholders' ability to use their labor, wealth, and creativity [70]. Any person granted rural land in the region shall be given a land holding certificate, on which the details of the land are registered by the Authority and his photograph is fixed [70]. However, previous studies focused on the effect of land registration on tenure security and land-management practice by comparing titled and untitled land holders at the kebele level. Moreover, Gedefaw et al. [71] focused on the effects of land certification on sustainable land management, particularly on terracing and manure use. Another study carried out by Mengesha et al. [25] investigated land certification effect on sustainable land management, especially on tree planting. One key exception is [72], who studied the effect of land certification on sustainable land-management practice in the dryland areas. Farming in dryland areas is risky due to lack of rainfall and unsustainable land-management practices. Still, no study on the contribution of land registration on sustainable land management has focused on the construction of water-harvesting systems in the dryland areas of Ethiopia generally, and in the Amhara region specifically. However, this study aimed to fill the existing research gap by investigating, with reasonable scientific justification, the changes brought about by individual households in terms of land tenure security and land management before and after registration in the dryland area of Ethiopia's East Gojjam Zone. Therefore, this study investigated the contribution of land registration on perceived tenure security of farmers, farmer confidence, women and marginalized groups, and sustainable land-management practice in dryland areas.

To achieve this objective, the following research questions were formulated:

a. Does land registration improve the sense of tenure security of farm households?
b. Does land registration improve the holding rights of women and marginalized groups in the study area?
c. Is there a change of crop productivity after the land registration process?
d. Does land registration improve perceived tenure security. If yes, what are the influencing factors?
e. Does land registration improve land-management practices such as water-harvesting system. If yes, what are the influencing factors?

1.2. Conceptual Framework

Figure 1 depicts a conceptual framework for contribution of land registration. The federal and regional laws provide the foundation for land reform in the form of land registration. The land administrations that have been established are in charge of implementation, which is also dependent on donor support and budget allocations for the activities. The effect of land registration on perceived tenure security, farmer confidence, marginalized groups, and sustainable land-management practice is also influenced by the initial conditions in farming households where reforms are implemented. The effects will be determined by factors such as individual and collectively owned resources and capabilities of households and communities, traditional norms, market exposure, other government policies, and agro-climatic conditions.

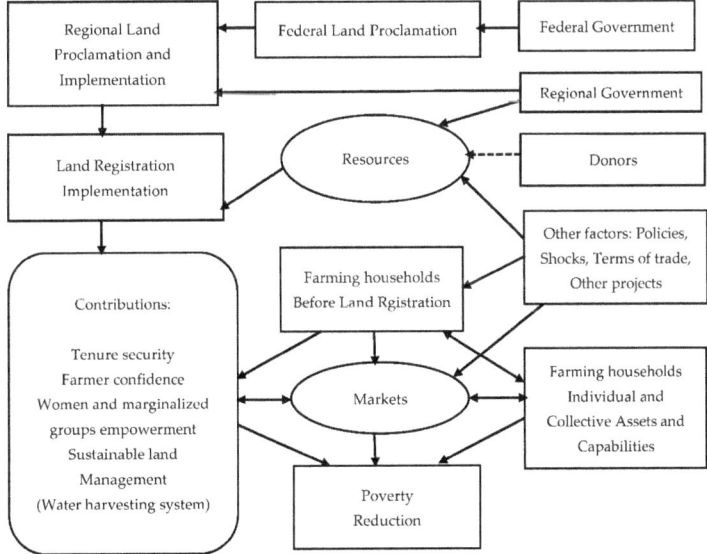

Figure 1. Conceptual Framework.

2. Materials and Methods

2.1. Study Area

East Gojjam Zone is one of eleven zones in Amhara National Regional State, Ethiopia. It is divided into 20 districts, 16 of which are rural and 4 of which are town administration districts. The Zone encompasses an area of approximately 14,004.47 square kilometers. The Oromia Region borders it on the south, West Gojjam Zone on the west, South Gondar Zone on the north, and South Wollo Zone on the east. The bend of the Abay River defines the Zone's northern, eastern, and southern boundaries. Mount Choke (also known as Mount Birhan) is its highest point, rising around 4100 m above mean sea level. East Gojjam Zone stretches from latitude 9°55′01″ to 11°14′12″ north and from longitude 37°29′37″ to 38°30′18″ east (see Figure 2).

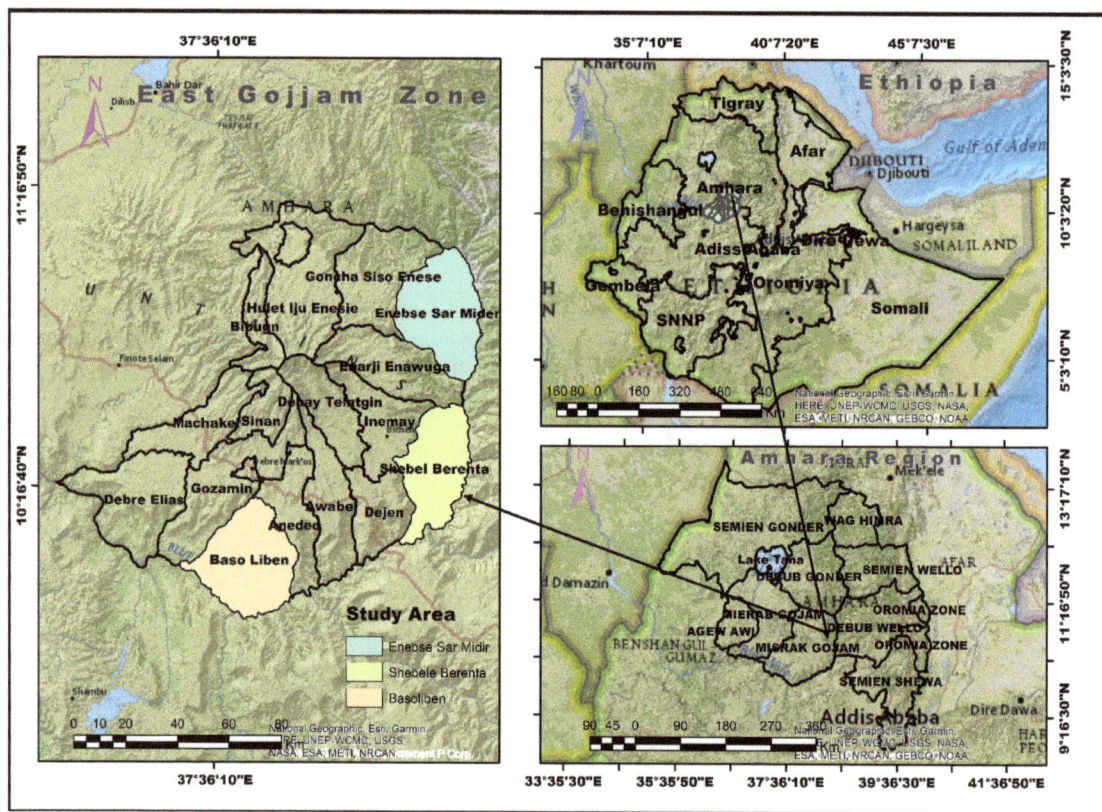

Figure 2. Study area map.

East Gojjam Zone is characterized by different landscapes such as mountains (Choke Mountain and Aba Mentous Mountain), plateaus (Yetnora, Awabal and Anaded, Gozamin, Debre Elias) and Gorges (Abay Gorge and Wamet). The study area is located between 759 and 4100 m above sea level. Different vegetation types have resulted from topographic variations combined with diverse climatic conditions, ranging from Afroalpine and sub-afroalpine vegetations to dry evergreen Montane Forest and Combretum Terminalia Woodland. The total population of the East Gojjam Zone is 2,153,937, of whom 1,066,716 are male and 1,087,221 are female. This zone has a population density of 153.80 people per square kilometer and the urban population accounts for 213,568 (9.92%) of the total population, while the remaining 1,940,369 (90.08%) are rural residents. This zone has a total of 506,520 households, with an average of 4.25 people per household and 492,486 housing units [73]. Because of the implementation of the land certification program, three representative districts, Enebse Sar Midir, Shebele Berenta, and Basoliben, were chosen to collect study data (Figure 2).

2.2. Methodology

2.2.1. Sampling Techniques

Three representative Districts (Enebse Sar Midir, Shebele Berenta and Basoliben) of Eastern Gojjam Zone were purposively selected for this study, due to land registration implementation history and dryland areas. The kebeles (lowest administrative structure) with the highest proportion of households registered were identified first in each sample district. Five kebeles were then chosen at random from these, and in total fifteen kebeles were selected from the three districts. With the assistance of kebele managers, administrators, and chairpersons of land administration committees, the names of female, male, and jointly certified households were identified on a separate slip of paper.

All 15,082 total households listed in fifteen selected kebeles were the sampling frame (N) (Table 1). To calculate the sample size (n), the statistical formula Cochran (1977) [74] was used. With a 95% confidence level and a 5% sampling error, the sample size (n) was calculated. As a result, the study's sample size (n) was 385 households. Probability proportional to size principle was used to assign a sample respondent from each kebele (Table 1). Finally, based on the number of respondents assigned to each sample category, the actual sample size was determined using a simple random sampling method. The detailed information about the sample districts, kebeles and sample size taken from each kebele is documented in Table 1.

Table 1. Total population of the sample kebeles and total sample size.

District	Kebeles	Total Population				Sample Size			
		Joint	Male	Female	Total	Joint	Male	Female	Total
Enebse Sar Midir	Gesese	695	156	336	1187	18	4	8	30
	Yetefet	479	124	160	763	12	3	4	19
	Ambalaye	430	111	226	767	11	3	6	20
	Segenet	795	336	268	1399	20	9	7	36
	Leule	666	158	289	1113	17	4	7	28
Shebele Berenta	Qarema	682	278	424	1384	17	7	11	35
	Abera	626	324	363	1313	16	8	9	33
	Beneyana Seqela	789	257	456	1502	20	7	11	38
	Gebsit	575	209	347	1131	15	5	9	29
	Yejuna Bayelie	548	293	351	1192	14	7	9	30
Basoliben	Korke	721	139	131	991	18	4	3	25
	Yeduge	343	57	88	488	9	1	2	12
	Anejeme	427	86	166	679	11	2	4	17
	Yelaminje	392	46	52	490	10	1	1	12
	Dendo	502	84	97	683	13	2	2	17
Total		8670	2658	3754	15,082	221	68	96	385

2.2.2. Data Collection Technique

Household surveys (HHS) conducted from September 2022 to October 2022 were the primary source of data. To collect primary data for the field interviews, both closed and open-ended structured questionnaires were used. Structured questionnaires were developed, tested, and adjusted to fit their intended purpose. Farmers were asked before and after land registration about their perceptions of land holding rights and land management activities. Four data collectors with a minimum of bachelor's degree (Undergraduate) in related fields of land administration and land management were employed for data collection. These enumerators were first trained in data collection techniques, study objectives, questionnaire management, and interviewing techniques. Face-to-face interviews were required because many of the respondents were illiterate. To avoid language barriers, one expert in land administration and management translated the questions from English into Amharic (local language). Official supporting letters written by each district office to the kebeles helped to enable data collection at kebele level.

The questionnaire was designed to gather information about respondents' personal and socioeconomic characteristics, as well as the effect of land registration on sustainable land management. Furthermore, each question was thoroughly explained and clarified to them with adequate explanation. The questionnaire was pre-tested by administering it to selected respondents at Korke kebele. On the fourth day of the exercise, enumerators were given the opportunity to make suggestions and remarks that could help them handle the interview. Based on the results obtained from the pre-test, necessary modifications were made to the questionnaire. Variables identified in the survey are documented in Table 2.

Table 2. Variables used in the study.

Variables	Definition and Values Used
Female	Female respondent (=1 female, =0 otherwise)
Male	Male respondent (=1 male, =0 otherwise)
Joint	Joint (Male and Female) respondent (=1 male and female jointly, =0 otherwise)
Age	Age of the respondents (=0, 1, 2, 3, ... , n)
Education	Educational level of the respondents (=1 literate, =0 illiterate)
Household size	Total household size (=0, 1, 2, 3, ... , n)
Land holding size	Total land holding size in hectare (=0, 1, 2, 3, ... , n)
Distance	Average distance farm to settlement in minutes (walking) (=0, 1, 2, 3, ... , n)
Land redistribution	Affected by land redistribution of 1997 (=1 yes, =0 no)
Expropriation	Fear of loss of land due to the expropriation by government at any time (=1 yes, =0 no)
Perceived tenure security	Fear of loss of farmland due to redistribution within the next five years (=1 (do not fear) yes, =0 otherwise)
Credit beneficiary	Credit beneficiary of the respondents (=1 yes, =0 no)
Training and advice	Training and advice on land management (=1 yes, =0 no)
Livestock holding	(Total livestock holding size) =0, 1, 2, 3, ... , n (in tropical livestock units)
Land-management practice	Application of land management (at least one) practices of parcel (=1 yes, =0 no)
Water harvesting construction	Application of water harvesting construction (=1 yes, =0 no)
Crop yields	Crop yields for major crop types (=0, 1, 2, 3, ... , n in quintal)

Data from household surveys were supplemented with qualitative data from direct field observations, focus group discussions (FGD), and key informant discussions. To supplement the quantitative data, FGD were held in each kebele. The FGD participants were chosen based on their knowledge of and experience with land-management practices. These people have lived in the kebeles for a long time and formed the kebele's Land Administration and Certification Committee (LACC). LACC members include elders, female-headed households, youth, and disabled people, as well as Development Agents and Kebele Managers. There were nine group discussions (three in each kebele). With the help of the "Kebele Land Administration Officer", each FGD had 10 to 12 participants. The focus of the discussion was on local-level entities dealing with land-related issues, the effect of land certification on land management, and other issues.

Youth, women, and the elderly were among the community representatives chosen for focus group discussions. A few members of the kebele Land Administration and Certification Committee, development agents, kebele leaders, and district experts were among the key informants. Interviews with representatives of the Environmental Protection and Land Administration Authority, the Bureau of Agriculture and Rural Development, and local authorities were conducted to better understand experts' perceptions of land registration and its intended objectives. Secondary data were gathered by reviewing several reports at the kebele, district, zonal, and regional levels. In addition, five federal, six regional, nine zonal, and four district experts participated in panels and discussion forums. These professionals work in rural land administration and land management offices, as well as other related fields. The professionals' discussion focused on accomplishments, bottlenecks, and recommendations for sustainable land management.

2.2.3. Data Analysis Techniques

Most of the data were analyzed quantitatively, and the analysis was supplemented by a qualitative analysis. Descriptive statistics such as percentages, mean, standard deviation, chi-square, and t-test were used to describe the socioeconomic characteristics of respondents. For a more detailed data analysis, a binary logit regression model was used. To determine the effect of other factors on selected variables, the following formula was used:

$$\ln\left[\frac{P_x}{(1-P_x)}\right] = \sum_{i=1}^{n} \beta_i X_i + U_i$$

where P_x is the probability for an observed set of variables that the event occurs, β_i is the ith coefficient to be estimated, X_i is the ith explanatory variable and U_i is a random error term.

2.3. Model Specification

For selected discrete and continuous variables, the presence of multi-collinearity and association was investigated. To identify multi-collinearity between continuous variables, the variance inflation factor (VIF) method was used [75]. A contingency coefficient test was used to evaluate associations between dummy variables. The variables found to be highly correlated with one or more of the other continuous or discrete variables (VIF > 10) were excluded from further analysis.

2.4. Description of Dependent Variables

Perceived tenure security, a dependent variable, shows whether respondents anticipate losing farmland because of redistribution within the next five years. For respondents who do not expect future redistribution, this binary variable has a value of 1, and for those who do, it has a value of 0. To investigate the influence of land registration on perceived tenure security, the following model was used:

Perceived tenure security = β0 + β1FEMALE + β2JOINT + β3EDU + β4AGE + β5HHSIZE + β6LANDHOLD + β7LANDRED + β8EXPROPRIATION + β9LH + Ui

where β0 to β9 is the coefficient to be estimated, FEMALE, JOINT, EDU, AGE, HHSIZE, LANDHOLD, LANDRED, EXPROPRIATION, and LH are explanatory variables, and Ui is a random error term.

Water harvesting construction, a dependent variable, has a value of 1 if the plot received water-harvesting system application, and 0 otherwise. The explanatory variables, on the other hand, are either continuous or binary. The influence of land registration on the water-harvesting system on plot j by household i was specified as follows:

$$WHC = \beta 0 + \beta 1 FEMALE + \beta 2 JOINT + \beta 3 EDU + \beta 4 AGE + \beta 5 LANDHOLD + \beta 6 DIST + \beta 8 CREDBENEF + \beta 9 TRAINING + \beta 10 LH + Ui$$

where β0 to β10 is the coefficient to be estimated, FEMALE, JOINT, EDU, AGE, LANDHOLD, DIST, CREDBENEF, TRAINING, and LH are explanatory variables and Ui is a random error term.

3. Results

3.1. Household Characteristics

Land registered in the names of females, males, and joint (male and female) accounted for 23%, 17%, and 60% of the sampled households, respectively. Concerning the educational attainment, 36% of the households sampled were illiterate, 42% could read and/or write, and 22% had completed grade five. Table 3 contains detailed information about the household characteristics.

Table 3. Land holding right and education of households.

Variables	Number	Percent
Land holding right		
Female	89	23.0
Male	64	17.0
Joint	232	60.0
Total	385	100.0
Educational level		
Illiterate	140	36.0
Read and/or write	162	42.0
Grade five and above	83	22.0
Total	385	100.0

Amhara Land Administration and Use (ALAU) Proclamation No. 133/2005 under Article 24 (2) states that where the land is a holding of a husband and a wife in common, the holding certificate shall be prepared by the name of both spouses [70]. As a result, joint titled implies implementation of land proclamation.

The age structure of households revealed that the mean age was 47 years. Furthermore, the average household size was 6.2 persons. Looking at the differences between respondent households, the largest family size was 12 and the smallest was one. There was an average difference of 0.03 ha in land holding size before and after land registration (see Table 4). The average difference between the number of farm plots before and after land registration was only 0.31.

Table 4. Summary of age of the household head, household size and land holdings (N = 385).

Variables	Mean Difference	St. Difference
Age	47	11.09
Household size	6.2	2.35
Landholding before and after land registration (in ha)	0.03	0.22
Plot number before and after land registration	0.31	0.38

3.2. Characteristic of Farm Plot

Farm plot characterization was assumed to demonstrate differences in the fertility status of farm plots before and after registration. Based on farmer perception, farms' fertility status could be classified as fertile, moderately fertile, or poorly fertile.

The findings in Table 5 clearly show that land registration had no effect on the level of farmland fertility. Out of all the households surveyed, 258 have an average of 0.60 ha of farmland on a flat slope, 126 have an average of 0.36 ha on a moderate slope, and 163 have an average of 0.40 ha on a steep slope. The cultivation of crops on steep slopes suggests that the study area lacks land use planning and a consequence of demand for land and food production exceeding supply of suitable land, pushing farmers to use marginal land.

Table 5. Fertility status of farm plots of sample households.

Farm Status	Frequency Number (%)	Before Registration		After Registration		
		Mean	St. dev.	Number (%)	Mean	St. dev.
Fertile	266 (88)	0.59	0.52	272 (90)	0.60	0.53
Moderate	193 (63)	0.44	0.34	194 (64)	0.45	0.35
Poor	131 (43)	0.37	0.26	127 (41)	0.39	0.27

3.3. Households' Confidence on Land Registration

According to Table 6, the last land redistribution affected 23% of the sampled households interviewed, either positively or negatively. However, approximately 26% of households are concerned about land redistribution over the next five years and losing their farms. Furthermore, 22% of households are concerned that the government will seize their farm plot at any time. Focus group participants proved that at each kebele, farmers living around town administration were highly frustrated with the expropriation of their farms.

Table 6. Households' confidence on land redistribution after registration (N = 385).

Name of the Variable	Yes		No	
	Number	Percent	Number	Percent
Households affected in 1997 land redistribution	90	23	295	77
Fear of land redistribution and farm loss in the next five years	102	26	283	74
Fear of government land expropriation at any time	83	22	302	78

3.4. The Effect of Land Registration on Women and Marginalized Groups

Table 7 shows that approximately 70% and 85% of households knew landless households in their village before and after land registration, respectively. The statistical test reveals a significant ($p < 0.01$) difference between the number of landless households in marginalized social groups before and after land registration. The qualitative information gathered from household surveys and focus group discussions showed that land registration protects the land rights of women and other marginal societies more than youths.

Table 7. Household Landlessness before and after land registration (N = 385).

Variable Name	Before Land Registration (Percent)		After Land Registration (Percent)		Chi-Square
	Yes	No	Yes	No	
Landless households	70.0	30.0	85.0	15.0	1.37 ***
Women	52.7	47.3	37.0	63.0	94.73 ***
Disabled	41.7	58.3	30.0	70.0	1.21 ***
Youth	60.3	39.7	84.7	15.3	63.57 ***
Poorest of poor	51.7	48.3	30.3	69.7	79.06 ***

*** = Significant at $p < 0.01$.

The chi-square test reveals a significant difference ($p < 0.01$) before and after land registration in the case of women's stronger land-holding rights in jointly led households. Furthermore, according to focus group participants and key informants, women have full rights to share the land equally during divorce; no one takes the land of women and other marginal societies.

3.5. Effect of Land Registration on Land-Management Practices

Table 8 clearly demonstrates that, with $p < 0.01$, approximately 80% and 95% of the households participated in at least one type of land management practice before and after land registration, respectively. Land management practices considered in this study included terracing, tree planting, compost application, manure application and the construction of water-harvesting structures (WHS). There was a significant difference between before and after land registration for each type of land management practice ($p < 0.01$).

Table 8. Land management practices in (%).

Factors	Before Land Registration		After Land Registration		Chi-Square
	Yes	No	Yes	No	
Land-management application	80	20	95	5	47.65 ***
Terracing	75.0	25.0	92.3	7.7	62.89 ***
Planting of tree	45.3	54.7	50.7	49.3	2.89 ***
Compost use	40.3	59.7	65.0	35.0	2.24 ***
Manure use	70.3	29.7	83.0	17.0	1.34 ***
Water-harvesting structure	15.7	84.3	25.3	74.7	97.73 ***

*** = significant at $p < 0.01$.

3.6. Effect of Land Registration on Crop Productivity

Except for wheat, there was no significant difference in major crop yields between 2021/22 (after land registration) and 2005/06 (before land registration) ($p < 0.1$). The average difference between wheat production in 2021/22 and 2005/06 is 0.89 quintals per household (Table 9). This finding indicates that there is no significant improvement in major crop yield per household following land registration, but rather a decrease. This could be due to changes in rainfall and other factors.

Table 9. Major crops produced in the year 2021/22 and 2005/06 (quintal/household).

Crops	Respondents	Difference in Mean	Difference in Std. Deviation	t-Test
Maize produced in (2021/22–2005/06)	66	0.51	3.74	−1.22
Wheat produced in (2021/22–2005/06)	182	0.89	7.15	1.74 *
Teff produced in (2021/22–2005/06)	150	0.31	4.33	1.26
Other crops produced in (2021/22–2005/06)	125	0.18	5.15	0.28

* = significant at $p < 0.1$.

3.7. Results of the Logit Model

To evaluate the relationships between dummy variables, a contingency coefficient test was used. The model is included for analysis because, as is evident from the results in Appendix A, there are no problems with multi-collinearity between the variables, and the contingency coefficient test result is very good.

3.7.1. Influencing Factors of Land Tenure Security

Three of the eight independent variables entered the model, namely education (significant at $p < 0.05$), land holding size (significant at $p < 0.01$), and land redistribution (significant at $p < 0.01$), were significantly and positively influencing households' fear of future land redistribution and loss of farmland (see Table 10).

Table 10. Factors influencing land registration on perceived tenure security (N = 385).

Explanatory Variables Name	B	Z-Value	$p > |Z|$	Marginal Effect
Female	−0.4596	−0.87	0.324	−0.0887
Joint	−0.0850	−0.13	0.719	−0.0136
Age	−0.0121	−0.28	0.755	−0.0015
Education	0.2562	2.02 **	0.035	0.0649
Household Size	−0.0257	−0.23	0.714	−0.0138
Land Holding size	0.6740	2.83 ***	0.004	0.1647
Land Redistribution	3.6758	7.14 ***	0.000	0.9216
Livestock holding	0.0081	0.22	0.732	0.0122
Constant	−2.3673	−3.43	0.001	
Log likeihood		−65.42575		
Chi squared		174.52		
Pseudo R²		0.5592		

*** and ** designate significance at $p < 1\%$ and $p < 5\%$, respectively: B (coefficients).

3.7.2. Factors Influencing Construction of Water-Harvesting Systems

Before land registration, as shown in Table 11, household size, livestock holding, and distance all had a significant effect on construction of water-harvesting systems at $p < 0.05$, $p < 0.01$, and $p < 0.01$, respectively. After land registration, age and household size influenced the construction of water-harvesting system with a significant difference of $p < 0.05$, whereas distance, livestock holding, and training and advice influenced the construction of water-harvesting systems with a significant difference of $p < 0.01$. The descriptive statistics also revealed that after land registration, the construction of water-harvesting systems increased by 59%.

Table 11. Influencing factors of land registration on construction of water-harvesting systems (WHS) (N = 385).

Variables	Before Land Registration			After Land Registration		
	B	Z-Value	Marginal Effect	B	Z-Value	Marginal Effect
Female	−0.3220	−0.74	−0.0523	−0.1490	−0.51	−0.0474
Joint	−0.3724	−1.27	−0.0740	−0.3360	−1.24	−0.1077
Age	−0.0235	−1.34	−0.0015	−0.0156	−2.07 **	−0.0051
Education	−0.1213	−1.13	−0.0216	−0.1285	−1.35	−0.0411
Household size	0.1180	2.27 **	0.0221	0.0892	2.01 **	0.0276
Land holding size	−0.0674	−0.31	−0.0135	−0.2760	−1.34	−0.0758
Training and advice	0.2694	0.78	0.0442	0.8953	2.69 ***	0.2089
Livestock holding	0.0987	2.50 ***	0.0183	0.1127	2.68 ***	0.0349
Credit beneficiary before LC	−0.1124	−0.52	−0.0176	-	-	-
Credit beneficiary after LC	-	-	-	0.1214	0.61	0.0479
Distance	−0.0075	−2.49 ***	−0.0014	−0.0078	−3.33 ***	−0.0027
Constant	−0.9554	−1.55		−0.5520	−1.04	
Log likelihood		−104.6043			−156.23964	
Chi squared		25.30			44.47	
Pseudo R²		0.1025			0.1126	

*** and ** indicate significance at $p < 1\%$ and $p < 5\%$, respectively: B (coefficients).

4. Discussion

4.1. Confidence of Households in Land Tenure Security

The last land redistribution affected 23% of the households in 1997 (Table 6). According to Deininger et al. [76], land redistribution affected 9% of Ethiopian farmers between 1991 and 1998, 18% in Tigray region, and 21% in the Amhara. Approximately 26% of households are concerned about land redistribution and losing their farm plots in the next five years, while 22% are concerned that the government will take their farms at any time (see Table 6). Thus, there is still concern about land redistribution, as found in Tigray, where 44% of farmers expect land redistribution and believe they will lose farms [77].

Furthermore, a previous study has found that 27% of respondents are confident that land redistribution will not happen in the future, while 9% believe it will occur within the next five years [78]. Given the aim of land certification, a small number of households are concerned about land redistribution over the next five years, and the government must address those households properly if land management is to improve.

4.2. The Effect of Land Registration on Women and Marginalized Groups

Land registration aimed to protect the land rights of marginalized groups such as the elderly, disabled, and women. After land registration, women, the disabled, and the poorest of the poor experienced less landlessness, whereas youths experienced an increase. This result was consistent with previous studies and discovered that 8.5% of farm holders are younger than 24 years old, indicating that landlessness is a significant issue in the Amhara region, especially for young people who have difficulty accessing land. This could be due to a lack of farmland, and land law prioritizes youths as one of society's most marginalized groups [61]. For instance, revised Amhara Region Land Administration and Use (ARLAU) Proclamation No. 133/2006 Article 9 (2) supports land holding in priority order for orphans, the disabled, women, and young people who join the new life of independence.

Women now have more land ownership rights after receiving land certification. Land registration has been shown in studies to promote gender equality, increase women's tenure security, and enhance land-management practice participation; [66,79,80] supported this conclusion. According to similar studies, the land certificates promote gender equality and encourage women to the field work [79]. Furthermore, this finding is in line with results found in Amhara Region pilot and non-pilot districts [81] and in Southern Ethiopia, who discovered that certification improved women's tenure security [82]. According to studies, the majority of households (85%) believe that land certification will improve women's status and provide incentives for land rental [66]. Finally, the land registration program promoted gender equality in Worja kebele in the Southern Nations Nationalities and Peoples region and 90% in Beresa kebele in Oromia [80].

4.3. Land Registration Effect on Land-Management Practices and Crop Productivity

Following land registration, household participation in land-management practices improved (see Table 8). This result is consistent with research carried out in Tigray, where 85.2% of households engaged in various sustainable land-management practices following land titling, compared to 34.1% growth prior to titling [83]. Similar studies discovered that a sizable majority of households in Ethiopia believed that registration of rural land increased incentives for spending on planting trees (88%), building structures for soil and water conservation (86%), and managing common property resources sustainably (66%) [66]. Likewise, land registration has strong implications for household participation in sustainable land management initiatives at the community level [84].

Most of the land tenure regularization programs predict an increase in land-based investments such as soil and land management infrastructure due to land registration and certification [85]. Deininger et al. [29] found positive and a statistically significant marginal effect of the land certification on the repairs and new investments in land management with an estimated average treatment effect of 30%. Land management incentives promote the positive impact of the land registration program [86]. In order to increase investments in land-related projects for sustainable land management, certificates are issued [29]. According to a similar report, 77.5% of Worja kebele farmers in the Southern Nation Nationalities and Peoples region and 70% of Beresa kebele farmers in Oromia completely agree that land registration increases investments in soil and land management [80]. In the same manner, reports indicate that 96.7% of farmers in pilot areas and 77.5% of farmers in non-pilot areas in the Amhara have participated in land management activities [81]. Additionally, studies conducted in Damot-Gale District, Southern Ethiopia revealed that the majority (62%) of the respondents indicated that they are practising land management due to a certificate, i.e., land certificate increases the perception of farmers in land management practices [87]. Finally, results from Melesse and Bulte [56] substantiate that land-certified households are more likely to adopt land management strategies than the uncertified ones. The participants of focus group discussion clearly indicated that land registration addressed issues of persistent gender inequality. As a result, registration improved decision-making in relation to land-management practices, and increased women's land rights. Studies show that the registration process made women more willing to work in the field and apply appropriate land-management practices [80].

Major crop yields decreased following land registration, except for wheat. The average difference in wheat production before and after land registration per household decreased by 0.89 quintals. This result demonstrates that, rather than improving significantly after land registration, major crop yield per household decreased. This might be brought on by changes in rainfall and other factors affecting crop growth. Because frequent droughts, the recent emergence of insect pests, and other factors have an impact on farmland productivity, crop productivity did not increase solely because of land registration in dryland areas. This outcome is consistent with earlier findings, according to which 50% of households in non-pilot districts and 63.3% of households in pilot districts of the Amhara region both agreed that land registration had no impact on farmland productivity [81]. On the contrary, studies have shown that improved land-management practices following registration have been associated with increased crop yields [71].

4.4. Factors Affecting Perceived Tenure Security

4.4.1. Education

The educational level of the respondents has a significant and positive effect on the fear of future land redistribution (see Table 10). Respondents who have higher levels of education are more likely to engage in off-farm activities and find alternative employment opportunities. As a result, there is a greater fear of losing farmland because the government could take over the land at any time. This survey result is consistent with the Amhara Rural Land Administration and Use (ARLAU) Proclamation No. 133/2006, which states in Article 12 (1a) that any land holder of a right to use the land may lose that right if he engages in non-farming activities and makes a living from these [70]. As a result, households are concerned that as education levels rise, so will the likelihood of non-agricultural activity, which may not be enough to meet individuals' basic needs but will result in farmland loss. On the contrary, this finding contradicts Pender et al. [88], in which the findings reported that education is likely to increase households' opportunities for salary employment off-farm and may increase their ability to start up various nonfarm activities. In addition, this may increase households' access to credit as well as their cash income, thus helping to finance purchases of physical capital and purchased inputs.

4.4.2. Landholding Size

There is a positive and significant correlation between respondents' total land holding size and their fear of future land redistribution (Table 10). Households with large land holdings are more concerned about land redistribution and the loss of farmland. According to participants in the focus group discussion, as food insecurity and crime rise, an increasing number of landless young people are threatening their farms. Consequently, land could be redistributed from elderly people who own large farms to landless youth. The findings of this study are supported by investigations that farm households with relatively larger farms feel more insecure than those with relatively less land, and farm tenure security in Ethiopia is inversely related to farm size [71,89].

4.4.3. Land Redistribution

The fear of land redistribution is significantly and positively associated with households affected by land redistribution in 1997 (Table 10). Fear of land redistribution is high, and it is even higher within the next five years than it is beyond (Table 5). Focus group discussion participants reported that land redistribution had occurred frequently in recent days. They have no idea what will happen in the future because it is dependent on the government and its policies. Even the land policy gives reason for concern, stating that land redistribution may be possible if the land is required for irrigation projects. Another concern is that the government could be replaced, and the legislation would not be properly implemented [71].

4.5. Factors Influencing Construction of Water-Harvesting System

4.5.1. Distance

The construction of a water-harvesting structure is negatively impacted by the distance between a farm plot and the settlement. After land registration, the average distance of a farm plot from the settlement increased by one minute, while the construction of a water-harvesting structure decreased by 0.27%. Farmland owners who live close to residential areas are more likely to build a water-harvesting system than those who live far from the settlement (see Table 11). This is since households prefer nearby farm plots over distant plots. According to studies, managing close farmland takes less time and energy, so longer walking distances between farmland and settlement areas reduce farmland cultivation adoption [90,91].

4.5.2. Training and Advice

The construction of a water-harvesting system has a significant and positive relationship with households that received land-management training and advice (see Table 11). A unit increase in training and advice from agricultural extension services increased the construction of a water-harvesting structure by 20.8% after the land was registered. This significant increase was caused by the provision of enough knowledge and instruction on sustainable land-management techniques. Focus group participants reported that households that received more training were motivated to build water-harvesting systems for their farmlands. Participants acknowledged that extension services offered to them were more likely to persuade them to make such land investments than development workers. Previous research has found that farmers who receive training are more likely to adopt, use, and implement land-management practices [91]. Farmers' attitudes and abilities in land management will improve because of increased access to training, as well as their knowing of the advantages and limitations of soil conservation. Additionally, training enhances one's capacity to understand and use specific knowledge about land management activities. A previous study confirmed that training had an impact on the adoption of land-management applications [90–93]. Numerous studies have examined the connection between farmers' training and their use of sustainable land-management techniques [94]. The current study's findings also showed that after the land registration process, the impact of training was increased.

4.5.3. Livestock Holding

The total number of livestock holdings and the decision to construct a water-harvesting system were significantly and positively correlated (see Table 11). Following land registration, the construction of water-harvesting systems increased by 3.4% as the total number of livestock increased by one unit. Small family sizes, a labor shortage, and a high livestock population are the most likely causes. There is a chance of selling livestock and converting to human labor. It is then possible to build water-harvesting systems using the human labor force gained from family members. Key informant participants confirmed that when farm households have a shortage of human labor, livestock sales are used to purchase labor for the building of water-harvesting systems. The results of this study have also been supported by earlier studies [90,93,95,96]. Additionally, livestock is a significant source of farm income that enables farmers to invest in land management strategies and purchase agricultural inputs. Moreover, it serves as non-human labor to construct structures for soil conservation [97]. According to earlier research, the quantity of livestock is a sign of financial stability, which improves the efficiency of land management [95,96].

4.5.4. Household Size

After land registration, the size of a household has a positive and significant influence on the construction of a water-harvesting structure. Construction of water-harvesting structures increased by 2.7% as household size increased by one member after certification (Table 11).

4.5.5. Age

After land registration, the construction of a water-harvesting structure is negatively impacted by the household head's age. As a result, after land registration, an increase of one year in the household's age resulted in a 0.51% decrease in the construction of water-harvesting structures (Table 11). Older farmers have larger land holdings than younger farmers, and they may lose land due to redistribution. So, older farmers were less invested in land management [98]. Greater family labor indicates a greater potential for labor-intensive investments such as water-harvesting construction. Larger households will be able to provide the labor needed to maintain conservation structures [98].

5. Conclusions and Recommendations

Land registrations are critical issues in Ethiopia's land administration system for improving land tenure security. To that end, Ethiopia has had a land registration and certification program in place since 1998, and the Amhara region has had one since 2002, with the goal of registering all land holdings and issuing land certificates to enhance farmers' land rights' security. In theory, land certification stimulates economic growth by providing incentives to increase agricultural production. Secure land rights are essential for economic development.

Because tenure insecurity is a problem in African countries, efforts should be made to provide land rights to people, and particularly to women and marginal groups. Appropriate land rights are considered a starting point for the empowerment of the poor. Land registrations are currently applied in Ethiopia to provide land tenure security. This contributes to the advancement of sustainable land-management practices. As a result, developing countries can learn from this success and emphasize tenure rights for their country's sustainable development.

Even though land registration has a significant impact on long-term land management, Ethiopia lacks a comprehensive land use policy. Land use regulation is not given much weight in the current rural land administration system. Land use rights are given less attention in rural land administration and land use proclamations. The legal framework is primarily concerned with issues of land administration. Of course, land ownership and tenure security are fundamental components of sustainable land-management practices, and they are a good place to begin. Nonetheless, in order to enforce sustainable land management, a land use policy for proper land use practices should be established. Otherwise, there will be no solution to the land degradation and deforestation problem. This, in turn, could be a threat for agricultural production and exacerbate the country's poverty situation.

In this study, land registration significantly improved farmers' perceptions and confidence in land tenure security, even though 13.7% of households remain concerned about future land redistribution and expropriation by the government at any time. Fear of land expropriation by the government emerged from the foundation of new town administration and that is holding rural kebeles.

The state still owns all land in Ethiopia, even though farmers feel more secure regarding competing claims to their land from neighbors and relatives. This policy continues to create insecurity, especially when local officials suggest that the government might seize the land if it is not used properly. Such claims by local officials have caused confusion among smallholders about the benefits of land registration on tenure security, sparking a debate about whether land registration must be accompanied by land ownership in order to realize secure use rights.

The logit model results revealed that education, land holding size, and households affected by last land redistribution were found to significantly and positively aggravate households' fear of future land redistribution and farmland losses. Except for youths, land registration effectively protects the land use rights of women and other marginalized groups in society. As a result of this, youths have raised the issue of land right immediately following land registration in the study area. Women's land holding rights were found to be stronger after land registration, with a significant difference of $p < 0.01$ between before and after land registration.

Regarding the effect of land registration on land management on cultivated land, household land management participation improved after land registration in the study area. However, the average distance of a farm plot from the settlement had a negative impact on the construction of water-harvesting system in the study area, whereas access to agricultural extension training and advice, as well as livestock holding, had a positive impact on the construction of water-harvesting systems. Nonetheless, steep slope areas in the study area are still used for crop cultivation. With the exception of wheat, which was significant at the $p < 0.1$ level, there was no significant difference in major crop yield per household after and before land registration. This result shows that there is no significant improvement in major crop yield per household after land registration, but rather a decrease. This could be due to variations in rainfall and other crop growth factors. As a result, crop yield did not improve solely through land registration in dryland areas because farmland productivity was affected by the occurrence of recurrent drought and other factors.

The findings also provide important policy implications and suggest that policymakers both at governmental and non-governmental agencies engaged in sustainable land management among rural agricultural households that aim to boost agricultural development should consider land registration as an important prerequisite. The evidence shows that if farm households are given more secure property rights on their land, they would be encouraged to increase their investments in sustainable land management. Thus, policymakers in Ethiopia should consider land registration as a matter of priority to ensure the success of sustainable land management programs and to promote the development of modern agriculture. Tenure security by ensuring the probability of benefiting from their investment in the long term. Thus, tenure security can also serve as an incentive mechanism for the success of sustainable land management.

The possible recommendations were that governmental and non-governmental offices should work together to raise awareness about the duties and responsibilities that land registration entails. Meanwhile, the government should look for clear policies, such as small-scale enterprise and urban agriculture, to address the issues of landless youths and farmers whose lands have been encroached upon by town administration expansions into rural kebeles. Furthermore, the government should strengthen the implementation of the society's land registration processes. However, to address youth landlessness, intensive farming practices should be promoted, which will increase labor needs and thus engage youths.

Funding: This research received no external funding.

Data Availability Statement: The data presented in this study are available on request from the corresponding author. The data are not publicly available due to restrictions e.g., privacy or ethical.

Acknowledgments: The author thanks respondents and land administration experts for their collaboration during survey interviews.

Conflicts of Interest: The authors declare no conflict of interest.

Appendix A

Severity of multi-collinearity between independent variables was assessed prior to estimating the logit model by calculating the variance inflation factor (VIF). VIF < 10 specifies that there is no multi-collinearity.

Table A1. Multi-collinearity test for perceived tenure security.

Continuous Independent Variables	VIF
Age	1.112
Household size	1.184
Land holding size	1.158
Livestock holding	1.165

VIF is variance inflation factor; source: survey, 2022.

Table A2. Multi-collinearity test for construction of water-harvesting system.

| Explanatory Variables | VIF | |
	Before Registration	After Registration
Age	1.158	1.123
Household size	1.214	1.187
Land holding size	1.159	1.151
Distance	1.022	1.021
Livestock holding	1.143	1.135

VIF is variance inflation factor; source: survey, 2022.

Table A3. Contingency coefficients for perceived tenure security.

	Female	Joint	Education	Land Redistribution	Expropriation
Female	1				
Joint	0.568	1			
Education	0.086	0.087	1		
land redistribution	0.023	0.023	0.032	1	
Expropriation	0.011	0.045	0.084	0.038	1

References

1. Barrows, R.; Roth, M. Land tenure and investment in african agriculture: Theory and evidence. *J. Mod. Afr. Stud.* **1990**, *28*, 265–297. [CrossRef]
2. Huenchuleo, C.; Barkmann, J.; Villalobos, P. Social psychology predictors for the adoption of soil conservation measures in Central Chile. *Land Degrad. Dev.* **2012**, *23*, 483–495. [CrossRef]
3. Cerdà, A.; Lavee, H.; Romero-Díaz, A.; Hooke, J.; Montanarella, L. Preface: Soil erosion and degradation in mediterranean type ecosystems. *Land Degrad. Dev.* **2010**, *21*, 71–74. [CrossRef]
4. Zhao, G.; Mu, X.; Wen, Z.; Wang, F.; Gao, P. Soil erosion, conservation, and eco-environment changes in the loess plateau of china. *Land Degrad. Dev.* **2013**, *24*, 499–510. [CrossRef]
5. Wang, X.; Wang, G.; Lang, L.; Hua, T.; Wang, H. Aeolian transport and sandy desertification in semiarid China: A wind tunnel approach. *Land Degrad. Dev.* **2013**, *24*, 605–612. [CrossRef]
6. Fernández-Calviño, D.; Garrido-Rodríguez, B.; López-Periago, J.E.; Paradelo, M.; Arias-Estévez, M. Spatial distribution of copper fractions in a vineyard soil. *Land Degrad. Dev.* **2013**, *24*, 556–563. [CrossRef]
7. García-Orenes, F.; Cerdà, A.; Mataix-Solera, J.; Guerrero, C.; Bodí, M.B.; Arcenegui, V.; Zornoza, R.; Sempere, J.G. Effects of agricultural management on surface soil properties and soil-water losses in eastern Spain. *Soil Tillage Res.* **2009**, *106*, 117–123. [CrossRef]
8. Abu Hammad, A.; Tumeizi, A. Land degradation: Socioeconomic and environmental causes and consequences in the Eastern Mediterranean. *Land Degrad. Dev.* **2012**, *23*, 216–226. [CrossRef]
9. Al-Awadhi, J.M. A Case Assessment of the Mechanisms Involved in Human-Induced Land Degradation in Northeastern Kuwait. *Land Degrad. Dev.* **2013**, *24*, 2–11. [CrossRef]

10. Lanckriet, S.; Derudder, B.; Naudts, J.; Bauer, H.; Deckers, J.; Haile, M.; Nyssen, J. A Political Ecology Perspective of Land Degradation in the North Ethiopian Highlands. *Land Degrad. Dev.* **2015**, *26*, 521–530. [CrossRef]
11. Lemenih, M.; Kassa, H.; Kassie, G.T.; Abebaw, D.; Teka, W. Resettlement and woodland management problems and options: A case study from North-Western Ethiopia. *Land Degrad. Dev.* **2014**, *25*, 305–318. [CrossRef]
12. Mekuria, W.; Aynekulu, E. Exclosure land management for restoration of the soils in degraded communal grazing lands in Northern Ethiopia. *Land Degrad. Dev.* **2013**, *24*, 528–538. [CrossRef]
13. Tesfaye, A.; Negatu, W.; Brouwer, R.; van der Zaag, P. Understanding soil conservation decision of farmers in the Gedeb understanding soil conservation decision of farmers in the Gedeb watershed, Ethiopia. *Land Degrad. Dev.* **2014**, *25*, 71–79. [CrossRef]
14. Yami, M.; Mekuria, W.; Hauser, M. The effectiveness of village bylaws in sustainable management of community-managed exclosures in Northern Ethiopia. *Sustain. Sci.* **2013**, *8*, 73–86. [CrossRef]
15. Glover, E.K.; Elsiddig, E.A. The causes and consequences of environmental changes in Gedaref, Sudan. *Land Degrad. Dev.* **2012**, *23*, 339–349. [CrossRef]
16. Omuto, C.T.; Balint, Z.; Alim, M.S. A Framework for national assessment of land degradation in the drylands: A case study of Somalia. *Land Degrad. Dev.* **2014**, *25*, 105–119. [CrossRef]
17. Bennett, J.E.; Palmer, A.R.; Blackett, M.A. Range degradation and land tenure change: Insights from a "released" communal area of Eastern Cape Province, South Africa. *Land Degrad. Dev.* **2012**, *23*, 557–568. [CrossRef]
18. Place, F. Land Tenure and Agricultural Productivity in Africa: A Comparative Analysis of the Economics Literature and Recent Policy Strategies and Reforms. *World Dev.* **2009**, *37*, 1326–1336. [CrossRef]
19. Requier-Desjardins, M.; Adhikari, B.S.S. Some notes on the economic assessment of land degradation. *Land Degrad. Dev.* **2011**, *22*, 285–298. [CrossRef]
20. De Janvry, A.; Gordillo, G.; Sadoulet, E.; Platteau, J.-P. In Access to land, rural poverty and public action. *J. Dev. Econ.* **2001**, *70*, 235–238.
21. Mbow, C. Use it sustainably or lose it! The land stakes in SDGs for Sub-Saharan Africa. *Land* **2020**, *9*, 63. [CrossRef]
22. Tseng, T.W.J.; Robinson, B.E.; Bellemare, M.F.; Yishay, B.A.; Blackman, A.; Boucher, T.; Childress, M.; Holland, M.B.; Kroeger, T.; Linkow, B.; et al. Influence of land tenure interventions on human well-being and environmental outcomes. *Nat. Sustain.* **2021**, *4*, 242–251. [CrossRef]
23. Behaylu, A.; Bantider, A.; Tilahun, A.; Gashaw, T. The Role of Rural Land Registration and Certification Program. *Int. J. Agric. Ext. Rural Dev. Stud.* **2015**, *2*, 44–52.
24. Shawki, N. Norms and normative change in world politics: An analysis of land rights and the sustainable development goals. *Glob. Change Peace Secur.* **2016**, *28*, 249–269. [CrossRef]
25. Mengesha, A.K.; Mansberger, R.; Damyanovic, D. Impact of Land Certification on Sustainable Land Use Practices: Case of Gozamin District, Ethiopia. *Sustainability* **2019**, *11*, 5551. [CrossRef]
26. Biraro, M.; Zevenbergen, J.; Alemie, B.K. Good practices in updating land information systems that used unconventional approaches in systematic land registration. *Land* **2021**, *10*, 437. [CrossRef]
27. Besley, T. Property rights and investment incentives: Theory and evidence from Ghana. *J. Polit. Econ.* **1995**, *103*, 903–937. [CrossRef]
28. De Janvry, A.; Emerick, K.; Gonzalez-Navarro, M.; Sadoulet, E. Delinking land rights from land use: Certification and migration in Mexico. *Am. Econ. Rev.* **2015**, *105*, 3125–3149. [CrossRef]
29. Deininger, K.; Ali, D.A.; Alemu, T. Impacts of land certification on tenure security, investment, and land market participation: Evidence from Ethiopia. *Land Econ.* **2011**, *87*, 312–334. [CrossRef]
30. Goldstein, M.; Udry, C. The profits of power: Land rights and agricultural investment in Ghana. *J. Polit. Econ.* **2008**, *116*, 981–1022. [CrossRef]
31. Goldstein, M.; Houngbedji, K.; Kondylis, F.; O'Sullivan, M.; Selod, H. Formalization without certification? Experimental evidence on property rights and investment. *J. Dev. Econ.* **2018**, *132*, 57–74. [CrossRef]
32. Deininger, K.; Chamorro, J.S. Investment and equity effects of land regularisation: The case of Nicaragua. *Agric. Econ.* **2004**, *30*, 101–116. [CrossRef]
33. De Soto, H. *The Mystery of Capital: Why Capitalism Triumphs in the West and Fails Everywhere Else*; Basic Civitas Books: New York, NY, USA, 2003.
34. Feder, G.; Nishio, A. The benefits of land registration and titling: Economic and social perspectives. *Land Use Policy* **1999**, *15*, 25–43. [CrossRef]
35. Newman, C.; Tarp, F.; Van Den Broeck, K. Property rights and productivity: The case of joint land titling in Vietnam. *Land Econ.* **2015**, *91*, 91–105. [CrossRef]
36. Lawry, S.; Samii, C.; Hall, R.; Leopold, A.; Hornby, D.; Mtero, F. The impact of land property rights interventions on investment and agricultural productivity in developing countries: A systematic review. *J. Dev. Eff.* **2017**, *9*, 61–81. [CrossRef]
37. Berry, S. Building for the Future? Investment, Land Reform and the Contingencies of Ownership in Contemporary Ghana. *World Dev.* **2009**, *37*, 1370–1378. [CrossRef]
38. Abdulai, A.; Goetz, R. Time-Related Characteristics of Tenancy Contracts and Investment in Soil Conservation Practices. *Environ. Resour. Econ.* **2014**, *59*, 87–109. [CrossRef]

9. Peters, P.E. Challenges in Land Tenure and Land Reform in Africa: Anthropological Contributions. *World Dev.* **2009**, *37*, 1317–1325. [CrossRef]
10. Meinzen-Dick, R.; Mwangi, E. Cutting the web of interests: Pitfalls of formalizing property rights. *Land Use Policy* **2009**, *26*, 36–43. [CrossRef]
11. Ostrom, E. Polycentric systems for coping with collective action and global environmental change. *Glob. Environ. Change* **2010**, *20*, 550–557. [CrossRef]
12. Blackman, A.; Corral, L.; Lima, E.S.; Asner, G.P. Titling indigenous communities protects forests in the Peruvian Amazon. *Proc. Natl. Acad. Sci. USA* **2017**, *114*, 4123–4128. [CrossRef] [PubMed]
13. Damnyag, L.; Saastamoinen, O.; Appiah, M.; Pappinen, A. Role of tenure insecurity in deforestation in Ghana's high forest zone. *For. Policy Econ.* **2012**, *14*, 90–98. [CrossRef]
14. Abdulai, A.; Owusu, V.; Goetz, R. Land tenure differences and investment in land improvement measures: Theoretical and empirical analyses. *J. Dev. Econ.* **2011**, *96*, 66–78. [CrossRef]
15. Twerefou, D.K.; Osei-assibey, E.; Agyire-tettey, F. Land tenure security, investments and the environment in Ghana. *J. Dev. Agric. Econ.* **2011**, *3*, 261–273.
16. Kabubo-Mariara, J. Land conservation and tenure security in Kenya: Boserup's hypothesis revisited. *Ecol. Econ.* **2007**, *64*, 25–35. [CrossRef]
17. Owubah, C.E.; Le Master, D.C.; Bowker, J.M.; Lee, J.G. Forest tenure systems and sustainable forest management: The case of Ghana. *For. Ecol. Manag.* **2001**, *149*, 253–264. [CrossRef]
18. Fenske, J. Land tenure and investment incentives: Evidence from West Africa. *J. Dev. Econ.* **2011**, *95*, 137–156. [CrossRef]
19. Unruh, J.D. Carbon sequestration in Africa: The land tenure problem. *Glob. Environ. Change* **2008**, *18*, 700–707. [CrossRef]
20. Ghaffari, G.; Keesstra, S.; Ghodousi, J.; Ahmadi, H. SWAT-simulated hydrological impact of land-use change in the Zanjanrood Basin, Northwest Iran. *Hydrol. Process.* **2010**, *24*, 892–903. [CrossRef]
21. Holden, S.T.; Ghebru, H. Land tenure reforms, tenure security and food security in poor agrarian economies: Causal linkages and research gaps. *Glob. Food Sec.* **2016**, *10*, 21–28. [CrossRef]
22. Easterly, W. Institutions: Top down or bottom up? *Am. Econ. Rev.* **2008**, *98*, 95–99. [CrossRef]
23. Jacoby, H.G.; Minten, B. Is land titling in Sub-Saharan Africa cost-effective? Evidence from Madagascar. *World Bank Econ. Rev.* **2007**, *21*, 461–485. [CrossRef]
24. Headey, D.D.; Jayne, T.S. Adaptation to land constraints: Is Africa different? *Food Policy* **2014**, *48*, 18–33. [CrossRef]
25. Crewett, W.; Korf, B. Ethiopia: Reforming land tenure. *Rev. Afr. Political Econ.* **2008**, *35*, 203–220. [CrossRef]
26. Melesse, M.B.; Bulte, E. Does land registration and certification boost farm productivity? Evidence from Ethiopia. *Agric. Econ.* **2015**, *46*, 757–768. [CrossRef]
27. Deininger, K.; Jin, S. Tenure security and land-related investment: Evidence from Ethiopia. *Eur. Econ. Rev.* **2006**, *50*, 1245–1277. [CrossRef]
28. Holden, S.; Deininger, K.; Ghebru, H. Tenure insecurity, gender, low-cost land certification and land rental market participation in Ethiopia. *J. Dev. Stud.* **2011**, *47*, 31–47. [CrossRef]
29. Ali, D.A.; Dercon, S.; Gautam, M. Property rights in a very poor country: Tenure insecurity and investment in Ethiopia. *Agric. Econ.* **2011**, *42*, 75–86. [CrossRef]
30. Miller, D.C.; Muñoz-Mora, J.C.; Christiaensen, L. Prevalence, economic contribution, and determinants of trees on farms across Sub-Saharan Africa. *For. Policy Econ.* **2017**, *84*, 47–61. [CrossRef]
31. Adenew, B.; Abdi, F. Land Registration in Amhara Region, Ethiopia, Securing Land Rights in Africa. *Dep. Int. Dev.* **2005**, 1–34.
32. Wubneh, M. Policies and praxis of land acquisition, use, and development in Ethiopia. *Land Use Policy* **2018**, *73*, 170–183. [CrossRef]
33. Yami, M.; Snyder, K.A. After All, Land Belongs to the State: Examining the Benefits of Land Registration for Smallholders in Ethiopia. *Land Degrad. Dev.* **2016**, *27*, 465–478. [CrossRef]
34. Ege, S. Peasant Participation in Land Reform: The Amhara land redistribution of 1997. *Ethiop. Chall. Democr. Below.* **2002**, 71–86.
35. Benin, S.; Pender, J. Impacts of land redistribution on land management and productivity in the Ethiopian highlands. *Land Degrad. Dev.* **2001**, *12*, 555–568. [CrossRef]
36. Deininger, K.; Ali, D.A.; Holden, S.; Zevenbergen, J. Rural Land Certification in Ethiopia: Process, Initial Impact, and Implications for Other African Countries. *World Dev.* **2008**, *36*, 1786–1812.
37. Belay, A.A.; Abza, T.G. Protecting the Land Rights of Women through an Inclusive Land Registration System. *Afr. J. Land Policy Geospat. Sci.* **2020**, *3*, 2657–2664.
38. Holden, S.T.; Deininger, K.; Ghebru, H. Impacts of low-cost land certification on investment and productivity. *Am. J. Agric. Econ.* **2009**, *91*, 359–373. [CrossRef]
39. Holden, S.; Deininger, K.; Ghebru, H. Can Land Rregistration and Certification Reduce Land Border Conflicts? In *Center for Land Tenure*; Studies Working Paper 05/11; NMBU: Ås, Norway, 2011.
40. Amhara National Regional State (ANRS). *The Amhara National Regional State Rural Land Administration and Use System Implementation, Council of Regional Government Regulation*; Regulation No. 133/2006; Regional Zikre Hig Gazeta: Bahir Dar, Ethiopia, 2006.
41. Gedefaw, A.A.; Atzberger, C.; Seher, W.; Agegnehu, S.K.; Mansberger, R. Effects of land certification for rural farm households in ethiopia: Evidence from Gozamin District, Ethiopia. *Land* **2020**, *9*, 421. [CrossRef]

72. Tsegaye, A.; Adgo, E.G.; Selassie, Y. Impact of Land Certification on Sustainable Land Resource Management in Dryland Areas of Eastern Amhara Region, Ethiopia. *J. Agric. Sci.* **2012**, *4*, 261. [CrossRef]
73. CSA. *Summary and Statistical Report of the 2007 Population and Housing Census*; UNFPA: Addis Ababa, Ethiopia, 2008.
74. Cochran, W.G. *Sampling Techniques*, 3rd ed.; John Wiley & Sons: Hoboken, NJ, USA, 1977; Volume 3, ISBN 0-471-16240-X.
75. Gujarati, D.N. *Basic Econometrics*, 4th ed.; McGraw Hill: New York, NY, USA, 2003; ISBN 9780072335422.
76. Deininger, K.; Jin, S.; Adenew, B.; Gebre-selassie, S.; Nega, B. *Tenure Security and Land-Related Investment Evidence from Ethiopia. Policy Research Working*; World Bank Publications: Washington, DC, USA, 2003; p. 2991.
77. Hagos, F. Land Registration and Land Investment: The case of Tigray Region, Northern Ethiopia. *Ethiop. J. Econ.* **2012**, *21*, 19–47.
78. Deininger, K.; Jin, S.; Adenew, B.; Samuel, G.-S.; Mulat, D. *Market and Nonmarket Transfers of Land in Ethiopia and Nonfarm Development*; World Bank Publications: Washington, DC, USA, 2003; p. 36.
79. Stein, H.; Tefera, T. *Land Registration in Ethiopia: Early Impacts on Women*; United Nations Human Settlements Program (UN-HABITAT): Nairobi, Kenya, 2008; pp. 1–28.
80. Giri, S. The Effect of Rural Land Registration and Certification Programme on Farmers' Investments in Soil Conservation and Land Management in the Central Rift Valley of Ethiopia. Master's Thesis, Wageningen University, Wageningen, The Netherlands, 2010; p. 80.
81. Belay, A. The Effects of Land Certification in Securing Rural Land Rights Amhara Region, Ethiopia. A Thesis Presented to Real Estate Planning and Land Law. Ph.D. Thesis, KTH Real Estate and Construction Management, Stockholm, Sweden, 2010.
82. Holden, S.; Tewodros, T. *From Being Property of Men to Becoming Equal Owners? Early Impacts of Land Registration and Certification on Women in Southern Ethiopia. Final Research Report Prepared for UNHABITAT, Shelter Branch, Land Tenure and Property Administration Section*; Global Land Tool Network (GLTN): Hawassa, Ethiopia, 2008; pp. 1–110.
83. Dagnew, M.; Fitsum, H.; Nicck, C. Implications of Land titling on Tenure Security and Long-term Land investment: Case of Kilte Awela'elo woreda, Tigray, Ethiopia. In Proceedings of the Sustainable Land Management Research and Institutionalization of Future Collaborative Research, Mekelle, Ethiopia, 8–9 August 2009; pp. 264–274.
84. Adamie, B.A. Land property rights and household take-up of development programs: Evidence from land certification program in Ethiopia. *World Dev.* **2021**, *147*, 105626. [CrossRef]
85. Bizoza, A.R.; Opio-Omoding, J. Assessing the impacts of land tenure regularization: Evidence from Rwanda and Ethiopia. *Land Use Policy* **2021**, *100*, 104904. [CrossRef]
86. Xu, L.; Chen, S.; Tian, S. The Mechanism of Land Registration Program on Land Transfer in Rural China: Considering the Effects of Livelihood Security and Agricultural Management Incentives. *Land* **2022**, *11*, 1347. [CrossRef]
87. Dalacho, D. The Effects of Rural Land Registration and Certification on Land Management Practices in Damot-Gale District, Southern Ethiopia. *J. Resour. Dev. Manag.* **2022**, *85*, 26–48.
88. Pender, J.; Place, F.; Ehui, S. *Strategies for Sustainable Land Management in the East African Highlands*; International Food Policy Research Institute: Washington, DC, USA, 2006; ISBN 0896297578.
89. Holden, S.; Yohannes, H. Land redistribution, tenure insecurity, and intensity of production: A study of farm households in Southern Ethiopia. *Land Econ.* **2002**, *78*, 573–590. [CrossRef]
90. Kerse, B.L. Factors affecting adoption of soil and water conservation practices in the case of Damota watershed, Wolaita Zone, Southern Ethiopia. *Int. J. Agric. Sci. Res.* **2018**, *7*, 1–9.
91. Asfaw, D.; Neka, M. Factors affecting adoption of soil and water conservation practices: The case of Wereillu Woreda (District), South Wollo Zone, Amhara Region, Ethiopia. *Int. Soil Water Conserv. Res.* **2017**, *5*, 273–279. [CrossRef]
92. Wolka, K.; Negash, M. Farmers' Adoption of Soil and Water Conservation Technology: A Case Study of the Bokole and Toni Sub-Watersheds, Southern Ethiopia. *J. Sci. Dev.* **2014**, *2*, 35–48.
93. Meseret, D. Determinants of Farmers' Perception of soil and water Conservation Practices on Cultivated Land in Ankesha District, Ethiopia. *Agric. Sci. Eng. Technol. Res.* **2014**, *2*, 1–9.
94. Rezvanfar, A.; Samiee, A.; Faham, E. Analysis of Factors Affecting Adoption of Sustainable Soil Conservation Practices among Wheat Growers. *World Appl. Sci. J.* **2009**, *6*, 644–651.
95. Kebede, B.A.; Black, T.G.; Mideksa, D.F.; Nega, M.B. Soil and water conservation practices: Economic and environmental effects in Ethiopia. *Glob. J. Agric. Econ. Econ.* **2016**, *4*, 169–177.
96. Abebe, Z.D.; Sewnet, M.A. Adoption of soil conservation practices in North Achefer District, Northwest Ethiopia. *Chin. J. Popul. Resour. Environ.* **2014**, *12*, 261–268. [CrossRef]
97. Taye, A.A. *Caring for the Land: Best Practices in Soil and Water Conservation in Beressa Watershed, Highlands of Ethiopia*; Wageningen University and Research: Wageningen, The Netherlands, 2006.
98. Demeke, A.B. *Factors Influencing the Adoption of Soil Conservation Practices in Northwestern Ethiopia*; Institute of Rural Development University of Goettingen: Goettingen, Germany, 2003.

Disclaimer/Publisher's Note: The statements, opinions and data contained in all publications are solely those of the individual author(s) and contributor(s) and not of MDPI and/or the editor(s). MDPI and/or the editor(s) disclaim responsibility for any injury to people or property resulting from any ideas, methods, instructions or products referred to in the content.

Article

Forest Transition and Fuzzy Environments in Neoliberal Mexico

Cynthia Simmons [1,*], Marta Astier [2,*], Robert Walker [3], Jaime Fernando Navia-Antezana [4], Yan Gao [2], Yankuic Galván-Miyoshi [5] and Dan Klooster [6]

[1] Department of Geography, University of Florida, Gainesville, FL 32611, USA
[2] Centro de Investigaciones en Geografía Ambiental, Universidad Nacional Autónoma de México, Antigua Carretera a Pátzcuaro 8701, Col. San José de la Huerta, Morelia 58190, Michoacán, Mexico; ygao@ciga.unam.mx
[3] Center for Latin American Studies, University of Florida, Gainesville, FL 32611, USA; robertwalker@ufl.edu
[4] Grupo Interdisciplinario de Tecnología Rural Apropiuada (GIRA A.C.), Carretera Pátzcuaro a Erongarícuaro 28, Tzentzenguaro, Pátzcuaro 61613, Michoacán, Mexico; jnavia@gira.org.mx
[5] ECOTRUST, 6721 NW 9th Ave. Suite 200, Portland, OR 97209, USA; galvan@ecotrust.org
[6] Latin American Studies College of Arts & Sciences, University of Redlands, 1200 E. Colton Ave, Redlands, CA 9237, USA; daniel_klooster@redlands.edu
* Correspondence: cssimmons@ufl.edu (C.S.); mastier@ciga.unam.mx (M.A.)

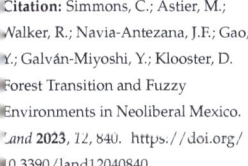

Abstract: Although deforestation remains a continuing threat to both the natural world and its resident human populations, a countervailing land cover dynamic has been observed in many nations. This process of landscape turnaround, the so-called forest transition, holds the potential of regenerating ecosystem services by sparing land from agricultural activities and abandoning it to forest succession. Here, we present a case study of a long-term process of forest transition that is ongoing in the Patzcuaro watershed of the state of Michoacán, Mexico. The research to be discussed comprises a remote sensing analysis designed to (1) capture the land cover impacts of a multidecadal process of trade liberalization (1996–2018); (2) ascertain the role that land tenure plays in land use dynamics affecting forest cover, and (3) resolve forest cover types into native forest, secondary vegetation, and "commodity" covers of fruit trees, in this case, avocado. Mexico presents a useful case for addressing these three design elements. Our analysis, undertaken for both private property and collective modes of resource management in five communities, reveals a forest transition annualized at 20 ha-yr^{-1}, or a gain of eight percent for the period. This translates into a relative rate of forest transition of 0.39%-yr^{-1} which is three times faster than what is occurring in the temperate biome on a national scale (0.07%-yr^{-1}). Most of the forest transition is occurring on private holdings and stems from field abandonment as farming systems intensify production with avocado plantations and cow–calf operations. As this study demonstrates, forest transitions are not occurring ubiquitously across nations but instead are highly localized occurrences driven by a myriad of distal and proximate factors involving disparate sets of stakeholders. Consequently, policy makers who are keen to expand forest transitions to fulfill their national climate action commitments under the Paris Agreement must first promote research into the complexity of landscapes and drivers of land change at regional and local scales.

Keywords: forest transition; land sparing; neoliberal policy; NAFTA; Mexico; deforestation

1. Introduction

The conversion of native forests is one of the most important processes contributing to the current global environmental crisis. It accounts for 38% of the carbon emissions from agriculture, which contributes ~25% of greenhouse gas (GHG) emissions to the atmosphere [1]. In addition, deforestation and forest degradation are the primary drivers of species extinctions on land and the loss of livelihood for the millions of people who rely on the ecosystem services provided by forests [2,3]. It should come as no surprise that measuring this environmental transformation and understanding its proximate and underlying causes now comprise a major interdisciplinary research area. Although deforestation remains a continuing threat to both the natural world and its resident human populations, a countervailing land cover dynamic has been observed in many nations. This process of landscape turnaround, the so-called forest transition (FT), holds the potential of regenerating ecosystem services by sparing land from agricultural activities and abandoning it to forest succession.

Specifically, FT provides a critical empirical foundation for greenhouse gas (GHG) emissions policy based on carbon sequestration and is therefore highly consistent with the goals of the United Nations Program on Reducing Emissions from Deforestation and Forest Degradation, or UN-REDD [4,5]. Recently, controlling anthropogenic greenhouse gas (GHG) release to the atmosphere with forest-based policy has been widely adopted by signatories to the Paris Climate Treaty, particularly by nations with large forests and dynamic agricultural economies, such as those found throughout the global south. FT, although first observed in the global north, is increasingly affecting forests wherever they are found. This article takes up the case of Mexico, which has set a goal of 0% deforestation by 2030 as part of its commitment to the Paris Climate Treaty. In particular, we present the results of a study addressing land cover change dynamics in an old agricultural frontier where FT was observed over twenty years ago. Our objective was to assess the continuity of this process given the significant changes in the institutional and economic environments affecting Mexico and the global economy more generally.

We pursue our objective as follows: First, we provide an overview of the manner in which FT has been conceptualized and touch on the literature that has extended its range of applications, both spatially and temporally. After this, we focus on Latin America and consider cases of FT occurring across the region at the present time. This sets the stage for the presentation of research on Mexico. For this, we present results from a remote sensing analysis combined with data collected from a survey of smallholders. Finally, we discuss our findings and draw a brief conclusion that questions to what degree FT can be regarded as an ecologically restorative process.

Before proceeding, we wish to address two terminological issues. The first concerns our use of the term "forest transition" as opposed to "land sparing." Although both reflect processes that for all intents and purposes are identical, we elected to use "forest transition" as our operative phrasing given its early usage as a term reflecting long-term patterns of land cover change tied to broad-scale global dynamics. This is appropriate given that the analysis we present covers a multidecadal period of nearly fifty years, during which time neoliberal reforms profoundly affected many national economies. The second issue involves FT's descriptive categories, a binary of highly aggregated land covers, one that exhibits exclusively positive environmental qualities (forest) and one that does not (non-forest). Such a sharp distinction is often absent from densely settled parts of the world such as Mexico, where human populations have long derived subsistence from agroecological matrices that include trees from the original biome as well as agricultural land uses. In these situations, land covers are environmentally "fuzzy" because they manifest varying intensities of ecosystem service provision. Our usage of the term "fuzzy" derives from the concept of the fuzzy set, whose members elicit variable degrees of membership. They cannot be construed as belonging to a binary scheme of categorization comprising covers that conserve the environment and its ecosystem services and covers that do not.

Mather [6] and Walker [7] use the terms forest areal transition and landscape turnaround, respectively, in referring to shifting trends in the national land covers of Europe, North America, and the Pacific Rim, from a period of forest decline to one of net expansion. Their original formulations rested on a hypothesized relationship between land cover dynamics and long-run structural changes in a national economy, with agricultural development driving deforestation, after which industrialization sparks rural-to-urban migration as farmers abandon their fields to take manufacturing jobs in cities. Intensifying this process of abandonment are technological changes that reduce the demand for farmland by raising agricultural productivity, which is further enhanced as communication networks reveal regional comparative advantages and production shifts from marginal to fertile areas. Technological changes also attenuate forest exploitation by providing new energy sources and building materials as substitutes for fuelwood and timber. Finally, as consumer preferences for natural amenities develop with rising incomes, state environmental policies reinforce the recovery of natural areas in the interest of conservation. The sum total of all these effects is that the pressure on forests is reduced and that formerly exploited lands are abandoned to forest succession, the foundational FT process [6,7].

Critics of the early formulation, which draws a correspondence between dynamics in a national economy and regional landscapes, point to its congruence with developmentalist thinking and the apparent restriction to an isomorphic pattern with necessary "stages", also known as the Rostow paradigm of growth [8]. This critique questions the alleged spontaneity of forest recovery in the face of industrial capitalism, arguing that FT necessarily requires state intervention in resource exploitation by the private sector and does not proceed by a natural succession of preordained stages [9]. It also points to the impact of institutions, trade, and historical contingency on FT, given that alleged necessary conditions do not always lead to the expected outcome [8,10–12]. Behavioral contingency plays a role as well, given the wide variety of skills individuals possess and the great diversity in their preferences for production and consumption.

More recent critiques call attention to scale and foundational issues of measurement and definition [13]. As for scale, there is no a priori reason to restrict FT to the nation-state given that processes of forest recovery occur not just at the macro-scale but also in highly localized settings [14–16]. In fact, aggregate deforestation might conceal regional FT as a function of the biome [15,17,18]. As for the issue of measurement, ecologists have pointed to the false dichotomy between conservation and agriculture inherent in FT theory, which is based on an agriculture–forest binary equating forests with conservation and agriculture with degradation [19,20]. Many landscapes are highly fragmented, with matrices combining agriculture and forest as opposed to distinct partitions into two categories, as noted at the outset. Further, the forest itself may provide agricultural values either by grazing or by the substitution of trees producing commodities for those of the original biome. In such landscapes, ecosystem services are more likely to be provided by a matrix of agroecological land use than by native forests. This framing, compatible with the whole landscape approach [21], recognizes a persistent rurality in which indigenous and other rural people do not abandon marginal lands but instead adapt their practices to the agroecological environment [22–24]. In these cases, small-scale sustainable agriculture can contribute effectively to an ecosystem services conservation strategy in the absence of a stylized FT.

This study presents research attentive to scale, both spatially and temporally, which accounts for the definitional issues raised by ecologists. With respect to spatial scale, we address complex, countervailing processes whereby aggregate deforestation conceals a disaggregate FT as a function of biome type [17]. In addition, the detail of our remote sensing analysis enables us to decompose the "forest" category into native forest and fruit tree plantations, which in our case possess contradictory implications for conservation. We are therefore able to address the binary framing of conventional FT theory with a "fuzzy" approach that translates forested and non-forested land covers into an agroecological matrix. In undertaking our analysis with a representative case study, we implement a multidecadal time span consistent with early FT conceptualization. Thus, the analysis period brackets long-run changes in Mexico's macro-economy driven by neoliberal reforms. Finally, the temporal scale implemented provides an opportunity to examine the impact of land tenure and property rights on localized FT, given that it covers changes in the institutional environment that have generated a complex pattern of private and communal holdings in the study area. Many have called attention to the role of property institutions in explaining dynamic land cover processes such as deforestation. Consequently, since land tenure affects land use decisions, FT, as an outcome of agricultural decision-making, must necessarily reflect the institutional context in which it occurs [10,25]. To date, little empirical work has documented how land rights impact restorative ecological processes such as FT.

1.1. Forest Dynamics in Latin America

Mather [6] and Walker [7] documented FT in much of the global north and argued there was no guarantee it would occur where nations were still experiencing high rates of deforestation, such as in the global south. Nevertheless, research has now documented FT in a wide variety of settings [25–28]. For example, Grau and Aide [27] point to 18 Latin American cases occurring between 1996 and 2008. Most of them—incipient and highly localized—involve relatively small areas, with half covering less than 5000 km^2 and only three exceeding 100,000 km^2. In addition to the reduced extent, the identified FTs are recorded for periods of short duration and do not necessarily reflect long-term dynamics associated with structural changes in an economy. Only two FTs appear to be occurring at a national scale, one in Puerto Rico and the other in El Salvador.

Puerto Rico is a commonwealth of the US, where manufacturing is heavily subsidized and emigration to the United States is unrestricted [7]. Here, annual rates of forest recovery from 1940 to 1990 have been estimated at between 9% [7] and 0.63% [29]. The other macro-scale FT is occurring in El Salvador, where civil war displaced a large part of the rural population in the 1980s [30]. Since 1990, forest covers have expanded by 7% in El Salvador's closed forests and 30% in its open forests [30]. The Brazilian Legal Amazon provides another relevant example given the area covered, ~5 million km^2. Here, Perz and Skole [31] reported declining rates of deforestation, with an increasing expansion of secondary forests for the period 1986–1992. Net secondary forest gain represents only 1% in the aggregate, but in old settlement areas of the lower basin, it reaches 25%. Finally, regional FT over a large area is occurring in the Atlantic Rainforest biome of Brazil, as documented by [17].

FTs in Puerto Rico, El Salvador, and Brazil reflect histories of agricultural land abandonment, where country-specific contingencies have managed to push and pull people from rural areas, with significant depopulation in the countryside [30,32]. Studies of Latin American FTs at more localized scales have emphasized the impact of changing livelihood strategies on household mobility [16,33,34]. For example, Rudel et al. [35] used remotely sensed and household data to investigate FT in the Peruvian Amazon. Schmook and Radel [36] took a similar approach to the southern Yucatan in Mexico. Results from both studies confirm an incipient FT in each case but diverge in their explanations. In Yucatán, out-migration and non-farm job opportunities appear to be driving a decline in deforestation. By contrast, out-migration does not appear to promote FT in the Peruvian Amazon, where household decisions about land use and livelihood diversification play key roles.

These Latin American examples suggest that landcover change dynamics sometimes, but not always, follow an FT narrative in which rural out-migration and forest recovery are linked to changes in national and global economies [15,37]. Our work provides an additional case study for Mexico, specifically in the highlands of Michoacán, where previous research has explored how landscape processes are linked to economic dynamics occurring at various scales, from global to local. Much of this work suggests an FT may be underway as agricultural abandonment leads to spontaneous forest regeneration [25,38], while other observations suggest drastic forest loss due to illegal logging and conversion to export-oriented avocado orchards [39].

1.2. Forest Transition: The Mexican Case

Mexico presents a useful case for addressing forest transition. Of particular note is the process of neoliberal reform that it has experienced since the 1980s with the General Agreement on Trade and Tariffs (GATT, 1985), which was followed by the North American Free Trade Agreement (NAFTA) in 1994 and, most recently, the United States–Mexico–Canada Agreement (USMCA) in 2020. These reforms affected land tenure and market access for smallholder farmers throughout Mexico. Land tenure was impacted by the ejidos, the publicly held properties created during the land reforms of the Mexican Revolution. Although private farm properties have long existed in Mexico, they expanded significantly following Amendment 27 to the Mexican Constitution in 1992, which allowed for the privatization of ejidal farm parcels under strict guidelines [40]. In addition to this, trade liberalization essentially thrust Mexico into the global economy by reducing tariffs on imports and eliminating subsidies to production from the parastatal sector [41,42].

Trade liberalization in Mexico and neoliberal reforms more generally have tracked an autonomous process of agricultural intensification, which began with the green revolution in the 1960s with new varieties of wheat and maize. Also in evidence is a multidecadal process of land cover dynamics. An examination of total forest change magnitudes at a national scale reveals overall declines in deforestation rates through much of the time period until 2013, when rates began to rise once again (see Figure 1). The rate of deforestation between 1985 and 1992, 0.53% yr^{-1}, declined to 0.11% yr^{-1} for the period 2002–2014. Of particular interest for this study are the forest cover dynamics revealed when the data are disaggregated by forest biomes. As the figure shows, deforestation rates in tropical humid biomes have steadily declined throughout the time period, and an apparent FT is underway in Mexico's temperate forests, which increased their extent from 341,805 km^2 in 2002 to 351,562 km^2 in 2018, for an annual rate of transition of 610 km^2 yr^{-1}. Unfortunately, deforestation rates in tropical dry biomes show a pronounced rise across the time period, a trend that parallels deforestation patterns across the global south [18,43,44]. The dramatic and continually high deforestation rates in tropical dry biomes clearly explain the total deforestation rate trends on a national scale.

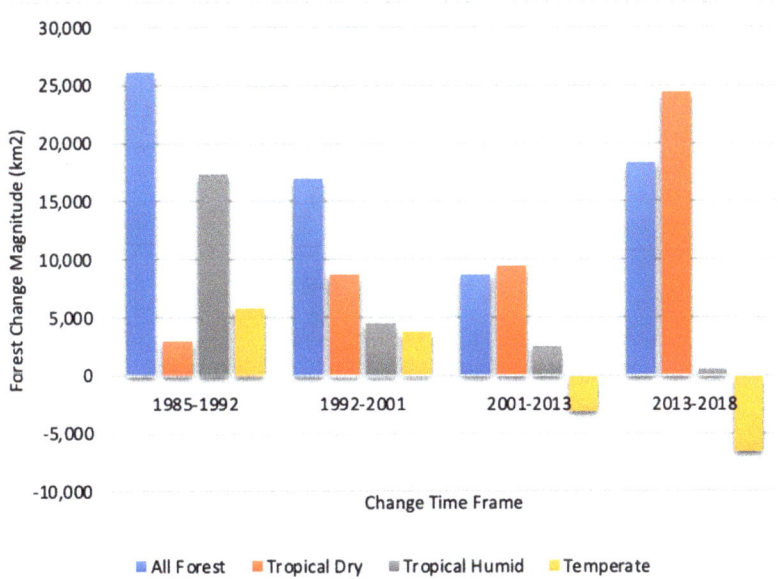

Figure 1. Forest change magnitudes 1985–2018 km^2 [45]. Five study sites in PLW, located in Michoacán state, central Mexico.

2. Methodology

2.1. Study Region: Physical and Social Characteristics

The present study addresses a specific case of a temperate FT, one occurring in the Patzcuaro Lake watershed (PLW) in the state of Michoacán, Mexico (see Figure 2). The PLW is part of the Purhépecha Indigenous Region, and, like many other parts of rural México, human settlement at this site is ancient, dating back more than 5000 years [46]. The watershed, covering ~1000 km² [47], drains into Pátzcuaro Lake, with a surface area of 130 km² and a 4.9 m mean depth [48]. The regional geomorphology is dominated by volcanic mountains, inter-mountainous valleys, piedmonts, and hills, with elevations ranging from 2035 m.a.s.l. at lake level to 3400 m.a.s.l., the altitude of the highest mountain [47]. The PLW possesses a mild, temperate climate, with an average annual temperature of 14.5 °C and an average rainfall of 1000 mm·yr^{-1} distributed mostly from June to October. Soils are of volcanic origin and are mostly fertile. Ecology is dominated by the temperate forest biome, consisting primarily of a mixed pine–oak forest. Fir trees (*Abies religiosa*) are found in the high elevations, while pine (*Pinus pseudos-trobus*, *P. michoacana*, and *P. montezumae*) and oak (*Quercus rugosa*, *Q. castanea*, and *Q. laurina*) grow on the wide slopes of the mountains. Lower elevations are covered by grasses and secondary herbaceous vegetation, presumably of anthropogenic origin [48].

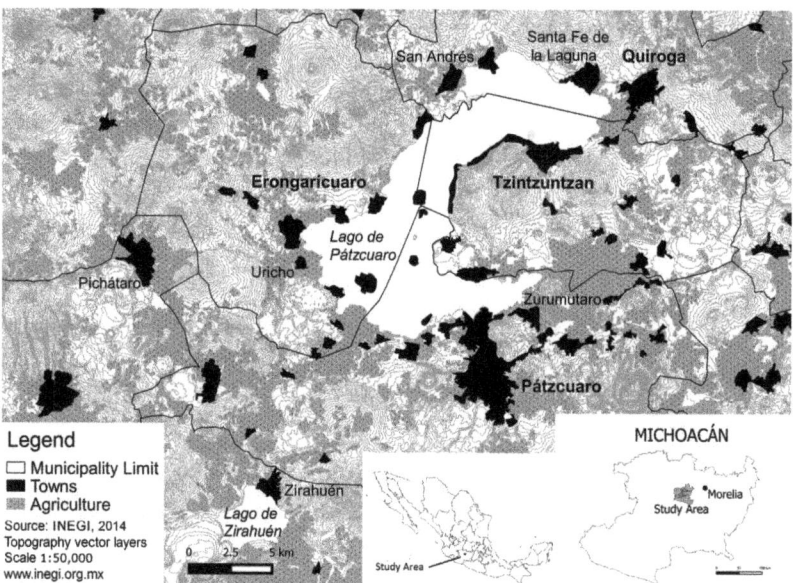

Figure 2. Map of the study area.

Agriculture occurs on small holdings (<5 ha) in the flat areas around the lake and in the valleys [47]. The main seasonal crops are maize, oats, wheat, legumes, and vegetables. Tree crops are also grown, avocado in particular, but maize is the most important crop, representing 71% of the planted area (8115 ha total in 2014), followed by oats, which are used to feed cattle [46]. PLW agriculture takes place within a complex spectrum of land tenure institutions, ranging from private property to those reflecting a more communal approach. These include Comunidades Agrarias (agrarian communities), which recognize long-standing collective tenure rights, and ejidos, which are collective land reform grants. Under both collective arrangements, individual land managers generally have strong usufruct rights to planted fields and orchards, while lands with native tree cover are shared by the community as a whole. The private properties, agrarian communities, and ejidos in our study area are mostly inhabited by Purépecha Indigenous peoples. The agrarian community (AC), also referred to as the Indigenous community in this part of Mexico, includes a higher density of collective social institutions for self-governance and ritual celebrations.

2.2. Prior Research

A substantial amount of research has addressed resource use and agriculture in the PLW. Klooster [25] used aerial photographs (from the years 1942, 1960, 1974, and 1990) and satellite imagery (2000) together with key informant interviews to investigate long-run landscape dynamics in the PLW, including one of the communities in the present study. This work suggested an overall expansion of forest cover, primarily on agricultural fields that were far from towns or on poor soils [25,49]. In a similar study on the neighboring Lake Cuitzeo Basin, Lopez et al. [38] used aerial photographs from 1975 and 2000 to show, for the entire 4000 km^2 watershed, a substantial expansion of scrubland areas following the abandonment of rainfed agriculture and a slight increase in the forest. Evidently, land abandonment occurred in 'relatively unproductive' areas, especially stony fields and hillsides with low yields. In several municipalities, however, avocado orchards replaced rain-fed agriculture.

These findings are generally consistent with the FT narrative stemming from agricultural abandonment [25,50]. Other land cover dynamics have been observed in the region. Although FT appears to be long-term and ongoing, deforestation to make way for avocado plantations is occurring in addition to logging, which is shifting species composition away from commercially valuable pine to less valuable oak by selective harvest [49]. Aside from outright resource exploitation, the PLW also provides examples of community forestry [39]. In an area approximately 50 km to the south of the PLW, Barsimantov and Navia Antezana [40] used Landsat TM and ETM images from 1990, 1996, 2002, and 2006 to conduct comparative landcover change analyses in four indigenous communities. Two of the four communities maintained effective communal organization, established legal community logging operations, and manifested drastically lower deforestation rates than the two others, which lacked effective communal organization. Here, communal forests were illegally logged, then extra-legally privatized by a transfer to non-community members who planted avocado. Most of the deforestation occurred during the early and mid-1990s, amid changes in communal tenure policy and a relaxation in log transportation rules.

2.3. Current Study

Our research comprises a remote sensing analysis designed to (1) capture the land cover impacts of a multidecadal process of trade liberalization; (2) ascertain the role that land tenure plays in land use dynamics affecting forest cover; and (3) resolve forest cover types into native and "commodity" covers of fruit trees, in this case, avocado. Our research complements earlier work by providing a quantitative assessment that disaggregates land change processes in the PLW into source land covers and by extending the analysis to the post-NAFTA period. The specific coverage area for the remote sensing analysis comprises five localities in the PLW, selected to ensure variation in land tenure regimes. These include communal lands in indigenous communities (San Andrés Tziróndaro, San Jerónimo Purenchécuaro) and ejidos (Erongarícuaro and La Zarzamora), in addition to private properties (also in Erongarícuaro) (see Figure 3). Ejido Zarzamora, Ejido Erongarícuaro, and the private properties under analysis are found in Municipio Erongarícuaro, while the two indigenous communities are located in Quiroga Municipio. We used remotely sensed data to produce land cover change matrices for each locality.

The remote sensing and GIS analysis involved change detection based on data from 1996 and 2018. For 1996, we created a land-use map by visual interpretation of black and white aerial photographs taken in 1996 at 2m spatial resolution. The aerial photographs were interpreted by an expert familiar with the study area and who had deep knowledge about the distribution of its land cover. In addition, information such as the form, shape, texture, and tone of the land cover classes was employed during the visual interpretation process. For 2018, we used very high spatial resolution images available in Google Earth, downloaded as an RGB photo and geometrically rectified, then projected to a common Universal Transverse Mercator projection (WGS 84, UTM 14N). The change analysis was carried out following an interdependent interpretation procedure developed by FAO [51], by which the first image is interpreted visually and then the second image is modified based on the interpretation. Accuracy assessment was not performed for the 1996 land-use map since there were no available high-resolution images in Google Earth to obtain ground truth data. The visual interpretation of aerial photographs by an expert was capable of producing land-use maps with high accuracy. By following this procedure, it was expected to be able to produce consistent change data. We superimposed the GE photo on the 1996 land cover map and visually identified the land cover areas that showed changes over the period; not all land use changed. The land cover categories implemented were forest (e.g., pine and oak), secondary vegetation (e.g., herbaceous vegetation, potential fallow, and dirty grazing pasture), annual crops (e.g., maize, legumes, vegetables), tree crops (e.g., avocado), and urban areas (built-up areas). Our change detection analysis produced transition matrices for the aggregate across all five localities as well as for each one. Net changes for each category of land use are given, as is the transition matrix for each pairwise permutation of the land use categories.

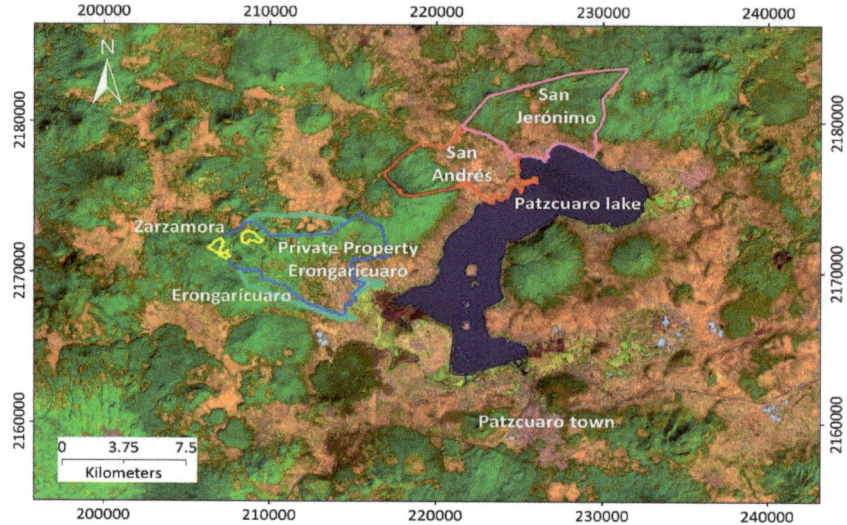

Figure 3. Five study sites in PLW, located in Michoacán state, central Mexico.

3. Results

Before presenting the results of the land cover change analysis, we provide background context by considering the socio-demographic conditions of the five localities (see Table 1). As is evident, the population of the agrarian communities (ACs) numbers in the thousands, considerably more than the relatively few people who live in the ejidos. A similar pattern emerges for the land area. In this case, both the land area for the ACs and the private holdings are larger than the ejidos. In general, the five localities have remained stable over time, even though the ejidos and AC of San Jerónimo are quite old. Although the ACs and ejidos have both benefited from government subsidies under PROCEDE, poverty remains entrenched.

Table 1. Characteristics of localities in study region.

	AC San Andrés	AC San Jerónimo	Ejido Erongarícuaro	Ejido Zarzamora	Private Property Erongarícuaro
Population:					
Total [a]	2418	1981	98	31	3639
Indigenous [a]	2418	1981	unknown	unknown	unknown

Table 1. Cont.

	AC San Andrés	AC San Jerónimo	Ejido Erongarícuaro	Ejido Zarzamora	Private Property Erongarícuaro
Poverty [a]	Medium	Very Low	Very Low	Very Low	Very Low
Area (ha):					
Initial Size [b]	2621.4	3146.8	648.0	117.7	NA
Current Size [c]	same	same	1045.0	same	NA
Resolution [b] Initial date	26 August 1997	22 September 1975	14 January 1918	3 February 1961	NA
PROCEDE [c] Current date	15 May 2006	24 September 2002	24 July 2002	30 June 1997	NA

Sources: [a] Consejo Nacional de Poblacion, Mexico [52,53]; [b] Presidential Resolution of Communal and Ejido [54]; [c] Program Certificación de Derechos Ejidales y Titulación de Solares—PROCEDE [54].

3.1. Changes in Land Use

Aggregate changes in land use across the five localities for the period reflect three processes, namely: (1) the reduction in the area dedicated to annual crop production, (2) the emergence of tree crops, specifically avocado, and (3) the abandonment of secondary vegetation to ecological succession (see Table 2a). The aggregate reduction in the area under annual crop production in the study region, in particular maize, is the dominant agricultural dynamic, with a decline of ~550 ha from ~3872 ha in 1996, or about 14 percent. For its part, avocado production expanded dramatically from a limited 5 ha dedicated to the crop in 1996 to more than 230 ha by 2018. During this same time period, private properties decreased the area under annual crops by 547 ha. The two indigenous communities manifest rather static dynamics, with farming based solidly on annual crops in both periods; neither has planted tree crops and only San Jerónimo shows any change in secondary vegetation, with a reduction of ~90 ha from ~350 ha in 1996. The annual crop area has remained stable in Ejido Erongarícuaro, where 8 ha of avocado were planted sometime after 1996. Tree crops were not planted in Ejido Zarzamora, which shows the abandonment of land under annual crops. The areas here are small relative to the other localities and therefore do not affect the aggregate dynamics.

Table 2. (a) The area of land use and land cover changes from 1996–2018. (b) Land use transition from 1996 to 2018.

a

Land Use Categories	Areas in 1996 (ha)	Areas in 2018 (ha)	Change Area (ha)
Annual crop	3872.19	3324.99	−547.2
Temperate forest	5448.3	5883.32	+435.02
Tree crops	5.51	228.87	+233.36
Secondary vegetation	956.05	771.96	−184.09
Urban zone	278.88	351.80	+72.92

b

		Land Use 2018 (ha)					
		Annual Crop	Forest	Tree Crops	Secondary Vegetation	Urban Areas	Total
Land Use 1996 (ha)	Annual crop	3293.47	280.58	101.95	125.05	71.14	3872.19
	Forest	9.35	5334.09	85.89	18.37	0.61	5448.30
	Tree crop	0.73		4.29		0.50	5.51
	Secondary vegetation	21.44	268.65	36.74	628.54	0.67	956.05
	Urban areas					278.88	278.88
	Total	3324.99	5883.32	228.87	771.96	351.80	

3.2. Forest Transition

In 1996, forest covered ~5448 ha of the five localities in the sample, and by 2018, 435 ha had been added, yielding an annualized transition of 20 ha-yr^{-1}, or a gain of eight percent for the period. This translates into a relative FT rate of 0.36%-yr^{-1}. Evidently, the local FT calculated for the PLW is three times faster than what is occurring in the temperate biome at the national scale in Mexico (0.11%-yr^{-1}).

About half of the study area's forest gain is accounted for by private properties (219 ha). This is followed by the gain of 169 ha in the indigenous community, San Jerónimo, in contrast to the forest cover stasis in San Andrés. Both ejidos show a transition, with an additional 28 ha of forest cover acquired by Ejido Erongarícuaro for the period and 16 ha by Ejido Zarzamora. The expansion of forest land on private properties is mostly accounted for by the suspension of annual crop production on 220 ha; 122 ha are contributed by succession from secondary vegetation. The sources of FT in San Jerónimo are reversed, with 123 ha coming from succession and 48 ha from the abandonment of crop fields. Hence, for these two localities, 268 ha of transition stem from field abandonment and 245 ha from vegetative succession. For the two ejidos, the scaling back of annual production accounts for most of the observed transition.

4. Discussion

The FT in the PLW, first documented by Klooster [25,49], has continued to the present, with the region's temperate forest gaining ground from the mid-twentieth century. Klooster [25] drew attention to the land-use component of FT, as advanced by Grainger, noting that field abandonments throughout the region were key to the region's land cover dynamics, especially the restoration of its temperate forests of pine and oak. The present study adds to the historical case presented by Klooster in documenting the contraction of annual crop production with the concomitant release of land to successional processes. We are able to show explicitly that most of the FT occurring across the five localities is attributable to the abandonment of land under annual crops. Such field abandonments arise for two primary reasons. Either 376 farmers abandon cornfields altogether to earn a living elsewhere or from off-farm activities, or they continue farming but with different crops and farming systems. We now consider each possibility. Our methodology cannot assess the quality of the remaining and recovering forests; the removal of the most commercially valuable species and individual trees is likely to happen [24,25].

It is certain that the study area's mobility patterns are consistent with farm closures. Although population dynamics are not observable at the locality level, insight can be gained with municipal data. Ejido Zarzamora, Ejido Erongarícuaro, and the private properties under analysis are found in Municipio Erongarícuaro, while the two indigenous communities—San Andrés and San Jerónimo—belong to Municipio Quiroga. Both Erongarícuaro and Quiroga grew during the period of NAFTA implementation (1990–2005), Erongarícuaro from 11,930 to 13,060 residents and Quiroga from 21,917 to 23,391. Dynamics affecting their rural and urban places differ, however. Of interest is Erongarícuaro's declining rural population (from 11,930 to 10,593). This suggests that people are leaving the rural economy for better opportunities elsewhere, which could explain the FT in each of the three localities found here (private property and two ejidos).

Alternatively, Quiroga experienced rural population growth between 1990 and 2005 (from 3788 to 5735). Although annual cropping persists in a reduced area, FT follows the succession of secondary vegetation. Moreover, the growth in the rural population hardly suggests that the restoration of the forest has been driven by outmigration. For the notable case of San Jerónimo, FT evidently arises for reasons other than farm closure, perhaps from changes in household production involving intensification or from the replacement of maize produced locally for human consumption with purchased grains [25]. FT dynamics may also include a shift from wood to natural gas for cooking, although rural people in the area tend to continue to use firewood even after adopting natural gas options [55].

Although we cannot clearly link migration dynamics to the landcover changes we detected, it is clear that smallholder farming strategies throughout the region, and Mexico more generally, have changed appreciably in the wake of NAFTA and closely related institutional reforms. Maize agriculture has long been important to the PLW and still represents a key component of household farming [56]. Nevertheless, our study frame brackets the period during which PLW farmers began their engagement with the maize–beef complex, a vertically integrated system of maize suppliers, feedlots, and supermarkets [46]. Now found in all regions, this highly integrated production chain, which exports beef to global markets, is concentrated near cereal-producing regions in temperate central Mexico such as *El Bajío*, comprising the States of Guanajuato, Jalisco, and the state of the study area, Michoacan.

The adjustments in farm-level production followed trade liberalization, NAFTA, in particular, which stimulated a structural change in the Mexican cattle economy, specifically a switch from the grass-fed system of the hacienda to a complex supply chain based on feedlots, cheap imported grains, and expanding global markets for beef. As a consequence, new opportunities for smallholders have opened as suppliers of calves and feeders. Although the mixed farming systems in the PLW have long incorporated cattle in their production portfolio, the survey data reported by Astier et al. [46] for two of the study area municipalities (Erongarícuaro and Quiroga) revealed a substantial herd expansion from 589 animals at the start of their individual operations to 1760 animals in 2019 for 130 smallholder households. At the same time, the average household herd size grew from 5.8 to 14.5 animals. In accompaniment has been a shift in the role that bovines play in the household economy, with the sale of feeders to the supply chain replacing reproduction as the primary reason for maintaining cattle.

Prima facie, this might be expected to have placed pressure on forests to convert to pasture. Unfortunately, our remote sensing analysis was unable to identify pasture as a distinct land cover category, which is probably moot given the presence of FT. The explanation evidently lies in the availability of cheap corn and the use of household grains to feed the animals [41]. This could explain why the amount of land used for grazing has diminished over time across the sample, from 1236 to 1084 ha; a decline tracked by the household average from 8.64 to 7.52 ha [41]. Earlier, we suggested that land abandonment in Municipio Erongarícuaro is attributable to farm closures and economic difficulties. An alternative explanation is that cheap grains bought by the sale of feeders have diminished the household demand for annual production. Following NAFTA, family cow–calf operations no longer produce maize for local consumption. Evidently, old maize fields have been converted to pasture and secondary forests in which cattle graze. Supporting this possibility, and mitigating our inability to identify pastures with remote sensing, is the fact that smallholders in the PLW do not graze their animals on well-formed pastures, but rather in forests and secondary vegetation (*agostadero*).

As for the possible impact of the 1992 amendments to Article 27 of the Mexican Constitution, we call attention to the sharp distinction between private properties on the one hand and ejidos and indigenous communities on the other, with respect to the production of avocado. The amendment undid basic principles of the Mexican revolution as codified in Article 27, which established the inalienability of *ejido* lands, effectively transforming them into communal holdings, much like Indigenous communities'. After 1992 (and with the implementation of NAFTA), *ejidatarios* were able to privatize holdings subject to regulations and legal protocols. Only then could land be used as collateral to guarantee the loans necessary to invest in avocado. This could explain why nearly all avocado plantations are found on private holdings. There are other possible explanations, the analysis of which lies beyond the scope of this paper. For example, private properties may have incidentally formed on lands with enhanced capabilities (soils and topography). Alternatively, communal controls on farm practices might have inhibited commercial prospects.

The extent of avocado agriculture appears to be growing. Orozco-Ramírez and Astier [56] measured its expansion in the PLW for two municipalities in the present study.

In Erongarícuaro, the avocado area grew from 8 hectares in 2003 to 211 hectares in 2014, and in Quiroga, from 5 to 73 hectares for the same period. In both cases, avocado areas grew at the expense of maize fields and the native forest of pine and oak. This rapid expansion is observable throughout the region. In the primary avocado-growing parts of Michoacán, associated land allocations increased from 3 to 34% of the total area, while fields for annual crops (mostly maize) shrank from 35 to 16%, and forests from 47 to 30% [57].

Our research documented a localized FT in the PLW, one that mirrors what is happening in the temperate forest biome across Mexico. This recovery of the forest area has occurred simultaneously with a significant structural change in regional agriculture. The old maize-oriented family farm appears to be giving way to more commercially oriented enterprises consisting of cow–calf operations, avocado plantations, or some combination of the two.

5. Conclusions

Economists and others often speculate about the link between resource management and land tenure. Bioeconomic theory has pointed to the superior performance of private property with respect to a variety of conservation outcomes [7]. It is argued that when resources are held in common, conservation-compatible incentives do not emerge. Although the theory has not been well articulated for the case of FT, it stands to reason that any kind of land cover change, to the extent it is driven by incentives, must bear some relationship to the land tenure situation. In the present case, the results are inconclusive. In absolute terms, private holdings dominate FT, affecting the native forest and accounting for about half of the new area captured by our analysis. However, one should not be too quick to conclude that FT is more likely to occur on private properties, which might be taken as the logical inference of bioeconomic theory and the incentives argument. Private property dominates in absolute terms, but its FT in relative terms (recovered native forest as a percentage of the land in private property) is 6%. For the case of Ejido Zarzamora, the relative FT is 20%; for Ejido Erongarícuaro, 5%; and for AC San Jerónimo, 5%. Although the AC of San Andrés shows no FT whatsoever, we note that our sample size forecloses an inferential conclusion, especially given the substantial variation in the relative rates across the communal properties.

FT research has called attention to the nature of the forest that returns to the landscape following the transition [7]. The hope is that the same ecosystem services, in the form of biodiversity maintenance and carbon sequestration, will be restored, but there is no guarantee. Our study reveals two confounding factors that weaken the prospects of such a beneficial outcome. The first is that Mexico's temperate FT involves a hidden land use, namely the grazing of cattle under closed canopies. It is difficult to imagine that this would not significantly impact both biodiversity and forest biomass. Further, absent sound forest management practices, woodcutting, and timber harvesting are likely to remove commercially valuable species from the forest, which could degrade its ability to provide the ecosystem services that local people value. Evidently, such exploitation in the PLW is pushing the forest to one dominated by oak.

The second confounding factor involves tree crop production, avocado, in particular. Tree crops with closed canopies are forests of a type, and it may be difficult to classify them as a separate land cover when remotely sensed data do not possess sufficient resolution. Even with adequate resolution and a sure classification, an analyst might naively interpret avocado plantations as a type of forest for the simple fact that avocado plants are trees. Is there any harm in this, one might naturally wonder. Would it be wrong to conclude that the encroachment of avocado plantations on abandoned maize fields represents a type of FT? We argue that an answer should depend on the ledger of ecosystem services before and after the plantation matures. In the case of avocado, environmental harms outweigh its production of ecosystem services.

Of prime concern in the case of avocado are irrigation and pollution [58]. Irrigation, in which well-capitalized interests typically invest, disturbs the hydrological balance, thereby affecting the microclimate and surface water run-off patterns [59,60]. The use of manure and synthetic fertilizers—also used by large-scale operations—have been shown to eutrophy receiving water bodies, which are also impaired by run-off with pesticide residues. The negative externalities of land-use change involving a switch from the temperate pine–oak forest of the PLW to avocado plantations are well documented and clearly militate against counting avocado trees as adding to the positive side of the ecological ledger [60]. It should not be forgotten that avocado agriculture also produces a valuable commodity that could potentially generate higher income. Nevertheless, assessing the extent to which an economic activity promotes human welfare requires an accounting of all associated costs. Our analysis has pointedly equated forest with native forest, which a priori eliminates the possibility of making a definitional error that would lead to an overestimation of the study area's FT.

Mexico's National Contribution to the Paris Climate Treaty addresses the agricultural sector and forest management. Like other nations with large extant forests, Mexico recognizes the ecosystem service of carbon sequestration and consequently has crafted a mitigation approach involving the principles of UN-REDD. As already mentioned, this involves the policy objective of achieving 0% net deforestation by 2030. Such an objective may not seem far-fetched given the onset of FT in its temperate forests. One should be cautious, however, in concluding that this is the best possible outcome for the environment as a whole. An agroecological matrix of land uses may provide more ecosystem services than a recovering "forest" that is grazed or a new one that generates valuable commodities. The present study does not suggest the FT occurring in the PLW offers no environmental benefit, given that we have excluded avocado trees from the "forest" category. Further, the FT we have measured appears quite robust. It has prevailed against the structural changes induced by trade liberalization, involving a partial shift from maize-oriented agriculture to cow–calf operations and avocado plantations. Only time will tell if these new and profitable activities will reverse the multidecadal FT presently occurring in this part of Mexico.

Author Contributions: Author Contributions: C.S., M.A. and R.W. conceived the paper; C.S., M.A., R.W., J.F.N.-A. conceptualized and designed the research; Y.G.-M. prepared and analyzed national data; J.F.N.-A. prepared demographic data; Y.G. prepared and analyzed regional and Land Use Change data; C.S., M.A., R.W. and D.K. wrote, reviewed and edited the paper; M.A. and R.W. administered the project and obtained the resources. All authors have read and agreed to the published version of the manuscript.

Funding: This research was funded by the National Science Foundation entitled International Trade Agreements, Globalization, Land Change and Agricultural Food Networks, 2016–2019 and also by UNAM-DGAPA-PAPIIT (No. IN-200319), Project titled Análisis de la dinámica de sistemas mixtos agricultura-ganadería y sus impactos en la distribución de los recursos productivos, la diversidad de maíz y la configuración de los paisajes rurales en el Centro-Oeste de México.

Data Availability Statement: Not Applicable.

Conflicts of Interest: The authors declare no conflict of interest.

References

1. Tubiello, F.N.; Salvatore, M.; Golec, R.D.C.; Ferrara, A.; Rossi, S.; Biancalani, R.; Federici, S.; Jacobs, H.; Flammini, A. *Agriculture, Forestry and Other Land Use Emissions by Sources and Removals by Sinks*; FAO Statistics Division: Rome, Italy, 2014.
2. Sala, O.E.; Stuart Chapin, F., III; Armesto, J.J.; Berlow, E.; Bloomfield, J.; Dirzo, R.; Huber-Sanwald, E.; Huenneke, L.F.; Jackson, R.B.; Kinzig, A.; et al. Global Biodiversity Scenarios for the Year 2100. *Science* **2000**, *287*, 1770–1774. [CrossRef] [PubMed]
3. Sunderlin, W.D.; Dewi, S.; Puntodewo, A.; Müller, D.; Angelsen, A.; Epprecht, M. Why Forests Are Important for Global Poverty Alleviation: A Spatial Explanation. *Ecol. Soc.* **2008**, *13*, 21. [CrossRef]
4. Angelsen, A.; Rudel, T.K. Designing and Implementing Effective REDD+ Policies: A Forest Transition Approach. *Rev. Environ. Econ. Policy* **2013**, *7*, 91–113. [CrossRef]
5. Barbier, E.B.; Delacote, P.; Wolfersberger, J. The Economic Analysis of the Forest Transition: A Review. *J. For. Econ.* **2017**, *27*, 10–17. [CrossRef]
6. Mather, A.S. The Forest Transition. *Area* **1992**, *24*, 367–379.
7. Walker, R. Deforestation and Economic Development. *Can. J. Reg. Sci.* **1993**, *16*, 481–497.

8. Perz, S.G. Grand Theory and Context-Specificity in the Study of Forest Dynamics: Forest Transition Theory and Other Directions. *Prof. Geogr.* **2007**, *59*, 105–114. [CrossRef]
9. Robbins, P.; Fraser, A. A Forest of Contradictions: Producing the Landscapes of the Scottish Highlands. *Antipode* **2003**, *35*, 95–118. [CrossRef]
10. Barbier, E.B.; Burgess, J.C.; Grainger, A. The Forest Transition: Towards a More Comprehensive Theoretical Framework. *Land Use Policy* **2010**, *27*, 98–107. [CrossRef]
11. Meyfroidt, P.; Rudel, T.K.; Lambin, E.F. Forest Transitions, Trade, and the Global Displacement of Land Use. *Proc. Natl. Acad. Sci. USA* **2010**, *107*, 20917–20922. [CrossRef]
12. Pfaff, A.; Walker, R. Regional Interdependence and Forest "Transitions": Substitute Deforestation Limits the Relevance of Local Reversals. *Land Use Policy* **2010**, *27*, 119–129. [CrossRef]
13. Mendoza-Ponce, A.; Corona-Núñez, R.O.; Galicia, L.; Kraxner, F. Identifying Hotspots of Land Use Cover Change under Socioeconomic and Climate Change Scenarios in Mexico. *Ambio* **2019**, *48*, 336–349. [CrossRef] [PubMed]
14. Walker, R. Forest Transition: Without Complexity, Without Scale. *Prof. Geogr.* **2008**, *60*, 136–140. [CrossRef]
15. Bonilla-Moheno, M.; Aide, T.M. Beyond Deforestation: Land Cover Transitions in Mexico. *Agric. Syst.* **2020**, *178*, 102734. [CrossRef]
16. Lorenzen, M.; Orozco-Ramírez, Q.; Ramírez-Santiago, R.; Garza, G.G. The Forest Transition as a Window of Opportunity to Change the Governance of Common-Pool Resources: The Case of Mexico's Mixteca Alta. *World Dev.* **2021**, *145*, 105516. [CrossRef]
17. Walker, R. The Scale of Forest Transition: Amazonia and the Atlantic Forests of Brazil. *Appl. Geogr.* **2012**, *32*, 12–20. [CrossRef]
18. Song, X.-P.; Hansen, M.C.; Stehman, S.V.; Potapov, P.V.; Tyukavina, A.; Vermote, E.F.; Townshend, J.R. Global Land Change from 1982 to 2016. *Nature* **2018**, *560*, 639–643. [CrossRef] [PubMed]
19. Mehrabi, Z.; Ellis, E.C.; Ramankutty, N. The Challenge of Feeding the World While Conserving Half the Planet. *Nat. Sustain.* **2018**, *1*, 409–412. [CrossRef]
20. Perfecto, I.; Vandermeer, J. The Agroecological Matrix as Alternative to the Land-Sparing/Agriculture Intensification Model. *Proc. Natl. Acad. Sci. USA* **2010**, *107*, 5786–5791. [CrossRef]
21. DeFries, R.; Rosenzweig, C. Toward a Whole-Landscape Approach for Sustainable Land Use in the Tropics. *Proc. Natl. Acad. Sci. USA* **2010**, *107*, 19627–19632. [CrossRef]
22. Bada, X.; Fox, J. Persistent Rurality in Mexico and 'the Right to Stay Home'. *J. Peasant. Stud.* **2022**, *49*, 29–53. [CrossRef]
23. Hecht, S.; Yang, A.L.; Basnett, B.S.; Padoch, C.; Peluso, N.L. *People in Motion, Forests in Transition: Trends in Migration, Urbanization, and Remittances and Their Effects on Tropical Forests*; CIFOR: Bogor, Indonesia, 2015; ISBN 978-602-387-013-4. [CrossRef]
24. Klooster, D. Environmental Certification of Forests: The Evolution of Environmental Governance in a Commodity Network. *J. Rural. Stud.* **2005**, *21*, 403–417. [CrossRef]
25. Klooster, D. Forest Transitions in Mexico: Institutions and Forests in a Globalized Countryside. *Prof. Geogr.* **2003**, *55*, 227–237. [CrossRef]
26. Aide, T.M.; Grau, H.R. Globalization, Migration, and Latin American Ecosystems. *Science* **2004**, *305*, 1915–1916. [CrossRef] [PubMed]
27. Grau, H.R.; Aide, M. Globalization and Land-Use Transitions in Latin America. *Ecol. Soc.* **2008**, *13*, 12. [CrossRef]
28. Ramírez-Delgado, J.P.; Christman, Z.; Schmook, B. Deforestation and Fragmentation of Seasonal Tropical Forests in the Southern Yucatán, Mexico (1990–2006). *Geocarto Int.* **2014**, *29*, 822–841. [CrossRef]
29. Rudel, T.K.; Perez-Lugo, M.; Zichal, H. When Fields Revert to Forest: Development and Spontaneous Reforestation in Post-War Puerto Rico. *Prof. Geogr.* **2000**, *52*, 386–397. [CrossRef]
30. Hecht, S.B.; Saatchi, S.S. Globalization and Forest Resurgence: Changes in Forest Cover in El Salvador. *BioScience* **2007**, *57*, 663–672. [CrossRef]
31. Perz, S.G.; Skole, D.L. Secondary Forest Expansion in the Brazilian Amazon and the Refinement of Forest Transition Theory. *Soc. Nat. Resour.* **2003**, *16*, 277–294. [CrossRef]
32. Hecht, S. The New Rurality: Globalization, Peasants and the Paradoxes of Landscapes. *Land Use Policy* **2010**, *27*, 161–169. [CrossRef]
33. Hernández-Aguilar, J.A.; Durán, E.; de Jong, W.; Velázquez, A.; Pérez-Verdín, G. Understanding Drivers of Local Forest Transition in Community Forests in Mixteca Alta, Oaxaca, Mexico. *For. Policy Econ.* **2021**, *131*, 102542. [CrossRef]
34. Vaca, R.A.; Golicher, D.J.; Cayuela, L.; Hewson, J.; Steininger, M. Evidence of Incipient Forest Transition in Southern Mexico. *PLoS ONE* **2012**, *7*, e42309. [CrossRef] [PubMed]
35. Rudel, T.K.; Bates, D.; Machinguiashi, R. A Tropical Forest Transition? Agricultural Change, Out-Migration, and Secondary Forests in the Ecuadorian Amazon. *Ann. Assoc. Am. Geogr.* **2002**, *92*, 87–102. [CrossRef]
36. Schmook, B.; Radel, C. International Labor Migration from a Tropical Development Frontier: Globalizing Households and an Incipient Forest Transition. *Hum. Ecol.* **2008**, *36*, 891–908. [CrossRef]
37. Mesa-Sierra, N.; de la Peña-Domene, M.; Campo, J.; Giardina, C.P. Restoring Mexican Tropical Dry Forests: A National Review. *Sustainability* **2022**, *14*, 3937. [CrossRef]
38. López, E.; Bocco, G.; Mendoza, M.; Velázquez, A.; Rogelio Aguirre-Rivera, J. Peasant Emigration and Land-Use Change at the Watershed Level: A GIS-Based Approach in Central Mexico. *Agric. Syst.* **2006**, *90*, 62–78. [CrossRef]

39. Barsimantov, J.; Navia-Antezana, J. Land Use and Land Tenure Change in Mexico's Avocado Production Region: Can Community Forestry Reduce Incentives to Deforest for High Value Crops? In Proceedings of the Twelfth Biennial Conference of the International Association for the Study of the Commons, Cheltenham, UK, 14–18 July 2008.
40. Barsimantov, J.; Navia Antezana, J. Forest Cover Change and Land Tenure Change in Mexico's Avocado Region: Is Community Forestry Related to Reduced Deforestation for High Value Crops? *Appl. Geogr.* **2012**, *32*, 844–853. [CrossRef]
41. Galvan-Miyoshi, Y.; Walker, R.; Warf, B. Land Change Regimes and the Evolution of the Maize-Cattle Complex in Neoliberal Mexico. *Land* **2015**, *4*, 754–777. [CrossRef]
42. Perramond, E.P. The Rise, Fall, and Reconfiguration of the Mexican Ejido*. *Geogr. Rev.* **2008**, *98*, 356–371. [CrossRef]
43. Siyum, Z.G. Tropical Dry Forest Dynamics in the Context of Climate Change: Syntheses of Drivers, Gaps, and Management Perspectives. *Ecol. Process.* **2020**, *9*, 25. [CrossRef]
44. Blackie, R.; Baldauf, C.; Gautier, D.; Gumbo, D.; Kassa, H.; Parthasarathy, N.; Paumgarten, F.; Sola, P.; Pulla, S.; Waeber, P.; et al. *Tropical Dry Forests: The State of Global Knowledge and Recommendations for Future Research*; CIFOR: Bogor, Indonesia, 2014.
45. Geografía (INEGI), I.N. de E. y Mapas. Uso de Suelo y Vegetación. Available online: https://www.inegi.org.mx/temas/usosuelo/ (accessed on 21 March 2023).
46. Astier, M.; Orozco-Ramírez, Q.; Walker, R.; Galván-Miyoshi, Y.; González-Esquivel, C.; Simmons, C.S. Post-NAFTA Changes in Peasant Land Use—The Case of the Pátzcuaro Lake Watershed Region in the Central-West México. *Land* **2020**, *9*, 75. [CrossRef]
47. Barrera-Bassols, N.; Alfred Zinck, J.; Van Ranst, E. Symbolism, Knowledge and Management of Soil and Land Resources in Indigenous Communities: Ethnopedology at Global, Regional and Local Scales. *CATENA* **2006**, *65*, 118–137. [CrossRef]
48. Torres, A.C. Lake Patzcuaro, Mexico: Watershed and Water Quality Deterioration in a Tropical High-Altitude Latin American Lake. *Lake Reserv. Manag.* **1993**, *8*, 37–47. [CrossRef]
49. Klooster, D. Beyond Deforestation: The Social Context of Forest Change in Two Indigenous Communities in Highland Mexico. *Conf. Lat. Am. Geogr. Yearb.* **2000**, *26*, 47–59.
50. Grainger, A. National Land Use Morphology: Patterns and Possibilities. *Geography* **1995**, *80*, 235–245.
51. Forest Resources Assessment 1990—Survey Tropical Forest Cover Studies of Change Processes. Available online: https://www.fao.org/3/w0015e/w0015e00.htm (accessed on 20 February 2023).
52. Población, C.N. de Índices de marginación 2020. Available online: http://www.gob.mx/conapo/documentos/indices-de-marginacion-2020-284372 (accessed on 20 February 2023).
53. Consejo Nacional de Población (CONAPO). Available online: http://www.conapo.gob.mx/work/models/CONAPO/Marginacion/Datos_Abiertos/Script_Indice_de_marginacion_por_localidad_2020.txt (accessed on 20 February 2023).
54. National Agrarian Registry, Mexican Government Registro Agrario Nacional-PHINA-Padrón e Historial de Núcleos Agrarios. Available online: https://phina.ran.gob.mx/consultaPhina.php (accessed on 20 February 2023).
55. Ruiz-Mercado, I.; Masera, O. Patterns of Stove Use in the Context of Fuel–Device Stacking: Rationale and Implications. *EcoHealth* **2015**, *12*, 42–56. [CrossRef]
56. Orozco-Ramírez, Q.; Astier, M. Socio-Economic and Environmental Changes Related to Maize Richness in Mexico's Central Highlands. *Agric. Hum. Values* **2017**, *34*, 377–391. [CrossRef]
57. Morales-Manila, L.M.; Reyes-González, A.; Cuevas-Garcia, G.; Onchi-Ramuco, M. *Inventario 2011 Del Cultivo Del Aguacate En El Estado de Michoacán*; Centro de Investigaciones en Geografía Ambiental, UNAM–COFUPRO: Morelia, Mexico, 2012.
58. Bravo-Espinosa, M.; Mendoza, M.E.; CarlóN Allende, T.; Medina, L.; Sáenz-Reyes, J.T.; Páez, R. Effects of Converting Forest to Avocado Orchards on Topsoil Properties in the Trans-Mexican Volcanic System, Mexico. *Land Degrad. Dev.* **2014**, *25*, 452–467. [CrossRef]
59. Chavez, A.G.T.; Morales-Chávez, R.; García-González, Y.; Francisco, A.; Rojas, G.T. T. Partición de La Precipitación En Cultivo de Aguacate y Bosque de Pino-Encino En Michoacan, Mexico. *Biológicas* **2020**, *21*, 1–18.
60. Quiroz Rivera, F. Comparación del Consumo Hídrico entre *Persea Americana* y *Pinus Pseudostrobus*. Ph.D. Thesis, Universidad Michoacana de San Nicolás de Hidalgo, Morelia, Mexico, 2019.

Disclaimer/Publisher's Note: The statements, opinions and data contained in all publications are solely those of the individual author(s) and contributor(s) and not of MDPI and/or the editor(s). MDPI and/or the editor(s) disclaim responsibility for any injury to people or property resulting from any ideas, methods, instructions or products referred to in the content.

Article

How Do Rising Farmland Costs Affect Fertilizer Use Efficiency? Evidence from Gansu and Jiangsu, China

Yuan Qi [1], Xin Chen [2], Jiaqing Zhang [1], Yaoyao Li [1] and Daolin Zhu [1,3,*]

[1] College of Land Science and Technology, China Agricultural University, Beijing 100193, China
[2] Department of Earth System Science, Ministry of Education Key Laboratory for Earth System Modeling, Institute for Global Change Studies, Tsinghua University, Beijing 100084, China
[3] Center for Land Policy and Law, Beijing 100193, China
* Correspondence: dlzhu@cau.edu.cn

Citation: Qi, Y.; Chen, X.; Zhang, J.; Li, Y.; Zhu, D. How Do Rising Farmland Costs Affect Fertilizer Use Efficiency? Evidence from Gansu and Jiangsu, China. *Land* 2022, *11*, 1730. https://doi.org/10.3390/land11101730

Academic Editors: Pengyu Hao and Le Yu

Received: 7 September 2022
Accepted: 30 September 2022
Published: 6 October 2022

Publisher's Note: MDPI stays neutral with regard to jurisdictional claims in published maps and institutional affiliations.

Copyright: © 2022 by the authors. Licensee MDPI, Basel, Switzerland. This article is an open access article distributed under the terms and conditions of the Creative Commons Attribution (CC BY) license (https://creativecommons.org/licenses/by/4.0/).

Abstract: As the farmland transfer market in China develops, moderate-scale operations increasingly grow but without much improvement in fertilizer use efficiency. This study theoretically analyzes the mechanism and effect of rising farmland costs on fertilizer use efficiency using multiple quadratic regression and mediating effects models. It empirically tests a micro-sample of 806 farmers in Gansu and Jiangsu provinces in China from two dimensions: the full samples and farmer heterogeneity. The results showed 0.544 as the average fertilizer use efficiency (hereinafter, *fe*) of farmers in Gansu and Jiangsu, highlighting the severe loss of *fe* caused by excessive fertilizer inputs. The multiple quadratic regression model further revealed an inverted U-shaped relationship between farmland costs and *fe*, with the U-shaped curve showing a remarkable inflection point at the USD 708/mu mark. When farmland costs are excessive (*cost* > CNY 708/mu), the increase in farmland costs inhibits the *fe*. An investigation of the corresponding impact mechanism for this scenario (i.e., *cost* > USD 708/mu) revealed that farmland costs directly suppress *fe* (−0.485) by distorting the fertilizer factor substitution effect and indirectly suppress *fe* (−0.037) by impeding the technology spillover effect of production specialization and production scale-up. We also found heterogeneity between two groups: ordinary farmers and new agricultural operators (e.g., large grain and family farmers), with the peak kernel density function of *fe* of new agricultural operators (0.85) being much higher than that of ordinary farmers (0.30). Moreover, the multiple quadratic regression between the groups revealed a lower inflection point for ordinary farmers (CNY 638/mu) than new agricultural operators (CNY 823/mu), highlighting that the *fe* of ordinary farmers was more likely to be inhibited by the excessive rise in farmland costs. To promote the sustainable development of China's agricultural production, we propose reducing the cost of farmland, promoting service-scale operations, and fostering new agricultural operators.

Keywords: farmland costs; fertilizer reduction; new agricultural operators; multiple quadratic regression model; mediating effect model

1. Introduction

Since the founding of "New China", especially since the country's reform and opening-up, China's agricultural production methods have evolved—from relying mainly on traditional agricultural resources (labor) to capitalizing on chemical agricultural resources (chemical fertilizers) [1]. The rapid growth of chemical fertilizer application has greatly supported the stable supply of agricultural products in China [2]. However, negative environmental externalities such as the surface pollution caused by heavy fertilizer application have been highlighted [3–5], including soil acidification, the eutrophication of water sources, and increased greenhouse gas emissions [6,7], which have primarily impeded China's sustainable agricultural development. In response, the Chinese government has actively launched a zero-growth fertilizer campaign and enacted a series of policies to promote green agricultural development in recent years. Under the guidance of ecological civilization, China's zero fertilizer growth target has also been largely achieved, but the total base remains high. In 2020, the total fertilizer application in China was at 52.5 million tons, a unit area application rate of 390 kg/ha [8]. This is much higher than the internationally accepted fertilizer application limit of 225 kg/ha. At this stage, China's grain supply and demand remain in tight balance, with significant structural contradictions. Thus, the promotion of chemical fertilizer reduction has become an important link affecting food security and sustainable agricultural development, becoming the focus of academia.

In recent years, scholars in China and abroad have paid attention to fertilizer reduction [9–11], especially regarding the impact mechanisms, with different focuses in different periods. Based on the changing characteristics of the external environment of agricultural production in China, the existing literature provides an in-depth analysis of the influential factors concerning fertilizer reduction in three stages: labor off-farm transfer [12], farmland transfer market development [13], and agricultural production service market development [14]. Initially, in the context of the off-farm transfer of agricultural labor, there was growing concern that the ongoing wave of off-farm employment would hurt the agricultural environment, as measured by fertilizer use intensity, because rational smallholders tend to hedge output risks by over-applying fertilizers [15]. Using panel data from two representative mountainous and plain areas in Sichuan and Henan, Zhang et al. [12] found that non-farm employment in mountainous (plain) areas has an inverted U-shaped (positive) relationship with fertilizer inputs. As the market for farmland transfer continues to develop, academics have turned to analyzing the impact of farmland transfer on farmers' green production behavior [13,16]. Starting from the relationship between farmland operation scale and fertilizer inputs [17,18], scholars have focused on the positive impact of farmland transfer and the resulting farmland scale operation on fertilizer reduction. Zou et al. [19] found a suppressive effect of land transfer on the fertilizer use intensity of large-scale farmers. The area transferred and nongrain crop planted were also negatively associated with fertilizer application intensity. In contrast, the fertilizer reduction effect of farmland transfer in smallholder production was weaker or not significant [20,21]. To compensate for the negative impact of agricultural labor shortage and small-scale decentralized operation on agricultural production [22], China has rapidly developed agricultural production services in recent years, and some scholars have started to focus on the contribution of agricultural production services to fertilizer reduction. They concluded that agricultural productive services promote specialized division of labor and precision production and are an important option for fertilizer use reduction [23,24].

In contrast to domestic research, international studies focus on the influence of market factors on farmers' fertilizer application, such as consumption preferences [25], fertilizer prices [26,27], agricultural prices [28], and government intervention [29,30]. Goetzke [25] argued from an agricultural market supply and demand perspective that consumer preference for green agricultural products drives the reduction of chemical fertilizer inputs. Through a combination of theory and empirical evidence, Banerjee [26] found that the relative prices of fertilizer and output have a negative effect on fertilizer application by farm households. Additionally, international researchers have also focused on the influence of farmers' intrinsic characteristics on fertilizer application, mainly in terms of individual farmers' characteristics [31], land operation scale [32], social networks [33], and farmers' moral codes [34], to analyze the influence of production motivation and behavior on fertilizer application due to differences in farmers' intrinsic characteristics. International and domestic research also share commonalities, for instance focusing on the impacts of external factors such as farmland market development [35] and agricultural production service market development [36] on fertilizer application, and both disagree on whether land trading can contribute to fertilizer reduction. Conley and Udry [35] found that the scale of operation through agricultural land trading could improve fertilizer application efficiency. However, Bambio and Agha [36] showed that the expansion of land trade and the scale of operation intensified the short-term production behavior of farmers, which was not conducive to fertilizer reduction.

The literature above has explored the impact mechanism of fertilizer reduction based on the stage characteristics of the external environment of agricultural production in China, and the internal characteristics and external factors of farmers, providing an important reference for this study. However, what has been overlooked is that along with the continuous development of the farmland transfer market, farmland prices have increased, and the high farmland cost has become an important feature of China's agricultural production environment [37,38]. According to a Rural Land Management Rights Transfer Market Insight Series Report, the average cost of farmland in China was CNY 11,000/hm^2/year in 2020, and the average annual growth rate of farmland cost in China has exceeded that of agricultural production costs (i.e., material, labor, and land costs). Moreover, the share of land cost in the total output of Chinese agricultural production is, on average, 31.35%, which is much higher than that of the United States (21.00%), Brazil (16.92%), or the European Union (8.81%) [39]. Furthermore, farmland transfer has not brought about a fundamental change in China's fragmented household management pattern, nor has the application of organic fertilizer increased with the advancement of China's farmland transfer market [2]. The underlying problem is that given the rising farmland costs in China, farmland prices must be kept at a reasonable range for farmland transfers to lead to fertilizer reduction with excessive farmland costs, the allocation structure and degree of use of production factors by agricultural producers change or even get distorted. However, few studies have focused on the poor fertilizer reduction effect of farmland transfer from the perspective of farmland cost. For a deeper investigation of the classic proposition of fertilizer reduction in Chinese agriculture, the impact of farmland cost on fertilizer use efficiency to achieve sustainable agricultural development must be explored.

Furthermore, in the current literature on the factors influencing fertilizer reduction, most scholars still characterize the degree of fertilizer application directly by the amount of fertilizer input per unit area. Notably, the marginal output of fertilizers will vary under different factor allocation structures and even utilization levels. The key to whether fertilizer input is excessive lies in the relative difference between the actual versus optimal input amount of fertilizer. Therefore, some scholars have used the Cobb–Douglas production function to characterize the degree of fertilizer application from the perspective of fertilizer use efficiency based on the principle that the marginal product value of fertilizers equals the price of fertilizers [40,41]. Hence, we also investigate fertilizer use efficiency to reveal its changing characteristics from the perspective of rising farmland costs and explore the mechanism of how rising farmland costs affect fertilizer use efficiency. The empirical test is conducted using survey data of 806 farmers in the Gansu and Jiangsu provinces, and multiple quadratic regression and mediating effect models from two dimensions: the full sample and farmer heterogeneity. It is important in curtailing the adverse effects of fertilizers on the environment such as soil acidification, eutrophication of water sources, and increased greenhouse gas emissions, among others.

The remainder of this paper is structured as follows. Section 2 discusses the structural effect of rising farmland costs on fertilizer use efficiency and the impact mechanism. Section 3 presents the study area, data resources, and basic model. Section 4 provides the empirical analysis and main findings. In Section 5, we discuss the limitations of the study and policy recommendations. Section 6 presents the conclusions.

2. Theoretical Analysis

2.1. Impact Pathways of Rising Farmland Costs on Fertilizer Use: Direct Versus Indirect Effects

With the transfer of surplus agricultural labor and the reform of China's farmland system, the development of the farmland transfer market has accelerated, leading to a continuous rise in the cost of farmland. This situation has profoundly changed farmers' behavior toward fertilizer application, which can be summarized as direct and indirect effects of the rise in farmland cost (Figure 1a).

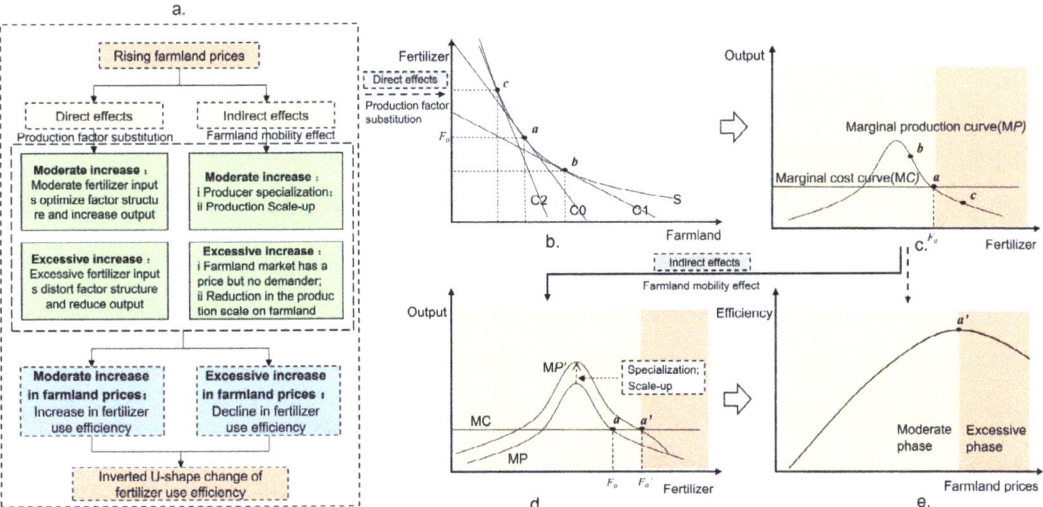

Figure 1. Mechanisms of the impact of rising farmland costs on fertilizer use efficiency. (**a**) illustrates the path of the impact of rising farmland costs on fertilizer use efficiency. (**b**) illustrates the substitution of fertilizers on farmland. (**c**) illustrates the effect of fertilizer input quantity on farmland output. (**d**) illustrates the effect of fertilizer input quantity on farmland output under the effect of indirect pathways. (**e**) illustrates the relationship between rising farmland costs and fertilizer use efficiency.

(1) Direct path based on the substitution effect of production factors

To maximize income, farmers restructure the input of each factor of production according to the changes in their relative prices, and the inputs of those that are relatively low-priced are increased to replace those that are relatively high-priced. In the case of fertilizer substitution for land, the substitution process is essentially influenced by the relative prices of production factors. The theory of induced technological innovation in agricultural development suggests that farmers will choose the appropriate technological innovation based on the relative prices of production factors. Moreover, fertilizer, as an agricultural production technology that involves biology and chemistry, is a land-substituting factor of production [42]. The increase in the farmland transfer price has caused the relative price of fertilizers to fall. The increased use of fertilizers, which is an input factor to promote crop growth, can partly replace farmland inputs that are higher-priced. Therefore, choosing to increase the amount of fertilizer applied per unit area to reduce the cost of farmland inputs is a rational choice for farmers.

(2) Indirect path based on the effect of farmland mobility

The farmland mobility effect includes producer specialization and the scale-up of production methods, and farmland transfer prices indirectly affect fertilizer application through the technological spillover effects on producers and the production scale. Examples are provided below.

a. Producer specialization. Producer specialization refers to the allocation of land from inefficient growers to more efficient growers. In a period of moderate price increases, the farmland market promotes the efficient mobility of farmland, thus forming an exit (entry) mechanism for "weak" ("strong") operators. Farmland flows into the hands of capable operators, who usually have better agricultural knowledge, management skills, and ability to solve the technical constraints faced by green agricultural production. They have a greater chance of acquiring green agricultural production technologies, and the modern production methods they bring implicitly promote the reduction of agricultural fertilizer production. However, excessive farmland prices prevent these management experts from effectively solving the financial constraints caused by extreme cost increases, and the farmland transfer market becomes "price-less." When farmland is costly, it is difficult for operators to apply green production technologies effectively, and the fertilizer-reducing effect of farmland mobility is hampered [20].

b. Scale-up of production methods. Profit maximization theory suggests that when the marginal output of an input factor of production is greater than the marginal cost, producers are incentivized to increase the input of that factor. In a period of moderate increase in farmland prices, producers will still choose to flow into farmland because of the economic incentive of scale payoffs in agricultural production. Additionally, several studies have shown that scale production can reduce the unit area cost of new technology use [43] and promote the diffusion of new agricultural technologies [44,45]. For example, in the promotion of green production technologies such as soil testing formula fertilization and water–fertilizer integration, having a certain area of farmland to meet the scale criteria for promoting green production technologies [18,46] can help promote chemical fertilizer reduction. However, when the price of farmland is too high, especially when the marginal cost of farmland is greater than the marginal payoff generated by scale benefits, producers become less willing to purchase farmland and even choose to abandon farmland [47]. In this situation, scale production becomes difficult, thus hindering the fertilizer reduction effect [2].

2.2. Characteristics of Changes in Fertilizer Use Efficiency under Different Scenarios of Rising Farmland Costs

Here, fertilizer use efficiency refers to the ratio of the actual amount of fertilizer input to the potential minimum amount of fertilizer input that can be achieved under a condition where the output and other input production factors remain constant [48]. To reveal further the changing characteristics of fertilizer use efficiency from the perspective of rising farmland costs, it is assumed that (i) only land and fertilizer are considered as input agricultural factors of production; (ii) to maximize profits, producers achieve cost minimization under equal yields by adjusting the combination relationship between these two elements; and (iii) point A is the optimal point of fertilizer application, where the marginal output of fertilizer is equal to the marginal cost of fertilizer.

(1) Moderate increase scenario

As shown in Figure 1b, when land prices change, the equal cost line between fertilizer and land shifts for producers to realize their original expected returns. Specifically, when the farmland price is low, producers tend to invest in more land. At this time, the iso-cost line C1 is tangent to the iso-return curve S at point b. However, owing to the excessive input of farmland area, the amount of fertilizer input is insufficient, resulting in low fertilizer use efficiency. As the farmland price rises moderately, the producer realizes the expected return by increasingly substituting fertilizer for land, and the equal-cost line moves from C1 to C0. At this time, the equal cost line C0 is tangent to the equal return curve S at point b. The structure of production factors is optimized, and the amount of fertilizer input is gradually matched with the farmland area. As shown in Figure 1c, the equilibrium point moves from point b to point a as the iso-cost line C1 shifts to C0, at which time the marginal output of fertilizer equals the marginal cost of fertilizer, and the efficiency of fertilizer use reaches its peak.

Additionally, the specialization of producers and the scale of agricultural production change the constraints of family agricultural production technology, causing a shift in the marginal output curve of fertilizers. As shown in Figure 1d, MP shifts upward to MP', facilitated by the farmland mobility effect. At this point, the fertilizer marginal output curve intersects with the marginal cost curve at point a'; the fertilizer use efficiency reaches the maximum, and the optimal input quantity of fertilizer expands to F_0'. Therefore, when the amount of fertilizer input resulting from the increase in farmland price is lower than F_0', the cost of farmland and fertilizer use efficiency show a positive relationship (Figure 1e).

(2) Excessive rise scenario

As shown in Figure 1b, as the farmland prices rise further, the equal cost line moves from C0 to C2, and the equilibrium point moves from point *a* to point *c*. At this point, the amount of fertilizer input is excessive and exceeds the degree of matching with the other factors of production. As shown in Figure 1c, the marginal output of fertilizers at point c is smaller than the marginal cost of fertilizers, and the amount of fertilizer input at this point exceeds the optimal amount of fertilizer input. Additionally, because of the excessive increase in farmland prices, the specialization of producers and the scale of agricultural production are inhibited, and the decrease in fertilizer use efficiency intensifies. Therefore, when the amount of fertilizer input resulting from the increase in farmland prices is greater than F_0', the cost of farmland and the efficiency of fertilizer use show a negative relationship (Figure 1e).

In summary, we propose the following hypothesis: the rising farmland cost—through the substitution effect of production factors, specialization of producers, and scale of agricultural production generated by the farmland mobility effect—leads the fertilizer use efficiency to exhibit an inverted U-shaped change with the rising farmland cost; in particular, the excessive rise in farmland cost hinders the improvement of fertilizer use efficiency.

3. Materials and Methods

3.1. Study Area

China is vast and varies greatly by region. We chose Gansu Province in northwestern China and Jiangsu Province in eastern China as the study areas to obtain representative results considering different farmland resource endowments, farmland prices, and fertilizer application rates (Figure 2), as explained in the following. First, Gansu is a typical inland mountainous region, where farmland resources are relatively poor; by contrast, Jiangsu is a flatter coastal province with high-quality farmland resources. Second, as the market development in Gansu Province is slow, the farmland transfer rent is relatively low—only 485 RMB/mu, while the farmland market in Jiangsu Province is well developed, and the farmland transfer rent is high—up to 920 RMB/mu. Moreover, in Gansu, where the fertilizer reduction policy is well implemented, fertilizer application is relatively low at approximately 800,000 tons. In Jiangsu Province, one of the major grain-producing provinces in China, the pollution is rather serious [40], and the fertilizer application is highly excessive from a nationwide perspective—up to 2.8 million tons. These facts are essential for determining the relationship between farmland cost and the fertilizer use efficiency in these regions.

3.2. Data Processing

Data were obtained from a survey of grain-growing farmers in the study area conducted in 2020. First, applying random sampling, 16 county-level administrative districts were selected as sample counties from these 2 provinces. Then, within each county, different types of towns were randomly selected considering factors such as resource endowment, geographical location, and economic development level. All villages within each selected town were divided into high- and low-income groups, with one of each income type being selected separately. Finally, in each village, no fewer than 10 households of different business types were randomly selected (large grain growers refer to households with more than 3.33 hectares of arable land; family farms refer to farmers identified and registered by the government as "family farms", and all others are ordinary farmers). Using this sampling strategy, we surveyed a total of 845 households (64 villages in 32 towns in 16 counties).

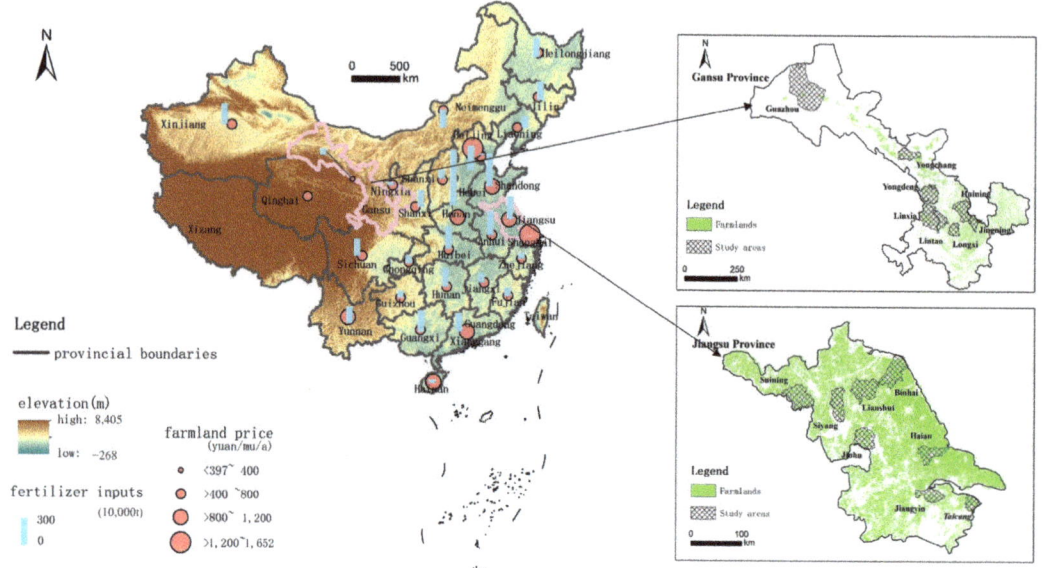

Figure 2. Profile of study areas. Note: Farmland price data were obtained from 41,452 sample sites surveyed by the Ministry of Natural Resources by letter; fertilizer input data were obtained from the Compilation of Information on Costs and Benefits of Agricultural Products in China.

For the specific survey of the data, we first trained the surveyors in a uniform manner and followed the steps of design, pre-survey, and finally revising the questionnaire. Secondly, in terms of survey content, we made detailed inquiries mainly about the production information of farmers, including basic data related to farmland transfer, fertilizer use, household characteristics, farmland characteristics, etc. However, since some farmers did not transfer their farmland, and some were not willing to disclose their specific agricultural production and operation information, they were excluded from our study. Finally, for the sake of data availability, we used 806 samples for this study by excluding samples who did not meet the needs of the study as well as outliers.

3.3. Methods

3.3.1. Measurement Method of Fertilizer Use Efficiency

Academics usually measure production efficiency or single-factor input efficiency through data envelopment analysis and stochastic frontier production function analysis (SFA) [49–51]. Considering that agricultural production is susceptible to stochastic factors such as weather and natural disasters, we use an SFA method that can effectively distinguish between stochastic perturbations and technical inefficiencies for estimation. The ratio of optimal quantity of the potential fertilizer input to the actual fertilizer input can be observed while keeping the technical efficiency fully effective. The specific steps are described as follows. First, we determined the formula for measuring fertilizer use efficiency (*fe*) based on the definition of *fe* in the theoretical analysis section. Second, we determined the production function under the condition of optimal amount of potential fertilizer inputs versus the production function under the condition of actual fertilizer inputs. We replaced the calculation of *fe* using polynomial substitution again. Finally, we substituted the actual output and input factor data of farmers into Equation (2) and used Stata to find out the α_1 and μi of each farmer, measuring the *fe* of each farmer in the calculation.

Specifically, we first characterized the fertilizer utilization efficiency with the following equation based on the definition:

$$fe_i = \frac{EFI_i}{AFI_i}, \quad (1)$$

where i denotes farmer; fe_i, fertilizer use efficiency; AFI_i, actual fertilizer input quantity; and EFI_i, optimal quantity of the potential fertilizer input.

Moreover, because we focus on measuring efficiency rather than examining the specific form of production, a simple C-D function is sufficient to support our study. Therefore, as with the majority of the literature (Shi et al., 2019), the C-D function is chosen as the specific form of the frontier production function, such that the efficiency frontier function can be expressed as follows:

$$lny_i = \alpha_0 + \alpha_1 lncf_i + \alpha_2 lnland_i + \alpha_3 lnlabor_i + \alpha_4 lnother_i + v_{1i} - \mu_i, \quad (2)$$

where y_i is the total output; and cf_i, $land_i$, $labor_i$, and $other_i$ represent fertilizer, land, labor, and other inputs, respectively. v_i is a random error term in the traditional sense, representing the random factors present in production (e.g., measurement errors and various uncontrollable random factors like weather and luck), which are assumed to follow a standard normal distribution. μ_i is a nonnegative technical efficiency term that reflects the deviation of the farmer's production from the frontier and is assumed to follow an exponential distribution with mean λ. It is also assumed that μ_i and v_i are independent of each other and both are uncorrelated with the explanatory variables.

In the selection of indicators, fertilizer input (in CNY), land operation area (in mu), farm labor (by person) and other agricultural production inputs other than fertilizers (specifically, land cost, labor cost, seedling fee, machine labor fee, etc.) are used to represent cf_i, $land_i$, $labor_i$, and $other_i$. y_i is characterized by agricultural production income (in CNY).

After estimating all the parameters of Equation (2) using the maximum likelihood method, the fertilizer use efficiency (fe) of the i_{th} farmer can be measured by Equation (3):

$$fe_i = \exp\left(\frac{-\mu_i}{\alpha_1}\right). \quad (3)$$

3.3.2. Multiple Quadratic Regression Model

Based on the theoretical analysis in part 2, we reasoned that the rising farmland cost has an inverted U-shaped effect on fertilizer use efficiency. To investigate the structural changes in the impact of rising farmland costs on fertilizer use efficiency, a multiple quadratic regression model is designed, and the quantitative relationship between farmland costs and fertilizer use efficiency for 806 farmers is examined using the following model:

$$fe_i = a \times cost_i + b \times cost_i^2 + \sum_{j=1}^{n} \partial_j X_{ij} + \beta_0 + v_{2i}, \quad (4)$$

where i denotes farm households; fe_i is fertilizer use efficiency; $cost_i$ is farmland cost, and $cost_i^2$ is the squared term of farmland cost. a, b, and β_0 are the parameters to be estimated. v_{2i} is a control for unobservables that vary by individual unit. X_{ij} is the control variable, and ∂_j is the regression coefficient of the control variable.

Specifically, X_{ij} is selected mainly based on the existing literature on the factors influencing fertilizer use efficiency [52,53], and other possible influential factors are controlled for in three dimensions: household head, household, and region. At the household head dimension, age and educational attainmentare included. Additionally, an age squared term is included in the regression of on fe, given the life cycle theory. Among the household characteristics variables, fertilizer application is a tedious and labor-intensive part of agricultural farming, and the labor supply status affects the efficiency of fertilizer use; thus, the proportion of off-farm workers is chosen. In addition, the share of nonfarm income reflects the importance that farmers attach to land. Farmers with a higher share of nonfarm income have more sloppy production practices, often resulting in less efficient fertilizer use.

Therefore, the share of nonfarm income in total household income is chosen. Among the regional control variables, conditions such as economic environment and social development among regions also have a large impact on fertilizer use efficiency, and two variables, county GDP per capita and urbanization rate, are introduced in the model. Table 1 shows the descriptive analysis of each variable.

Table 1. Descriptive statistics of variables.

Type of Variable	Representation	Definition	Mean	Standard Deviation
Dependent variable	fe	Efficiency of agricultural production scale	0.54	0.24
Variables of fe measurement	profit	Agricultural production income (yuan)	15575.97	30193.28
	land	Land operation area (mu)	12.48	20.11
	labor	Farm labor inputs (person)	1.87	0.73
	fertilizer	Fertilizer inputs (yuan)	2198.94	3543.08
	capital	Other agricultural production inputs other than fertilizers (yuan)	9115.15	17778.89
Independent variable	cost	Transaction price (yuan)	565.91	239.01
Intermediate variables	area	Area of farmland inflow (mu)	11.04	19.66
Peasant variable	age	Age (year)	41.28	11.13
	age^2	Age × age (year)	1827.84	873.76
	edu	1 = illiterate, 2 = secondary school, 3 = middle school, 4 = high school, 5 = university and above	2.12	1.23
Family variable	revenue	Non-farm income/total revenue	0.46	0.22
	off-farm	Number of non-farm workers/total number	0.38	0.24
Regional variable	pergdp	GDP/population (10,000 yuan/person)	129.99	193.46
	urban	Urban population/total population (%)	55.36	11.59

3.3.3. Intermediary Effect Model

According to the theoretical analysis, both the producer's specialization and production scale-up are based on the farmland transfer. When the area of land transfer is small, both effects hardly come into play. Therefore, to test whether the farmland cost indirectly affects the efficiency of fertilizer use through the effect of farmland transfer, we further introduce the intermediate variable of farmland inflow area to form the following model:

$$fe_i = a_1 * cost_i + \sum_{j=1}^{n} \partial_{2j} X_{ij} + v_{3i}. \tag{5}$$

$$area_i = b_1 * cost_i + \sum_{j=1}^{n} \partial_{3j} X_{ij} + v_{4i}. \tag{6}$$

$$fe_i = c_1 * cost_i + d_1 * area_i + \sum_{j=1}^{n} \partial_{4j} X_{ij} + v_{5i}. \tag{7}$$

where fe_i, $cost_i$, and X_{ij} are the same as in Equation (4), and $area$ is the inflow area of farmland. a_1, b_1, c_1, and d_1 are the coefficients to be estimated; ∂_{2j}, ∂_{3j}, and ∂_{4j} are the regression coefficients of each control variable, and v_{3i}, v_{4i}, and v_{5i} are the individual effects. To enhance the comparability of the effect sizes of each path, the natural logarithms of the variables *cost*, *pergdp*, and *area* are taken in the mediating effect model.

Additionally, we split the intermediary effect analysis into two sets of multivariate primary model mediation effects analysis by classifying the sample groups using the inflection point values of the linear model, allowing us us to satisfy the requirements of existing mediated effects models while using binary regression and mediation model methods.

3.4. Description of Variables

Table 1 shows the descriptive analysis of each variable.

Dependent variable. The average fertilizer use efficiency of 806 households was found to be 0.544 using Equations (1)–(3), which is in general agreement with the current academic research [48]. Among them, the fertilizer use efficiency of 51% of the farmers was even lower than the average.

Independent variable. We found that the farmland cost obtained from the survey was consistent with the price level of samples surveyed by the Ministry of Natural Resources by letter in Section 3.1. The average of farmland cost obtained from the research was RMB 566/mu/year, of which more than 45% of farmers spend more than RMB 566/mu/year when transferring farmland.

Intermediate variable. We found that the average size of farmland transferred by farmers was 0.69 ha, but about 75% of them transferred farmland below the mean level. This indicates that most of the farmers' willingness to transfer farmland has yet to be further stimulated by government policies.

Control variables. At the individual level, we found that the average age of farmers in the sample group was 41.28 years and that 85.11% of them had not attended high school. From the household perspective, the average percentage of household non-agricultural income was 46%, and at least one-third of them were engaged in non-agricultural work. At the regional level, the economic development level and urbanization level of the districts and counties where the sample is located were high, including an average GDP per capita of 1.3×10^6 and an urbanization level of 55.36%.

4. Results
4.1. Fertilizer Use Efficiency in the Study Provinces

According to Equations (1)–(3), the average fertilizer use efficiency (hereinafter, *fe*) of the 806 households is 0.544. This indicates that nearly half of the inefficient fertilizer inputs are used in agricultural production in the study provinces, highlighting the severe inefficient fertilizer use caused by excessive fertilizer inputs. The maximum reduction rate of fertilizer input is 45.60% if technical inefficiency can be eliminated while keeping the quantity of other factor inputs and output levels constant. Furthermore, calculating the *fe* by farmer type, we see from the peak of the kernel curve in Figure 3 that the *fe* peak for the new agricultural operators is higher than that for ordinary farmers. This result indicates that under the current level of agricultural production technology and the market environment of production factors, it is easier for new agricultural operators to improve *fe* than it is for ordinary farmers. Moreover, this phenomenon is consistent with the current theoretical perceptions in academia: new agricultural operators usually have better agricultural knowledge and management skills and are better able to solve the technical constraints faced by agricultural green production [7,13].

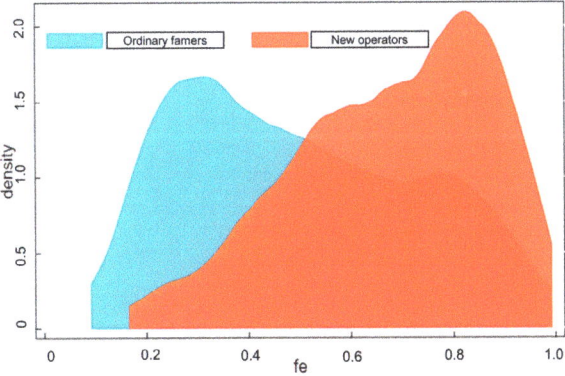

Figure 3. Kernel density distribution of farmers' *fe*.

4.2. The Effect of Rising Farmland Costs on Fertilizer Use Efficiency

Table 2 shows the regression results of the effect of rising farmland costs on fertilizer use efficiency. The table shows that both the primary term (*cost*) and squared term (*cost*^{2}) of farmland cost have a significant effect on fertilizer use efficiency (*fe*). The coefficient of the squared term is -3.72×10^{-7}, and the coefficient of the primary term is 5.27×10^{-3}. These results suggest the existence of a remarkable inflection value of the fitted function of the model, which indicates a nonlinear relationship between the farmland cost and the fertilizer use efficiency (Figure 4). Through the extreme value condition of a multivariate quadratic function in which $cost_0 = -2a/b$, the corresponding inflection point of farmland cost is calculated as RMB 708/mu. According to the inflection point measured by the model, the farmland cost is divided into two intervals: $cost \leq CNY\ 708/mu$ and $cost > CNY\ 708/mu$, representing two stages of moderate and excessive increases in farmland cost, respectively.

Table 2. Estimation results of multiple quadratic regression model. Note: *** indicates significance at the levels of 1%.

Variables	Coefficient	T Statistic
cost	5.27×10^{-3} ***	8.63
cost^2	-3.72×10^{-7} ***	-4.15
age	0.048 ***	13.20
age^2	-0.001 ***	-12.80
edu	0.051 ***	10.86
revenue	-0.167 ***	-6.87
off-farm	-0.256 ***	-10.89
pergdp	0.001	-0.35
urban	0.002 ***	4.28
constant	-0.582 ***	-7.87
inflection point	708	
R^2	0.702	
N	806	

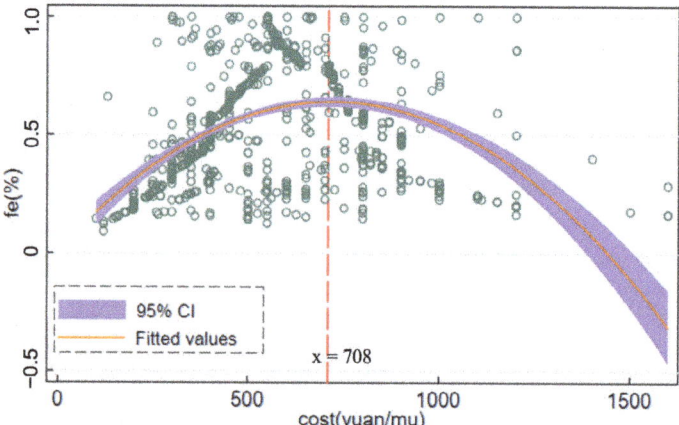

Figure 4. The inverse "U" shape relationship between *cost* and *fe*.

The inverted U-shaped effect of *cost* on *fe* is relatively consistent with the change curve of fertilizer use efficiency theoretically analyzed here, suggesting that the effect of farmland cost on fertilizer use efficiency changes from promotion to inhibition with the increase in farmland cost. Specifically, in the phase of moderate increase in farmland costs, the increase promotes the fertilizer use efficiency. This means that when the farmland cost is low, producers tend to buy relatively low-priced farmland, resulting in an input structure characterized by more farmland and less fertilizer among production factors, and the fertilizer use efficiency is at a low level owing to the relatively insufficient amount of fertilizer input. In the phase of moderate increase in farmland cost, producers increase the amount of fertilizer input and reduce the area of farmland input, as fertilizers act as a substitute production factor for farmland. Fertilizer use is gradually matched with the scale of farmland, technology level, and other factors; the input structure of production factors is optimized, and the efficiency of fertilizer use is continuously improved. In addition, the moderate rise in the farmland cost encourages market competition, leading to the gradual elimination of producers with poor operating ability and insufficient factor endowment from the market. As competent operators and large-scale households become the main agricultural producers, they improve at solving the technical constraints faced by agricultural green production, and fertilizer use efficiency is further improved. Therefore, the overall fertilizer use efficiency shows an increasing trend with the rising farmland cost.

However, as the farmland cost rises further to an excessive level, the impact on fertilizer use efficiency changes structurally to inhibit fertilizer use efficiency improvements. That is, when the farmland cost is relatively high, agricultural producers are financially constrained such that their willingness to purchase farmland and substitute fertilizers with relatively lower prices is reduced. The limited scale of farmland for production causes a structural distortion of less farmland and more fertilizer in production factor inputs, and fertilizer use efficiency is reduced because of excessive fertilizer inputs. In addition, an overpriced farmland destroys the original supply-and-demand balance in the market, causing the competitive mechanism of farmland market to fail. The production incentives for agricultural subjects—capable operators and large-scale households, the technical spillover of scale effect, and producer specialization—all suffer, further indirectly causing the reduction of fertilizer use efficiency.

From the regression results of the control variables, we see at the peasant level a significant inverted U-shaped relationship between *age* and *fe*, indicating that peasants of moderate age have richer management ability and production experience and can improve the efficiency of fertilizer use by optimizing the allocation structure of production factors. The significant positive correlation between *edu* and *fe* indicates that peasants who are relatively more educated have better agricultural knowledge and management skills and are more likely to improve fertilizer use efficiency through knowledge spillover and technology substitution. At the household level, *off-farm* significantly suppresses *fe*, indicating that off-farm labor leads to a reduction in agricultural labor, and growers tend to choose the more quantity, less frequency approach in fertilizer application to compensate for the lack of agricultural labor, which leads to inefficient fertilizer use. In addition, *revenue* significantly suppresses *fe*, suggesting that growers with a higher proportion of off-farm income have less funds dedicated to agricultural production; agricultural production is simpler and sloppy, and fertilizer efficiency tends to be lower. At the regional level, *pergdp*, *urban*, and *fe* are all positively correlated, but only *urban* passes the significance test. The results suggest that a higher level of urbanization may facilitate the effective flow of labor factors between agricultural production and nonagricultural work, resulting in an exit mechanism for weak business ability and an entry mechanism for strong business ability in agricultural production. Laborers who stay in rural areas for agricultural production usually have better agricultural knowledge and management skills and are better able to solve the technical constraints faced by green agricultural production, thus improving fertilizer use efficiency.

4.3. Analysis of the Impact Mechanisms: Direct and Indirect Effects

Based on the inflection point value of farmland cost, the sample group was first divided into moderate (*cost* $\leq$ 708) and excessive (*cost* > 708) groups. Further, the mediating effect model was used to group the effect mechanism for empirical analysis. The regression results all indicated (Table 3) that farmland cost directly and, through mediating variables, indirectly influenced fertilizer use efficiency. To illustrate the mechanism of the effect of cost on *fe* under the excessive farmland cost scenario, we present the following analysis for the excessive farmland cost group.

Table 3. Estimation results of intermediary effects model. Note: *, **, *** denote reaching 10%, 5% and 1% significance levels, respectively; T-statistics in parentheses.

Variable	Model 1	Model 2	Model 3
lncost	−0.522 *	−1.202 ***	−0.485 ***
	(−9.28)	(−3.51)	(−8.53)
lnarea	–	–	0.030 ***
			(2.830)
lnage	0.073 **	0.520 ***	0.057 *
	(2.47)	(2.90)	(1.93)
edu	0.069 ***	−0.014	0.069 ***
	(8.83)	(−0.30)	(9.02)
revenue	−0.168 ***	−0.070 ***	−0.166 ***
	(−3.82)	(−0.26)	(−3.83)
off-farm	−0.157 ***	0.061	−0.159 ***
	(−3.81)	(0.24)	(−3.92)
lnpergdp	0.015	0.122	0.012
	(0.11)	(1.21)	(0.72)
urban	0.001	−0.003	0.001
	(0.11)	(−0.31)	(0.17)
constants	3.576 ***	6.927 ***	3.384 *
	(8.36)	(2.42)	(7.93)
R^2	0.672	0.162	0.683
Sobel test	–	–	−0.037 ***
N	141	141	141

Indirect effects: As shown in Table 3, *cost* has a significant and negative coefficient on *fe* (Model 1). After adding the mediating variable *area* (Model 3), *cost* shows a significant negative effect on *fe*, while *area* also shows a significant positive effect on *fe*. The Sobel test results reveal that the mediating effect of *area* on *fe* is significant at −0.037 (Model 3). This indicates that in the high farmland cost scenario, it is impossible for the operators to effectively address the financial constraints caused by excessive cost increases. The farmland market is priced but not marketed, which makes it difficult for the operators to apply green production technology and indirectly inhibits fertilizer use efficiency. In addition, producers' willingness to purchase farmland decreases in the context of excessively rising farmland costs. The insufficient scale of farmland inputs in agricultural production makes it difficult to take advantage of the technological spillover effect of the scale effect of agricultural production, which further inhibits fertilizer use efficiency.

Direct effects: After deducting the mediating effect of *area* from the total effect, the remaining part then reflects the magnitude of the factor substitution effect. The parameter estimation results reported in Model 3, Table 3 show that in the excess group, after deducting the mediating effect of *area*, the effect of *cost* on *fe* remains significant at −0.485. This suggests that the high cost of farmland further increases the relative price of farmland and that fertilizers continue to play a substitution role for farmland. However, the limited scale of farmland inputs leads to the structural distortion of less farmland and more fertilizer, thus reducing fertilizer use efficiency due to excessive fertilizer inputs.

4.4. Estimations Results between Ordinary Farmers and New Agricultural Operators

After dividing the samples into groups of general farmers and new agricultural operators, the multiple quadratic regression of farmland cost on fertilizer use efficiency was rerun. The model estimation results (Table 4) show that both the primary and secondary terms of farmland cost in the two sample groups significantly affect the fertilizer use efficiency at the 1% level, and both have great inflection points, which are CNY 638/mu and CNY 823/mu. According to the two sets of inflection points, the farmland costs in the samples were classified into two stages: suitable and excessive. An inverted U-shaped relationship was found between *cost* and *fe* in both the general farmers' group and the group of new agricultural operators. This means that with the shift from moderate to excessive increase in farmland cost, the impact of farmland cost on fertilizer use efficiency structurally changes from promotion to an inhibition (Figure 5). In particular, the excessive increase in farmland cost hinders the improvement of fertilizer use efficiency, which further verifies our theoretical analysis and the empirical results for the full sample.

Table 4. Model estimation results of different growers Note: *** denotes reaching 1% significance levels.

Variables	Model 1 (Ordinary Farmers)		Model 2 (New Agricultrul Operators)	
	Coefficient	T Statistic	Coefficient	T Statistic
cost	8.10×10^{-3} ***	7.12	5.72×10^{-3} ***	8.63
$cost^2$	-6.35×10^{-7} ***	−7.00	-3.47×10^{-7} ***	−4.15
age	0.042 ***	11.21	0.066 ***	4.64
age^2	−0.001 ***	−10.82	−0.001 ***	−4.70
edu	0.054 ***	10.24	0.031 ***	3.19
revenue	−0.208 ***	−7.95	0.013	0.22
off-farm	−0.228 ***	−9.17	−0.273 ***	−4.40
pergdp	0.001	−0.87	0.001	0.98
urban	0.002 ***	4.31	0.001	0.27
constant	−0.547 ***	−7.24	−0.851 ***	−3.03
inflection point value	638		823	
R^2	0.719		0.599	
N	665		141	

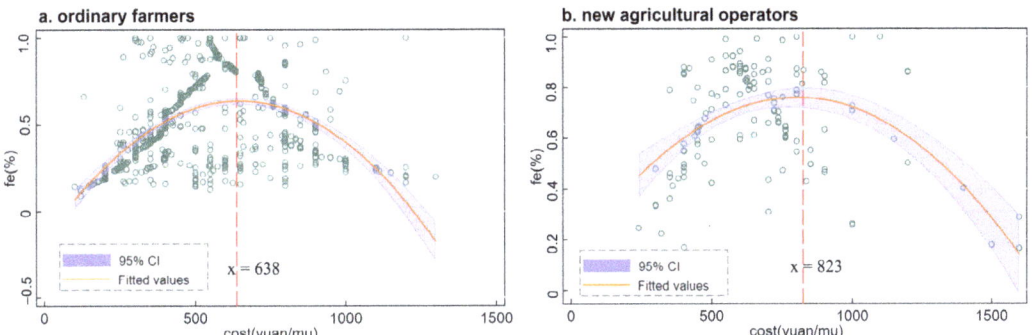

Figure 5. The scatter plots of *cost* and *fe* in ordinary farmers and new agricultural operators.

Notably, the inflection point of the ordinary farmers is smaller than that of the new agricultural operators. This indicates that the fertilizer use efficiency of ordinary farmers is more vulnerable to the excessive increase in farmland cost compared with that of new agricultural operators. The reason for this is that different types of producers have different factor allocation behaviors and factor endowments. Compared with ordinary farmers, new agricultural operators, such as large grain farmers and family farms, usually have better agricultural knowledge and management skills at the same factor input level and are better able to solve the financial constraints faced by agricultural production. With rising farmland costs, the new agricultural operators can maintain their profit margin in agricultural production in various ways such as technological substitution and factor structure optimization, which are more tolerant of the rising farmland costs. Therefore, fertilizer use efficiency is less affected by the farmland cost.

5. Discussion

5.1. Innovations and Outlooks for Future Research

High farmland costs have become an important feature of China's agricultural production environment. However, relatively little attention has been paid to the loss of fertilizer use efficiency caused by rising farmland cost. This study's contributions are summarized as follows: (1) We focus on fertilizer use efficiency from the perspective of rising farmland costs and identify key factors for the poor fertilizer reduction effect of farmland transfer; (2) the effect of rising farmland costs on fertilizer use efficiency is revealed, and the underlying mechanism and feasible strategies to promote fertilizer efficiency by reducing farmland costs are clarified; (3) by differentiating between ordinary farmers and new agricultural operators, this study reveals the differences in production behavior of different operators in the face of rising farmland costs, and provides a new policy logic for fertilizer reduction.

However, this study has its limitations. First, as revealed in previous studies, differences in crop type can lead to large variations in fertilizer use [19], and more fertilizers tend to be used in nongrain crops [7]. Moreover, another study suggested that rising farmland costs drive producers to prefer nongrain crops [54]. This suggests that fertilizer use efficiency and planted crop types may vary with the extent of rising farmland costs. Therefore, in the context of rising farmland costs in China, especially in the current environment of increasing cash crop acreage in China, there is an urgent need to further analyze the impact of rising farmland costs on fertilizer use efficiency from the perspective of changes in planted crop types in the future. Second, because of issues on the availability of microdata on farmers, some potential explanatory variables, such as soil conditions and degree of farmland use, were omitted in this study, and only cross-sectional microdata for 2019 were used. Additionally, there may be a mutual causal relationship between farmland cost and fertilizer use efficiency. As plots with higher fertilizer use efficiency tend to have higher output values that can further counteract farmland prices, this phenomenon should be further investigated. In future work, these issues can be adjusted and expanded through a combination of long time series and multiple cross-sectional data to provide a comprehensive understanding of farmland costs and fertilizer use efficiency. In addition, China is a large country with obvious geographical differences, and the selection of local cities in Gansu and Jiangsu may not provide sufficient evidence to reveal the whole picture and draw general conclusions. In future work, this study can be further improved by integrating more cities with different natural endowments and social development characteristics.

5.2. Policy Implications

This study shows that the high rise in farmland cost has become the main obstacle to sustainable agricultural production in China. Achieving the green development of Chinese agriculture requires both organic integration with the current farmland transfer market and the improvement of China's current fertilizer reduction policies.

First, the standardized management of the farmland transfer market should be strengthened, for example, through the development of land transfer information platforms, contract management, price evaluation, and other services to prevent excessive increases in farmland costs. The government should increase subsidies for the producers who use farmland, and especially for fertilizer reduction, establish a fertilizer reduction compensation system based on market price compensation, supplemented by government ecological compensation. In addition, a monitoring mechanism for farmland prices should be established, and fixed sample points should be selected nationwide to monitor farmland prices regularly.

Second, against the backdrop of the high farmland costs in China, the Chinese government should recognize the necessity of various forms of moderate-scale operations. Integrating smallholder production into the modern agricultural development track with the help of agricultural production services not only helps prevent the risk of high land costs but also ensures the sustainable development of agricultural scale operations. For example, the scale of agricultural services can realize contiguous scale economy by facilitating the mechanization of fertilizer application behavior to compensate for the indistinguishability of fertilizers.

Third, because new agricultural operators are more efficient in fertilizer use and less affected by excessive increases in farmland costs, the Chinese government should vigorously cultivate new agricultural operators in the future. For new agricultural business entities, the establishment of awards in lieu of subsidies can further encourage them to make long-term investments that are conducive to fertilizing the land and promote models and experiences of chemical fertilizer reduction technologies. Meanwhile, the government should encourage ordinary farmers to transform into new agricultural operators, for example, by increasing compensation for fertilizer reduction and actively guiding them to adopt new technologies (such as soil inspection and fertilizer application).

6. Conclusions

To identify the key factors that constrain farmland transfer in China and influence fertilizer reduction, in the context of high farmland costs in China, this study theoretically analyzed the mechanism and effect of rising farmland costs on fertilizer use efficiency. Accordingly, a multiple quadratic regression model and a mediating effects model were used to test empirically a microsample of 806 famers in Gansu and Jiangsu provinces of China and provide empirical support for the theoretical analysis in two dimensions: full samples and grower heterogeneity. We draw the following conclusions.

The lack of fertilizer use efficiency due to excessive fertilizer input is a serious problem in China, and we calculated the average fertilizer use efficiency of growers in Gansu Province and Jiangsu Province in China to be 0.544. The multiple quadratic regression model further revealed an inverted U-shaped relationship between farmland costs and fe, and the U-shaped curve showed a remarkable inflection point at the farmland cost of 708 Yuan/mu. The influence of rising farmland costs on fe showed a structural change from promotion to inhibition. Specifically, when the farmland costs were excessive ($cost > 708$ yuan/mu), the increase in farmland costs inhibited the fe. Then, we investigated the impact mechanism under the scenario of an excessive rise in the farmland costs ($cost > $ USD 708/mu). We found that farmland costs directly suppressed fe (-0.485) by distorting the fertilizer factor substitution effect and indirectly suppressed fe (-0.037) by impeding the technology spillover effect of production specialization and production scale-up. In addition, we found heterogeneity between ordinary farmers and new agricultural operators (e.g., large grain farmers, family farmers, etc.), with the peak kernel density function of fe of new agricultural operators (0.85) being much higher than that of ordinary farmers (0.30). Moreover, the multiple quadratic regression between the general farmers and new agricultural operators revealed that the inflection point value of the ordinary farmers (638 yuan/mu) was lower than that of the new agricultural operators (823 yuan/mu), indicating that the fe of ordinary farmers is more likely to be inhibited by the excessive rise in farmland costs. The reasons for this are that new agricultural operators usually have better agricultural knowledge and management skills, and they are better able to address the technical constraints faced by green agricultural production. In the face of rising farmland costs, they maintain profit margins in multiple ways such as through technological substitution and factor structure optimization. Therefore, new agricultural operators are more tolerant of the rising cost of farmland, and their fertilizer use efficiency is less affected by the cost of farmland.

Author Contributions: Conceptualization, Y.Q.; methodology, Y.Q. and X.C.; software, Y.Q.; validation, J.Z. and Y.L.; formal analysis, Q.Y; investigation, Y.Q. and J.Z.; data curation, Z.J; writing—original draft preparation, Y.Q.; writing—review and editing, X.C. and D.Z.; supervision, D.Z. All authors have read and agreed to the published version of the manuscript.

Funding: This research was funded by the National Natural Science Foundation of China (42171252).

Informed Consent Statement: Informed consent was obtained from all subjects involved in the study.

Data Availability Statement: Data cannot be disclosed because of the confidentiality agreement signed

Conflicts of Interest: The authors declare no conflict of interest.

References

1. Gao, J.J.; Shi, Q.H. The Shift of Agricultural production growth path in China: Based on the micro perspective of farm input. *J. Manag. World* **2021**, *37*, 124–134.
2. Li, B.; Shen, Y. Effects of land transfer quality on the application of organic fertilizer by large-scale farmers in China. *Land Use Policy* **2021**, *100*, 105124. [CrossRef]
3. Gong, H.; Li, J.; Liu, Z.; Zhang, Y.; Hou, R.; Ouyang, Z. Mitigated Greenhouse Gas Emissions in Cropping Systems by Organic Fertilizer and Tillage Management. *Land* **2022**, *11*, 1026. [CrossRef]
4. Hou, P.; Jiang, Y.; Yan, L.; Petropoulos, E.; Wang, J.; Xue, L.; Yang, L.; Chen, D. Effect of fertilization on nitrogen losses through surface runoffs in Chinese farmlands: A meta-analysis. *Sci. Total Environ.* **2021**, *793*, 148554. [CrossRef] [PubMed]
5. Zhang, Y.; Long, H.; Wang, M.Y.; Li, Y.; Jiang, T. The hidden mechanism of chemical fertiliser overuse in rural China. *Habitat Int.* **2020**, *102*, 102210. [CrossRef]
6. Wu, J.; Sha, C.; Wang, M.; Ye, C.; Li, P.; Huang, S. Effect of Organic Fertilizer on Soil Bacteria in Maize Fields. *Land* **2021**, *10*, 328. [CrossRef]
7. Zhang, L.; Luo, B.L. The logic of reduction in agriculture: An analytical framework. *Issues Agric. Econ.* **2022**, 15–26.
8. Huang, Y.Z.; Luo, X.F. Reduction and substitution of fertilizers: Farmer's technical strategy choice and influencing factors. *J. South China Agric. Univ.* **2020**, *19*, 77–87.
9. Epule, E.T.; Bryantz, C.R.; Akkari, C.; Daouda, O. Can organic fertilizers set the pace for a greener arable agricultural revolution in Africa? Analysis; synthesis and way forward. *Land Use Policy* **2015**, *47*, 179–187. [CrossRef]
10. Martey, E.; Kuwornu, J.K.M.; Adjebeng-Danquah, J. Estimating the effect of mineral fertilizer use on Land productivity and income: Evidence from Ghana. *Land Use Policy* **2019**, *85*, 463–475. [CrossRef]
11. Ren, S.; Song, C.; Ye, S.; Cheng, C.; Gao, P. The spatiotemporal variation in heavy metals in China's farmland soil over the past 20 years: A meta-analysis. *Sci. Total Environ.* **2022**, *806*, 150322. [CrossRef] [PubMed]
12. Zhang, Y.; Long, H.; Li, Y.; Ge, D.; Tu, S. How does off-farm work affect chemical fertilizer application? Evidence from China's mountainous and plain areas. *Land Use Policy* **2020**, *99*, 104848. [CrossRef]
13. Gao, L.; Zhang, W.; Mei, Y.; Sam, A.G.; Song, Y.; Jin, S. Do farmers adopt fewer conservation practices on rented land? Evidence from straw retention in China. *Land Use Policy* **2018**, *79*, 609–621. [CrossRef]
14. Zhang, M.L.; Chen, Z.J.; Wen, Z.L.; Zhang, Y.H. Research on the influence of socialized agricultural services on fertilizer reduction application—An analysis of the regulation effect based on factor allocation. *J. Agrotech. Econ.* **2022**, 1–21.
15. Chou, H.G.; Luan, H.; Li, J.; Wang, Y.J. Impact of risk aversion on farmers' fertilizer over-application behavior. *Chin. Rural Econ.* **2014**, 85–96.
16. Xie, J.; Yang, G.; Wang, G.; Song, Y.; Yang, F. How do different rural-land-consolidation modes shape farmers' ecological production behaviors? *Land Use Policy* **2021**, *109*, 105592. [CrossRef]
17. Cai, Y.P.; Du, Z.X. Analysis of ecological consciousness of production behavior of family farms and its influencing factors—An empirical test based on national family farm monitoring data. *Chin. Rural Econ.* **2016**, 33–45.
18. Wu, Y.; Xi, X.; Tang, X.; Luo, D.; Gu, B.; Lam, S.K.; Peter, M.; Vitousek; Chen, D. Policy distortions; farm size; and the overuse of agricultural chemicals in China. *Proc. Natl. Acad. Sci. USA* **2018**, *115*, 7010–7015. [CrossRef]
19. Zou, W.; Cui, Y.L.; Zhou, J.N. The impact of farmland transfer on farmers' fertilizer reduction: An analysis of transferability and security of land rights. *China Land Sci.* **2020**, *34*, 48–57.
20. Ju, X.; Gu, B.; Wu, Y.; Galloway, J.N. Reducing China's fertilizer use by increasing farm size. *Glob. Environ. Chang.* **2016**, *41*, 26–32. [CrossRef]
21. Liu, X.Y.; Zhang, D.; Xu, Z.G. Does grain scale farmers also overuse fertilizer?—Based on the heterogeneity of large-sized farmers and small-sized farmers. *J. Agrotech. Econ.* **2020**, *9*, 117–129.
22. Lu, H.; Chen, Y.J.; Hu, H.; Gen, X.H. Can agricultural socialized services promote farmers to adopt pro-environment agricultural technologies? *J. Agrotech. Econ.* **2021**, *3*, 36–49.
23. Ji, M.F. Agricultural productive service industry: The third dynamic energy in the history of China's agricultural modernization. *Issues Agric. Econ.* **2018**, 9–15.
24. Zhang, L.; Luo, B.L. Agricultural downsizing: The logic of scale in farming and its evidence. *Chin. Rural Econ.* **2020**, 81–99.
25. Goetzke, B.; Nitzko, S.; Spiller, A. Consumption of organic and functional food a matter of well-being and health. *Appetite* **2014**, *77*, 96–105. [CrossRef]
26. Brunelle, T.; Dumas, P.; Souty, F.; Dorina, B.; Nadaud, F. Evaluating the impact of rising fertilizer prices on crop yields. *Agric. Econ.* **2015**, *46*, 653–666. [CrossRef]
27. Yanggen, D.; Kelly, V.A.; Reardon, T.; Naseem, A. *Incentives for Fertilizer Use in Sub-Saharan Africa: A Review of Empirical Evidence on Fertilizer Response and Profitability*; Michigan State University, Department of Agricultural, Food and Resource Economics: East Lansing, MI, USA, 1998.
28. Takeshima, H.; Adhikari, R.P.; Shivakoti, S.; Kaphle, B.D.; Kumar, A. Heterogeneous returns to chemical fertilizer at the intensive margins: Insights from Nepal. *Food Policy* **2017**, *69*, 97–109. [CrossRef]

29. Haghjou, M.; Hayati, B.; Choleki, D.M. Identification offactors affecting adoption of soil conservation practices by some rainfed farmers in Iran. *J. Agric. Sci. Technol.* **2014**, *16*, 957–967.
30. Takeshima, H.; Nkonya, E. Government fertilizer subsidy and commercial sector fertilizer demand: Evidence from the federal market stabilization program (FMSP) in Nigeria. *Food Policy* **2014**, *47*, 1–12. [CrossRef]
31. Schreinemachers, P.; Chen, H.T.; Nguyen, T.T.L.; Borarin, B.; Bouapao, L.; Gautam, S.; Le, N.T.; Pinn, T.; Vilaysone, P.; Srinivasan, R. Too much to handle? Pesticide dependence of smallholder vegetable farmers in Southeast Asia. *Sci. Total Environ.* **2017**, *593–594*, 470–477. [CrossRef]
32. Khanna, M. Sequential adoption of site–specific technologies and its implication for nitrogen productivity: A double selectivity model. *Am. J. Agric. Econ.* **2001**, *83*, 35–51. [CrossRef]
33. Genius, M.; Koundouri, P.; Nauges, C. Information transmission in irrigation technology adoption and diffusion: Social learning, extension services and spatial effects. *Am. J. Agric. Econ.* **2014**, *96*, 328–344. [CrossRef]
34. Moslem, S.; Hamidreza, G. Application of the extended theory of planned behavior to predict Iranian farmers' intention for safe use of chemical fertilizers. *J. Clean. Prod.* **2020**, *263*, 12151.
35. Conley, T.G.; Udry, C.R. Learning about a new technology: Pineapple in Ghana. *Am. Econ. Rev.* **2005**, *100*, 35–69. [CrossRef]
36. Bambio, Y.; Agha, S.B. Land tenure security and investment: Does strength of land right really matter in rural Burkina Faso? *World Dev.* **2018**, *111*, 130–147. [CrossRef]
37. Lai, M.C.; Wu, P.I.; Liou, J.L.; Chen, Y.; Chen, H. The impact of promoting renewable energy in Taiwan—How much hail is added to snow in farmland prices? *J. Clean. Prod.* **2019**, *241*, 118519. [CrossRef]
38. Luo, B.L. The key; difficulty and direction of agricultural supply-side reform. *Rural Econ.* **2017**, 1–10.
39. Du, T.; Zhu, D.L. Studies on the spatio-temporal evolution and macro-mechanism of land circuation price in China. *Resour. Sci.* **2018**, *40*, 2202–2212.
40. Kong, F.B.; Guo, Q.L.; Pan, D. Evaluation of the extent of over-fertilization of grain crops in China and spatial and temporal variation. *Econ. Geogr.* **2018**, *38*, 201–210+240.
41. Zou, W.; Zhang, X.Y. Effects of land management scale on fertilizer use efficiency: Taking Jiangsu as an example. *Resour. Sci.* **2019**, *41*, 1240–1249.
42. Hu, H.; Yang, Y.B. Study on fertilizer application by farming households from the perspective of factor substitution—Based on data of farming households in fixed observation sites in rural areas nationwide. *J. Agrotech. Econ.* **2015**, *239*, 84–91.
43. Song, H.N.; Luan, J.D.; Zhang, S.Y.; Jiang, J.Y. Land fragmentation; production diversification and agricultural production technical efficiency—Empirical analysis based on stochastic frontier production function and mediation model. *J. Agrotech. Econ.* **2021**, 18–29.
44. Hei, X.R. Reflections on the scale of agricultural operations in China. *Issues Agric. Econ.* **2016**, *37*, 4–15.
45. Pan, D.; Kong, F.; Zhang, N.; Ying, R. Knowledge training and the change of fertilizer use intensity: Evidence from wheat farmers in China. *J. Environ. Manag.* **2017**, *197*, 130–139. [CrossRef] [PubMed]
46. Zhao, C.; Kong, X.Z.; Chou, H.G. Does the expansion of farm size contribute to the reduction of chemical fertilizers?—Empirical analysis based on 1274 family farms in China. *J. Agrotech. Econ.* **2021**, *40*, 110–121.
47. Xu, Y.; Huang, X.; Bao, H.; Ju, X.; Zhong, T.; Chen, Z.; Zhou, Y. Rural land rights reform and agro-environmental sustainability: Empirical evidence from China. *Land Use Policy* **2018**, *74*, 73–87. [CrossRef]
48. Shi, C.L.; Zhang, Y.; Guo, Y.; Zhu, J.F. The impact of land fragmentation on farmer's chemical fertilizer use efficiency. *J. Nat. Resour.* **2019**, *34*, 2687–2700.
49. Gai, Y.; Qiao, Y.; Deng, H.; Wang, Y. Investigating the eco-efficiency of China's textile industry based on a firm-level analysis. *Sci. Total Environ.* **2022**, *833*, 155075. [CrossRef]
50. Lampach, N.; To-The, N.; Nguyen-Anh, T. Technical efficiency and the adoption of multiple agricultural technologies in the mountainous areas of Northern Vietnam. *Land Use Policy* **2021**, *103*, 105289. [CrossRef]
51. Yang, J.; Lin, Y. Driving factors of total-factor substitution efficiency of chemical fertilizer input and related environmental regulation policy: A case study of Zhejiang Province. *Environ. Pollut.* **2020**, *263*, 114541. [CrossRef]
52. Ma, L.; Feng, S.; Reidsma, P.; Qu, F.; Heerink, N. Identifying entry points to improve fertilizer use efficiency in Taihu Basin; China. *Land Use Policy* **2014**, *37*, 52–59. [CrossRef]
53. Wang, P.P.; Han, Y.J.; Zhang, Y. Characteristics of change and influencing factors of the technical efficiency of chemical fertilizer use for agricultural production in China. *Resour. Sci.* **2020**, *42*, 1764–1776. [CrossRef]
54. Zhu, D.L. The economic mechanism and governance path of "non-food" arable land. *China Land* **2021**, 9–11.

 *land*

Review

Ill Fares the Land: Confronting Unsustainability in the U.K. Food System through Political Agroecology and Degrowth

Mark Tilzey

Centre for Agroecology, Water and Resilience, Coventry University, Coventry CV8 3LG, UK; ab7997@coventry.ac.uk

Abstract: The U.K. food system exhibits strong unsustainability indicators across multiple dimensions, both in terms of food and nutritional insecurity and in terms of adverse climate change, biodiversity, and physical resource impacts. These indices of an unsustainable and inequitable social metabolism are the result of capitalist agriculture and society in general and, more specifically, of neoliberal and austerity policies adopted with vigour since the global financial crisis. The causal, capitalistic, and, latterly, more neoliberal bases of the U.K. food system are delineated in the first section of the paper. These bases are then detailed in terms of their impacts in exacerbating climate change, biodiversity (and resource) decline and loss, and food and nutritional insecurity. The political narratives and policy frameworks available to dissemble, mitigate, or, more rarely, to address (resolve) these impacts are then delineated. It is argued that the only policy framework available that strongly integrates food security (social equity) with ecological sustainability is political agroecology and an accompanying degrowth strategy. The final section of the paper details what political agroecology and degrowth might entail for the U.K. food system.

Keywords: U.K. food system; capitalism; neoliberalism; climate change; biodiversity; food insecurity; political agroecology; food sovereignty; degrowth

Citation: Tilzey, M. Ill Fares the Land: Confronting Unsustainability in the U.K. Food System through Political Agroecology and Degrowth. *Land* **2024**, *13*, 594. https://doi.org/10.3390/land13050594

Academic Editor: Hossein Azadi

Received: 13 February 2024
Revised: 26 April 2024
Accepted: 27 April 2024
Published: 29 April 2024

Copyright: © 2024 by the author. Licensee MDPI, Basel, Switzerland. This article is an open access article distributed under the terms and conditions of the Creative Commons Attribution (CC BY) license (https://creativecommons.org/licenses/by/4.0/).

1. Introduction

A recent and authoritative assessment of the U.K. food system [1] concluded that it currently fails the test of sustainability on multiple criteria—food security, dietary quality and equality, greenhouse gas (GHG) emissions, biodiversity and soil conservation, and so on. The author (Lang) makes an incontrovertible case for the inherent and growing fragility of the U.K. food system across these dimensions, especially, perhaps, in relation to the historical and current reliance of the system upon the U.K.'s ability to buy 'food from nowhere' according to the principle of 'cheapness', without asking necessary questions concerning the ecological and social sustainability of the food so produced. Lang's detailing of this fragility is comprehensive; his analysis and critique are undertaken from what he terms a 'multi-criteria' or what might be called a 'systems-theoretical', perspective. This has the advantage, in contrast to neoclassical economic ('hegemonic') theory, of bringing into play all dimensions of the food system, placing on an equal footing the 'economic', 'political', 'socio-cultural', and 'ecological' variables involved. While this approach is essential as a starting point, it becomes less convincing the more questions are asked concerning causality underlying the dynamics of the system. In other words, the question of 'structural causality', or the principal causal driver(s), underpinning and propelling the system seems to be lacking. This is unfortunate because structural causality, following the principles of critical realism, enables us to make sense of the way in which the above variables interact in a causal hierarchy. Thus, critical realism synthesises multiple determinations, identifies the underlying real mechanisms, and connects them to actual and empirical aspects of the *explanandum*, the phenomenon to be explained [2–4]. This missed opportunity on Lang's part to drill down analytically into structural causality has an important impact on his normative proposals for rendering the U.K. system more food-secure and resilient. Notably absent amongst these is a critique of capitalism, surely the *primum mobile* of a system that is predicated on market hegemony, exponential growth and consumption, social inequality, and the externalisation of ecological and social costs—all features central to the contradictions of the *capitalist* U.K. food system. Lang, however, appears to assert that the search for a more food secure U.K. should be deployed as a test of the strength of British capitalism [1], as if the two aims were somehow entirely compatible and natural bedfellows.

This paper, while covering territory similar to Lang's book [1], develops a rather different line of analysis, with concomitantly rather different normative conclusions. This is not to suggest that the two analyses and their respective normative recommendations are incompatible but rather to aver that the political and ecological transformation of the U.K. food system required to render it sustainable is likely to be rather deeper and more comprehensive than Lang allows. Given the deep ecological and social contradictions of the U.K. food system that we will delineate in this paper (contradictions integral to the unsustainability of the U.K. capitalist economy as a whole), it appears appropriate, therefore, to build on the more radical analyses and normative proposals for transformative change set out in a series of recent publications addressing political ecology, agroecology, food sovereignty, and degrowth [4–7]. In line with these radical analyses, a recent paper by Guerrero Lara et al. [8] identifies and synthesises a useful research agenda for critical agrarian studies in relation to the degrowth problematic, and we propose, in the present paper, to engage with this agenda in relation to the U.K. food system (while noting that significant elements of this agenda have in fact been parsed in the aforementioned publications, albeit not necessarily in relation specifically to 'degrowth'). These authors identify four areas in need of further research and development in relation to agri-food studies and degrowth: degrowth conceptualisations, theorisation of transformations towards sustainability, the political economy of degrowth agri-food systems, and rurality and degrowth.

Concerning the first area, degrowth conceptualisations, Guerrero Lara et al. point to the relative analytical neglect of the ecological conditions and the energy/material throughput of proposed alternative agri-food systems by comparison to the more common focus on the social principles of degrowth. Research that identifies and quantifies possible changes in social metabolism and nutrition, they suggest, can serve as a 'reality check' in relation to assertions regarding the potential of alternative agri-food models to reduce energetic/material throughput (ecological sustainability) whilst sustaining or enhancing human nutritional standards and social well-being/equity (social sustainability). More specifically, they ask, 'what is the social metabolic space of possibilities for the reduction of material and energy throughput in agri-food initiatives from food production to consumption to make them "thermodynamically efficient" rather than striving for more economically efficient modes of consumption and production' (p. 1583). It is also necessary, of course, to gauge the biophysical indices of the current (capitalistic) social metabolism in order to assess their degree of (un)sustainability and, therefore, the level of transformation required to render the social metabolism 'thermodynamically efficient'. Building on previously published work [4,9,10], these tasks we undertake in the present paper in sections addressing the climate change, overseas land footprint, and biodiversity impacts of the U.K. food system and the outlining of a policy framework for agroecology, food sovereignty, and degrowth.

Within this first area (degrowth conceptualisations), Guerrero Lara et al. also point to the need for degrowth research in agri-food systems to engage more with literature addressing policy and social movements. They suggest that few studies have investigated the role of policies such as those governing trade and agriculture as factors influencing the degrowth transformation of agri-food systems. While this unfortunately neglects significant studies that have addressed the constraints on transformations to sustainable food systems (implying or specifying degrowth) imposed by varying configurations of capitalism [11–15], the role of policy and its relationship to social movements (that is, the relationship between 'systemic' and 'anti-systemic' agents) in the dynamics of transformative change are discussed, in varying degrees, in all sections of the present paper. Work specifically addressing the relationship between 'systemic' ('hegemonic' and 'sub-hegemonic') and 'anti-systemic' ('alter-hegemonic' and 'counter-hegemonic') agency in relation to agri-food systems has also been presented in detail elsewhere [4,10,15,16].

In respect of the second area identified by Guerrero Lara et al., the theorisation of transformations towards sustainability, these authors assert, correctly, that research has commonly lacked a consideration of capitalism, has usually related to 'Western' countries, and, by the same token, has been of limited applicability to 'non-Western' societies. While this criticism is certainly well-founded as a generality, it is important to note work that has attempted to address these lacunae [4,10,12,15–18] since this forms an essential backdrop to the current paper. Guerrero Lara et al. highlight three themes within this area that would, they maintain, benefit from greater research focus: learning from critical perspectives within sustainability transformation scholarship, investigating the multiplicity of change agents beyond grassroots initiatives, and bringing in the more-than-human dimension.

Inter alia, they suggest the following:

- Research addressing peasant/indigenous movements and decoloniality should move centre stage—here, the following publications should be noted [4,9,10,15,16,18,19], and the present paper will build on this work by specifying the 'ecological imperialism' of the U.K. food system;
- Research should ask what role multiple agents of change might play in a degrowth transformation and how their political agendas might intersect or conflict—here, complementing prior work addressing political (class) positionalities and discourses [4,10–12,15,16], the present paper will delineate these positionalities and discourses and argue that dominant class interests in the U.K., in upholding capitalism in various forms, impose strong constraints on a degrowth agenda;
- More research is needed to identify, critique, and theorise the roles that state and non-state and systemic or anti-systemic actors may have in promoting or inhibiting degrowth transformation of agri-food systems—here, again, Guerrero Lara et al. appear to have overlooked work in precisely this area [4,10,15–17], and in the present paper, we build on this theorisation of the state in relation to systemic and anti-systemic class interests to explore the dynamics of the U.K. food system;
- Finally, in bringing in the more-than-human dimension, this paper will continue the theme developed elsewhere [4,9,10,15,20] that ecological sustainability should be a fundamental desideratum of an enduring and stable social metabolism. Moreover, the paper will address explicitly the relationship between the U.K. food system and biodiversity conservation.

Turning to the third area highlighted by Guerrero Lara et al., the political economy of degrowth in agri-food systems by recentring capitalism, these authors rightly argue that the transformation to a degrowth society cannot be envisaged without conflict in a growth-dependent capitalist system. Inter alia, the authors point importantly to the need for more exploration of how the mechanisms of capitalist institutions impede the success of degrowth agri-food initiatives, how they may be contested, and what alternatives to capitalism may be sought. For instance, farmers confront *structural* constraints imposed by a capitalist regime of private landownership (absolute property rights), pushing them to cultivate in a productivist manner and largely prohibiting agroecological initiatives towards degrowth through lack of access to land. The authors ask, crucially, 'in a society predicated on private property ownership, what elements need to be unmade as part of a degrowth transformation to ensure the decommodification of land and prioritise the use value of land over its exchange value? How can the degrowth movement pursue large-scale land decommodification?' (p. 1588). In raising the issue of structural politico-economic constraints, the authors point to the need to situate agri-food degrowth initiatives within the wider context of capitalist food regimes and within the context of close intersectoral linkages within a state-defined economy as a whole. The present paper addresses these concerns throughout but especially in sections addressing the dynamics of the U.K. food system, contested policy discourses, and a policy framework for agroecology, food sovereignty, and degrowth. Again, these sections represent an outgrowth of work published elsewhere addressing these themes [4,10,12,15,17,19].

The last area highlighted as requiring further research by Guerrero Lara et al. is that of rurality and degrowth. They point out that most agri-food degrowth research in the global North is concentrated in urban and peri-urban areas—this is due to the concentration here of the greatest economic precarity and consequently of oppositional movements. The countryside in the global North has largely lost its peasantry and is the home of relatively prosperous, conservative, and property-owning farmers, constituting a rather uncongenial environment for oppositional movements [4,10,15,16,19,21]. As Guerrero Lara et al. ask, 'how can degrowth speak to large-scale [or indeed to any market-dependent family] farmers who have been formed and shaped by the capitalist economy's ruthless paradigm of continuous growth?' (p. 1589). Despite widespread antipathy amongst the family farm constituency to ideas associated with agri-food degrowth, many smaller market-dependent farms appear unlikely to survive as economically viable entities beyond the short term, especially in more neoliberally inclined economies such as the U.K. Given that such rising threats to family-farm livelihoods, arising especially from neoliberalisation, appear currently to be resulting in the political embrace of right-wing populism amongst this constituency, it is important to ask, 'how might degrowth help to effectively fight rural marginalisation and decline?' (p. 1589). In the present paper, we attempt to answer this question in the section describing a policy framework for agroecology, food sovereignty, and degrowth whilst acknowledging, in the previous section on contested agrarian discourses, the scale of the material and ideological task in securing alignment to such a politico-ecological transformation.

The present paper is structured as follows. We begin, in the first section of the paper, by delineating the causal, capitalistic, and, latterly, more neoliberal bases of the U.K. food system. In the next three sections, we then detail the empirical indices of unsustainability demonstrated by the U.K. food system, focusing on their impacts in exacerbating climate change and overseas land footprint, biodiversity (and natural resource) decline and loss, and food and nutritional insecurity. We argue that these indices are the result of capitalist agriculture and society in general and, more specifically, of neoliberal and austerity policies adopted enthusiastically by U.K. governments since the global financial crisis and, especially, since 2010. In the fifth section, the principal political narratives and policy frameworks being articulated either to obstruct action, to mitigate, or, more rarely, to address (resolve) these impacts are then delineated. These are defined as 'hegemonic' (neoliberal), 'quasi-hegemonic' (environmental neoliberal), 'sub-hegemonic' ('state–capitalism'), 'alter-hegemonic' (ecocentric), and 'counter-hegemonic' (agroecology-food sovereignty). It is argued that the only political narrative and policy framework available with the capacity to strongly integrate food security (social equity) with ecological sustainability is political agroecology as food sovereignty as part of a programme of degrowth. The final section of the paper outlines what political agroecology and degrowth might entail for U.K. food system.

2. The U.K. Food System and Causal Basis of Agrarian Growth and Unsustainability: From National Developmentalism, through 'Embedded' Neoliberalism, to 'Radical' Neoliberalism

2.1. National Developmentalism

The current fossil-fuel, agrochemical-based, and fully capitalist[1] configuration of British agriculture broadly achieved its current form during the post-Second World War period. This was a period characterised by Fordism and the development of sectoral articulation between agriculture and industry and social articulation between a fully proletarianized workforce in its role as both producer and consumer, in which Britain, following the severe food insecurity of the war and disruptions to food imports, sought to become self-sufficient in the production of principal food staples. This brought home to Britain the ecological contradictions of productivist agriculture, albeit now intensified through greater dependence of fossil fuels and agrochemicals, that had previously been externalised onto the spaces of export agriculture abroad during the period that Tilzey [4,17] denotes as the 'Second or Imperial' food regime. A massive acceleration in labour productivity and yields due to fossil fuel-powered mechanisation and agro-chemicalisation, with ecological costs in terms of GHG emissions, soil degradation, water pollution, and severe biodiversity loss effectively externalised [20], was structurally tied to a particular phase of capitalist development that we may term 'national developmentalism' [19,22]. As applied to the agriculture sector, we may refer to this as the 'Third or Political Productivist' food regime [17,20], a state-managed policy framework to which an acceleration of the processes of 'appropriationism' and substitutionism'[2] were pivotal. 'Political productivism', thus embodied in U.K. post-war policy and subsequently in the Common Agricultural Policy (CAP) of the European Union (EU) (to which the U.K. acceded in 1972, only to leave again, as part of 'Brexit', in 2021), was implemented by deploying the instruments of guaranteed prices, investment grants, input subsidies, state regulation of major commodity markets, and their insulation from overseas competition [9,20].

The result was to 'hothouse' agrarian capitalism through a policy framework in which higher net farm income could be secured only by means of productivity and yield increases. This acted as a massive incentive to cut costs through the substitution of machinery for labour, enlarging holdings, and borrowing money for land purchase and capital projects, all dependent centrally on increased fossil fuel and agrochemical consumption. This, in turn, created indebtedness, further reinforcing the imperative to cut costs and increase output [24], enabled by the 'cheap' 'ecological surplus' afforded by fossil carbon extracted primarily (before North Sea oil) from the global South under the aegis of imperially installed autocratic regimes. The appropriation of such 'ecological surplus' (principally through Surplus Extraction 2 (see below)), enabled the fossil-fuel based capitalisation of agriculture, the final elimination of peasant-based production, and the 'secure' employment, as a 'labour aristocracy', of the resulting really subsumed proletariat in urban industry (facilitated also by the racialised super-exploitation of immigrant labour from the periphery as Surplus Extraction 1 (see below))—the result was the full instantiation, under Fordism, of the 'imperial mode of living' [12,25][3].

2.2. Embedded Neoliberalism

From the 1980s, however, the regime of 'political productivism' conceded gradually to a more neoliberal (or 'market productivist') regime of accumulation within the CAP, with commodity support measures giving way to direct payments supplemented by discretionary budgets for agri-environmental measures designed to mitigate the more egregious impacts of productivism on the ecosystems of rural areas [9,20]. Tilzey [4,17,26] nominates this new regime the 'Fourth or Neoliberal' food regime, since, like Bonanno and Wolf [27] and Otero [28], he considers the term 'neoliberal' to capture explicitly the central role of the state in re-regulating for and undergirding the strategies of certain fractions of capital, especially those with a transnational orientation. This trend towards neoliberalism, articulated by the EU and supported strongly by the U.K. as a then member state, reflected in no small part the greatly increased influence of transnational and neoliberally inclined class interests in defining and promoting a more globally and market-oriented agricultural policy [11–13]. 'Hegemonic' neoliberalism gradually gained ascendancy vis-à-vis those 'sub-hegemonic' (neo-mercantilist and social protectionist) class constituencies that had formed the bedrock of national developmentalist Fordism [11,12]. This neoliberal strategy was designed to stimulate the further expansion of productivism, now of a more market-oriented kind, and its increased integration into global agri-food circuits of capital. The progressive elimination of 'market distorting' commodity support in favour of (WTO-compatible) direct payments (Pillar 1 of the CAP) was complemented by the creation of Pillar 2 (rural development and agri-environmental monies), designed to afford some measure of continuing support to farmers marginalised in the neoliberalisation process, to provide countryside consumption spaces for the urban populace (while conserving a residual biodiversity and landscape resource), and to supply the new 'health conscious' and 'environmentally aware' middle-class, a 'reflexive' consumer with organic and/or locally produced food commodities. An (asymmetrical) bipolarity in policy became increasingly evident during the 1990s and the new millennium, therefore, between globalising norms of governance for market productivism on the one hand and regionalised or 're-territorialised' norms of governance for 'post-productivist' and 'multifunctional' agri-rural activities on the other. These latter 'ecologised' and 'localised' constituencies we nominate 'alter-hegemonic' interest groups [4,15].

Thus, while changes to Pillar 1 were designed to facilitate the progressive penetration of globalising market relations and international market dependency into EU and British agriculture, Pillar 2 budgets (far smaller than those allocated to Pillar 1), simultaneously, came to be disbursed on a competitive and selective basis and were (and remain) heavily constrained in their ability to counteract the overarching processes of neoliberal restructuring. Budgets for agri-environmental management likewise came to be defined and defended increasingly according to neoclassical 'public goods' criteria (see below), entailing more restrictive forms of subvention in line with WTO 'green box' disciplines and the accompanying requirement to minimise 'market distortion' [4,12,13]. Despite these clear shifts towards neoliberal governance norms, the EU insisted on the retention of significant direct supports for farmers within Pillar 1 of the CAP, the function of which was and remains to act as a WTO-compatible 'hidden subsidy' to Europe's farmers for reasons of political legitimacy, the continuation of which, in the face of global Southern opposition, comprised an important factor in the breakdown of the WTO Doha Development Round of free trade negotiations in 2008 [4]. Such continuing and generous direct subsidy within Pillar 1, together with the very existence of Pillar 2, indicate that the CAP cannot be described as unambiguously neoliberal. Rather, the Polanyian-derived term 'embedded neoliberalism' seems more apposite [12] since, however attenuated the 'agricultural welfare state' [29] might now appear by comparison to its Fordist heyday, the CAP continues to fulfil both the accumulation and legitimacy functions of the 'state'.

2.3. Radical' Neoliberalism

Indeed, it was the very retention of such legacies of 'sub-hegemonic', social democratic Fordism that incurred the ire of doctrinaire neoliberals (and their right-wing 'authoritarian populist' allies) in the U.K., especially amongst members of the Conservative Party and in state departments such as the Treasury and DEFRA (the Department of Environment, Food, and Rural Affairs). Harnessing the growing discontent of the British working classes, especially with globalisation and austerity following the financial crisis of 2008, and deploying the EU as a convenient scapegoat for these ills, right-wing 'authoritarian neoliberals' of the Conservative Party contrived to engineer a 'Brexit' departure from Europe under the guise of the national populist slogan 'taking back control', thereby disguising its actual Thatcherite agenda of untrammelled neoliberalism (modified by certain necessary, populist, but minimal concessions to its new proletarian constituency). The Conservative Party and DEFRA were able to portray Brexit as an unprecedented opportunity to address the ecological disbenefits of agricultural productivism (including, by implication, climate change impacts) since, according to neoclassical and neoliberal economic doctrine, these were the outcome not of capitalism but rather of the continuing market interventionism (statism) of the of the EU's CAP. Remove such interventionism, and the 'free play' of market forces would secure that axiom of neoclassical theory, 'optimal allocation of scarce resources'. With this in place, any environmental 'market failures' could then be made good with state subvention for 'public goods'.

The mainstream environmental movement in the U.K. has long held a similar view of the CAP and has, therefore, tended to be beguiled by neoliberal and neoclassical economic arguments for the freer play of 'market forces' as putatively the best means, with the added proviso of 'public goods' payments, to assure environmental sustainability (including climate change mitigation) [9]. The CAP has thus been a relatively easy target for the U.K.'s mainstream environmental conservation movements, precisely because such critiques sit comfortably with the new neoliberal economic agenda of the Conservative U.K. government and its intention to dismantle the 'distorting' influence of 'sub-hegemonic' market/direct support structures [9]. This agenda is now being enacted post Brexit with the phasing out in the U.K. of inherited CAP supports through Pillar 1 over the period up to 2028 and the intended restriction, thereafter, of state subvention in the agriculture sector to neoliberally configured and WTO-compatible 'public goods' payments. It is this 'hegemonic' discourse of neoliberalism and 'quasi-hegemonic' *environmental* neoliberalism (see below), propounded by the Conservative Party and the principal departments of state (and supported, albeit cautiously, by the main environmental NGOs), that is, in combination, the main determinant of U.K. climate change mitigation, agri-environmental, and food security policy. Below, we examine this discourse, together with other 'sub-hegemonic', 'alter-hegemonic', and 'counter-hegemonic' discourses in relation to agrarian climate change, biodiversity loss mitigation, and food security policy. Before doing so, we lay out the current climate change, biodiversity, and food in/security impacts of the U.K. alimentary system.

3. Current Climate Change Impacts and Overseas Land Footprint of the U.K. Food System

3.1. U.K. Food System Climate Change Impacts and Land Footprint

In the first study of its kind, Audsley et al. [30] employed a detailed inventory of emissions developed from the life cycle analysis of a wide range of foods and processes comprising three parts: primary production to the regional distribution centre (RDC), from the RDC to consumption (through retail and cooking), and land-use change (LUC). On this basis, they estimated that the supply of food and drink to the U.K. results in a direct emission equivalent of 152 $MtCO_2$ (million tonnes of carbon dioxide equivalent, a figure including other GHGs, notably methane [CH_4] and nitrous oxide [N_2O]). A further 101 $MtCO_2e$ from LUC, mainly overseas, largely due to deforestation and forest degradation in Latin America and Southeast Asia, is attributable to U.K. food consumption. Total U.K. economy consumption emissions are estimated to be about 748 $MtCO_2e$ (excluding LUC, which, if added, sums to 849 $MtCO_2e$). This means that direct emissions from the U.K. food system comprise about twenty percent of the currently estimated consumption emissions of the U.K. economy, a figure that rises to thirty percent if LUC emissions are added [30,31].

Of direct emissions (excluding LUC), fifty-eight percent arise from animal production and products, which, however, account for only thirty percent of consumer energy intake [30–32]. Overall, about twenty percent of direct U.K. food chain emissions occur outside the U.K. However, if LUC is taken into account, this figure increases dramatically to around fifty percent, meaning that about half of total U.K. food system emissions arises outside the country. Audsley et al. conclude that the direct and indirect (that is, LUC) effect of the supply of the food to the U.K. as a contributor to global land-use change pressures is a significant factor in U.K. consumption emissions. Their study also attributes a large proportion (seventy-five percent) of LUC emissions to ruminant meat production, primarily through the production of soya feed for beef and dairy, and to lesser extent through direct beef exports, sourced overwhelmingly from Latin America on areas formerly characterised by biodiverse, high carbon sequestration biomes.

Primary production, that is, production of food commodities, accounts for fifty-six percent of direct emissions (excluding LUC), with nearly half comprising N_2O from agricultural soils through the application of synthetic fertilisers and CH_4 from enteric fermentation from ruminant livestock (primarily cattle and sheep). The source of the other half or so is dominated by CO_2 emissions from fossil energy used in the manufacture of agricultural inputs, such as energy use in highly mechanised farming, commodity storage, and some processing. Beyond primary production, energy use in processing, manufacture, transport, retail, and food preparation/cooking accounts for thirty-seven percent of all direct emissions. However, if we again factor in LUC, this contributes, as noted, around 100 $MtCO_2e$, most of which comprises CO_2 emissions through direct destruction or degradation of high carbon sequestering biomes in the global South. Again, this indicates that LUC, located primarily in the highly biodiverse and high carbon sequestering biomes of the global South (Latin America (soya and beef) and Southeast Asia (palm oil)) comprises the single largest contributor to U.K. food system GHG emissions, with emissions from primary production and then RDC to consumption (through retail and cooking) representing the second and third most important sources of U.K. food system GHG emissions [30]. This represents very considerable 'carbon leakage' (the displacement of GHG emissions to states outside those where product consumption occurs) in the U.K. food system.

3.2. Impacts Increased by Meat- and Dairy-Heavy Diets

GHG emissions in primary production and in LUC are exacerbated by meat- and dairy-heavy diets characteristic of the 'imperial diet' since meat production is energetically less efficient than producing plant foods directly for human consumption. Moreover, ruminants produce the potent GHG CH_4 as a by-product of digestion. GHG emissions are exacerbated still further when ruminants are fed grains produced on the basis of fossil energy-dependent mechanisation and agrochemicals, the latter producing the most potent GHG N_2O, some 265 times as virulent as CO_2. The worst emissions occur when primary production as specified above is preceded by LUC that entails the destruction or degradation of high carbon sequestering and climate stabilising native vegetation (it is known that trees emit natural aerosols that act as water vapour condensation nuclei, which then both cool the air within and around forests and increase local rainfall, significantly mitigating climate change *in addition* to their role as carbon sequestrators [33]). Sadly, much of the feed grown in the global South and destined to feed ruminants in the U.K. is produced under the 'worst case' scenario described above. In this regard, de Ruiter et al. [34] endeavoured to detail the total agricultural land footprint associated with the U.K. food supply, differentiating between the impacts of feed versus food. Thus, thirty-eight percent, or 22,630 Kt, of the total U.K. crop supply (that is, domestic production and imports minus exports) in 2010 was used for animal feed. Eighty-seven percent of all barley is destined for animal feed, while about ninety-three percent of all soya beans (the great bulk from Latin America) is also used for feed. About fifty-five percent of the total cropland footprint for U.K. feed, or about 2619 kha, was located overseas in 1987, and this increased to sixty-four percent, or 3293 kha, in 2010.

De Ruiter et al. [34] find that the total cropland footprint of U.K. food supply increased between 1987 and 2010. Thus, the cropland footprint for both feed and food in 2010 was 8833 kha compared to 8406 kha in 1987. This suggests increased carbon intensity of production since grassland-fed livestock are less carbon-intensive than grain-fed livestock. These crops have been sourced increasingly from abroad, both in respect of crops for feed and for food. Thus, in 1987, the domestic share of the cropland footprint for feed and food was about forty-five and forty-two percent, respectively, and this share decreased to thirty-six and thirty-eight percent, respectively, in 2010. Thus, sixty-four percent of feed crops and sixty-two percent of food crops were imported into the U.K. in 2010, representing increasing 'carbon leakage', especially in the case of feed since this has tended to displace the more productive sector of grass-grown domestic livestock production in the U.K. In other words, the main exporters to the U.K. of ruminant feed products have intensified their production, while the U.K. has seen a commensurate extensification (de-intensification) of grass-based ruminant production. This suggests that some environmental gains through extensification of grass-based ruminant production in the U.K. have been secured at the expense of 'carbon leakage' overseas, especially to Latin America, the primary source of U.K. soya. The U.K. has also witnessed a slight decline in red meat (ruminant) consumption, but this reduction has been matched by an increase in poultry and pork consumption, both fed almost exclusively on crop feed. While these livestock types produce less GHGs in production than ruminant equivalents, they are even more dependent on the importation of feed crops, especially soya, thus sustaining 'carbon leakage' to the global South [34].

The land required overseas to meet the U.K.'s annual demand for soya between 2016 and 2018 was on average 1.7 Mha, or an area similar to that of Wales [WWF and RSPB 2020]. Sixty-five percent of the soya land footprint was located in Argentina, Brazil, and Paraguay during this period, all countries classified by the authors as high or very high risk.[4] Fifty-six percent of the U.K.'s soya imports between 2016 and 2018 were in the form of soymeal, a prime ingredient of animal feed and increasingly associated with high protein diets. The authors' data indicate that at least seventy-five percent of all imported soya is either embedded in imported meat, eggs, and dairy or is used for animal feed [35].

Again, in 2019 [36], U.K. imports of soya amounted to over 2000 Kt, forty-two percent of which came from Argentina, twenty-seven percent from Brazil, and eight percent from Paraguay. Some forty percent of this amount (1000 Kt) came without any sourcing requirements at all concerning sustainable production in relation to deforestation/conversion certification criteria. A further twenty-three percent, however, derived from quite dubious 'book and claim' certification, whereby production, which may well be unsustainable, is supposedly 'offset' by environmental 'credits' purchased elsewhere. Only nine percent of soya production outside 'low risk' countries (such as the USA and Canada, which have historically destroyed their native vegetation) can be unequivocally attributed to sustainable production on the basis of 'physical certification'. Thus, some sixty-three percent of the seventy-seven percent of U.K. soya imports from Latin America is likely to be closely linked to climate change enhancing LUC through the destruction and degradation of biodiverse carbon sinks, in addition to the soil-degrading (CO_2-releasing) and fossil fuel-dependent character of the capital-intensive production process itself. To this it is necessary to add the fossil fuel consumption entailed in transportation and processing of soya.

3.3. Ecological Imperialism

The very considerable 'carbon leakage' of the U.K. food system (some fifty percent of GHG emissions) implies that the U.K. is engaged in a relation of 'ecological imperialism' with supplier states. This is because the latter, with the complicity of agro-exporting oligarchies in the (semi-) peripheral state–capital nexus, are subordinated to the consumption demands of this 'core' state, entailing the displacement of peasant/indigenous populations and with them the potential for sustainable agroecological food production [10]. This process of displacement generates 'ecological surplus', through 'appropriation by dispossession'[5], on the basis of the 'mining' of socio-natural 'capital' nurtured by those non-capitalist populations in the form of diverse semi-natural biomes and organically cultivated soils (principally through Surplus Extraction Mechanism 2; see below). It also entails the externalisation of costs onto the supplier country in the form of compromised food security and increased precarity for displaced populations, degraded ecosystems and soils, and the exacerbation of regionalised climate change impacts in the form of droughts, heat waves, and destructive storms [10]. All these impacts contribute to the global climate crisis through the destruction of carbon sinks and the displacement of carbon sequestering agroecosystems by fossil fuel-powered agri-business. Such externalised costs, a measure of the 'spatio-temporal fix' of ecological imperialism, should, of course, be borne by the U.K. itself or, better, not generated at all, the latter feasible only by recourse to agroecological production as part of degrowth with equity (see below).

Within the U.K. itself, agriculture, in 2018, produced only fifty-three percent (by value) of the food consumed in the country, a decline in 'self-sufficiency' in relation to the heyday of 'political productivism' and reflecting the trend towards neoliberalisation in the food system [1]. Land distribution is highly inequitable, with less than one percent of the population owning half the land in England, for example, with landownership even more concentrated in Scotland [37–39]. There were 217,000 farm holdings in the U.K. in 2017, with around twenty percent of these comprising 'very large' holdings (over hundred ha in area) and using the majority of land (seventy-six percent), while forty-eight percent of farms are 'small' (less than twenty ha), farming just four percent of land [40,41]. Production output is also very unevenly distributed—thus, in England in 2017, a small number of large farms (only seven percent) produced fifty-five percent of output by value on only thirty percent of farmed area [40].

These larger farms are highly capitalised and usually capitalist enterprises (although commonly centred around family labour), and achieve very high productivity (the ratio of labour input to output), but only with massive quantities of climate change-inducing fossil fuel, synthetic fertiliser, and agrochemicals, together with extraordinarily expensive equipment and infrastructure, again dependent on fossil fuel. Agrochemicals derive, of course, from oil, while immense amounts of fuel and electricity are required to synthesise artificial fertilisers from natural gas. Most fertiliser is made from ammonia (NH_3), which is itself produced in factories, whereby nitrogen from the atmosphere is synthesised with hydrogen atoms extracted from fossil fuels at high temperature and under high pressure. Being highly soluble, synthetic fertiliser that is not taken up by plants is washed into surface or groundwaters, causing huge eutrophication problems exacerbated by the loss of soil structure and organic soil content through the persistent application of artificial fertilisers themselves (with oxidation of soil organic matter itself being a significant source of CO_2 emissions). The excess synthetic fertiliser that is not leached into water is converted by bacteria into nitrous oxide (N_2O), a GHG that, as noted, is many times more potent than CO_2.

Once the raw ingredients have left the farm, another long chain of energy consumption begins, comprising processing, packaging, transport, retail, cold storage, cooking in homes and restaurants, and, lastly, waste disposal [30]. Of these 'direct emissions' or supply chain emissions (primary production to RDC and RDC to consumption but excluding LUC), primary production generates fifty-six percent, with CH_4 and N_2O accounting for more than half of these [31]. CH_4 is a more potent GHG than CO_2 (although much shorter-lived) and is produced by ruminant livestock, especially when fed on grains rather than grass, while manure from these animals also releases CH_4 and N_2O. The ecological inefficiencies and destructiveness of raising livestock fed with agro-industrially produced grains (especially soya, which embodies LUC impacts as documented above) are thus immense. Overall, sixty percent of the grain grown in the U.K. is fed to animals [41,42], while some eighty-five percent of agricultural land is devoted directly or indirectly to livestock production [32,34,43]. As much land overseas is used to support the U.K. grain-based livestock production system as is used in the U.K. itself [31,34].

3.4. Surplus Extraction Mechanisms

We can translate these highly unsustainable parameters of the U.K. food system into three Surplus Extraction Mechanisms, all subsumed within the category of 'imperialist rent', that is, the above average or extra profits realised as a result of inequality between the North and South in the global capitalist system [44]. The three main drivers of the U.K. economy, namely banking/services (finance capital), mining, and fossil fuel extraction, are all predicated on 'imperialist rent'. First, U.K. finance capital has invested in labour-intensive and polluting industries that have re-located to the global South during the neoliberal era especially, and here, 'imperialist rent' is founded on the super-exploitation of labour power through Surplus Extraction 1. This is founded on the huge wage differentials that exist between the global South and global North, despite comparable productivity levels of labour power between the two. This differential is key to the super-exploitation of labour power in production in the South and the differential location of high consumption in the global North, a phenomenon known as 'labour arbitrage' [45]. This surplus extraction mechanism occurs primarily through the export of industrial commodities from the South to the North. This transfer of surplus value from global South to North helps to support the levels of affluence in the latter, on which meat- and dairy-rich diets, with their adverse climate impacts, are predicated.

Second, the key mining and fossil fuel extraction sectors of the U.K. economy realise 'imperialist rent' primarily through Surplus Extraction 2. This is undertaken principally through capital-intensive extractive processes with little use of human labour power, reliant on the 'ecological surplus' embodied in energy-dense fossil fuels and the 'socio-ecological capital' built up by non-capitalist social systems and extracted through 'appropriation by dispossession' (land grabbing). Imperialist rent is founded, inter alia, on the failure to enforce in the global South norms of environmental regulation, rehabilitation, and social compensation that would be required in the North. Super-profits thereby generated through both Surplus Extraction 1 and 2 afford the transfer of wealth to the U.K. that enables the 'imperial mode of living', the background affluence that undergirds, inter alia, the adoption of meat- and dairy-heavy diets with their ecological inefficiencies and climate change impacts. Super-exploitation of global Southern labour power and environments, or the supply of 'cheaps', implies not only more disposable wealth in the U.K. (as in the North generally) but more consumption since the cheaper the commodity, the more of it will be consumed, varying 'elasticities of demand' notwithstanding. Super-exploited labour and environments in the global South enable the continued formation of 'labour aristocracies' and specialisation in finance/services and high end manufactures in the U.K., with the country, despite increasing wealth differentials, maintaining status as the fifth or sixth richest state globally, as measured by GDP [31]. The historical and sustained full agrarian transition of the U.K. population out of agriculture (with only 426,000 people, or 1.5 percent of the U.K. labour force, remaining in agricultural production [41]) implies that agriculture is highly capitalised and resource intensive in order to supply the high consumption demands of its generally wealthy, non-agrarian, population.

The climate change impacts of such high capitalisation and fossil fuel dependency of U.K. agriculture are exacerbated by grain-fed meat and dairy-oriented diets of the U.K. population as enumerated above. The ecological inefficiencies and climate change impacts of these 'imperial' diets are, as we have seen, to a considerable extent externalised onto the global South through 'carbon leakage', such that around half of GHG emissions associated with the U.K. food system are generated overseas. The GHG emissions thus externalised onto the global South through direct production and through land-use change (LUC), principally to supply feed for livestock consumption, operate largely through Surplus Extraction 2.

Additionally, land devoted to livestock grazing in the U.K., especially on poorer soils in the uplands, could be used to a much greater degree than at present for growing timber and fuelwood. (These areas are currently dominated by sheep production, but consumption of sheep meat contributes only minimally to the food security or calorific value of food consumed in the U.K.) Not only does this compromise potentials for greater CO_2 absorption in the U.K. through woodland expansion, but it also displaces timber and fuelwood production overseas to areas where it may involve LUC from forest to agriculture. Thus, the land required overseas to supply the U.K.'s demand for timber has increased threefold since 2011 (from 2.8 to 8.4 Mha), an area greater than the size of Scotland [35]. Around one-fifth of the U.K.'s overseas land footprint was located between 2016 and 2018 in high risk countries (as defined earlier), including Brazil, China, and Russia. Fuelwood is used primarily for energy generation, and demand in the U.K. has increased from an average of twenty-two percent of total imports in 2011 to thirty-two percent in the period 2016–2018. This increase is likely to be linked to policies designed to increase the share of renewable sources in the U.K.'s energy matrix. Although well intended, these policies fail sufficiently to assess the carbon impacts of biofuels [35].

Finally, the U.K. food system (like the U.K. economy as a whole) emits far more GHGs than are sequestered in the U.K. We have suggested that, in total, U.K. GHG emissions are around 850 $MtCO_2e$ [30]. Net emissions in 2017 were 460 $MtCO_2e$ [46], implying that the U.K. emits some 460$MtCO_2e$ more GHGs than it sequesters. (This net emissions figure does not, however, include 'carbon leakage' due to overseas GHG emissions associated with the U.K. economy, while much of the carbon supposedly sequestered by U.K. vegetation does not in reality occur because most of this is cancelled out by the oxidation of degrading peatlands [46]. Consequently, net emissions are likely to be considerably higher in reality than the above figure suggests.) This indicates, in turn, that the U.K. is reliant upon extra-territorial GHG sinks to mitigate the climate change impacts flowing from its GHG emissions. These it receives in large measure gratis, thus significantly lowering (externalising) the costs to the U.K. that would be borne were the country to sequester its own GHG emissions. Since such carbon sinks, in terrestrial terms, are located differentially in the global South (in the form of differentially intact ecosystems and agroecosystems sustained by indigenous and peasant production), this represents a further mechanism of surplus extraction from the South to the U.K. (and the global North more generally) since the South is bearing the cost of sequestration that should be borne by the U.K. The sad irony here, of course, is that Surplus Extraction Mechanisms 1 and 2, through destruction and degradation of these ecosystemic sinks, are actively undermining the capacity of the global South to continue to sequester GHGs, not only compromising Surplus Extraction 3 for the U.K. (and the North) but, more importantly, posing dire consequences for the future of the Earth in terms of accelerated climate change.

4. Biodiversity Impacts of the U.K. Food System
4.1. Generic Issues and Structural Causality Underlying Biodiversity Decline

It is important to recall at the outset that a large proportion of the most valued biodiversity habitats and landscapes in the U.K., as in Europe as a whole, has arisen from pre-industrial agrarian management of the, consequently, *semi-natural* environment over a very long period [47]. Thus, much of the biodiversity resource of the U.K. depends for its survival, and a fortiori for its flourishing, upon the continuation or re-adoption of low-input farming systems and practices. This comprises a co-evolutionary relationship between pre-industrial farming and the sustainable management of semi-natural habitats within a framework of *land sharing* rather than *land sparing*. The latter comprises a dichotomous framework in which *de-naturalised* agriculture is given over to productivism, while the residue of non-farmed landscape is abandoned to *de-socialised* re-wilding. The co-evolutionary relationship between farming and the semi-natural environment began seriously to erode with the rise of capitalism and industrial agricultural techniques, a trend that, as we have detailed, accelerated out of all recognition following the Second World War. The post-war period has witnessed steep declines in the area of semi-natural habitat and in the number and range of characteristic farmland bird, invertebrate, and plant species. Survival of these habitats and species now occurs (increasingly scarcely) *despite* rather than (as before) *due to* agricultural practices [20]. (We focus in this section on biodiversity impacts of productivist agriculture in the U.K., while acknowledging, as detailed in the previous section, the significant adverse impacts of the U.K. food system on biodiversity overseas, especially in the global South.)

As with climate change impacts, this massive acceleration in the rate of biodiversity loss and decline may be attributed *structurally* to the impacts of capitalism as a general tendency and, more specifically, to a particular model of capitalism that we have identified as 'political productivism' (and its tendential neoliberal successor 'market productivism') [20]. The environmental impacts of productivism can be enumerated as a series of generic issues:

- Loss and fragmentation of semi-natural 'infield', traditionally grazed habitats through agricultural 'improvement' (application of synthetic fertilisers and herbicides) or conversion of these to arable land;
- Overgrazing of semi-natural habitats, primarily in the uplands;
- Loss or mismanagement of 'interstitial' habitats, for example, hedgerows, field margins, ditches, ponds, etc.;
- Drainage or drying out of wetland habitats due to water over-abstraction;
- Pollution and eutrophication of surface and groundwaters leading to loss or degradation of aquatic ecosystems;
- Loss of crop rotations and arable–pasture mosaics leading to severe reduction in characteristic farmland species;
- Shift from spring-sown to autumn-sown cereals leading to loss of nesting sites for characteristic farmland bird species;[6]

- Generalised application of pesticides leading to loss of arable weed species, invertebrates, and thereby food sources for other wildlife groups;
- Generalised application of synthetic fertiliser leading to the loss of degradation of semi-natural vegetation, decline in the organic content and structure of soils, and eutrophication of ground and surface waters [20,47–52].

These generic issues can be linked causally to the essential features of the capitalist and productivist impulses embodied in 'appropriationism' and 'substitutionism', a relationship that we can nominate as *structural causality* [4].

As a result of these generic or structural impacts of capitalist productivism, semi-natural 'infield' habitats have been pushed to the margins of agrochemically based agriculture, subsisting as a residual resource peripheral to most farming systems [20]. Only in the uplands, where physical constraints have prohibited the widespread application of synthetic fertilisers and pesticides, do semi-natural habitats still comprise integral elements of farming systems [48]. In much of the lowlands, however, semi-natural 'infield' habitats survive typically as mere fragments within an otherwise ecologically impoverished farming landscape. Even 'common' species characteristic of more productive farmland (that is, traditionally farmed 'artificial' infield habitats such as arable and grass leys) have exhibited alarming declines over the course of the productivist era [49]. Freshwater habitats continue to suffer decline and loss through nutrient pollution and water abstraction from agrochemically based farming practices. Rivers in catchments where large-scale chicken/egg production is located (birds fed principally on imported soya) suffer increasing eutrophication and ecological degradation through the spreading of manure on land in quantities that the soil cannot absorb. Surplus nutrients, principally phosphates, are then washed into adjacent watercourses. In the uplands, habitat deterioration rather than outright loss has been the norm, the result most frequently of ecological overgrazing by livestock, principally sheep [50].

The current agricultural and environmental policy framework in the U.K. (these frameworks differ slightly, although are diverging increasingly, between the constituent countries' administrations to whom agricultural and environmental policy is devolved) affords a proportion of this residual resource a modicum of legal protection and conservation management by means of statutory regulation and/or environmental land management schemes (ELMS). As we shall see below, Brexit U.K. is introducing a new suite of ELMS following the discontinuation of CAP-derived schemes—these will be configured in a more purely neoliberal way than hitherto in conformity with the neoclassical economic theory of 'market failure' and the payment of so-called 'public money for public goods'. We will summarise the deficiencies of this theory and approach below. For the moment, it is sufficient to note that while environmental regulation, cross-compliance prescriptions attached to continuing direct farm payments (Basic Payment Scheme), and existing ELMS have slowed the decline of biodiversity in the farmed landscape (in other words, decline would have been worse without these sources of mitigation), alarming declines in the extent, quality, and numbers of both habitats and species continue to characterise the U.K. agricultural environment. In 2019, the *State of Nature* report indicated the following:

A wide range of changes in agricultural management in recent decades has led to greater food production but these changes have also had a dramatic impact on farmland biodiversity. For example, populations of farmland birds have more than halved on average since 1970, and similar declines have been seen in many other taxonomic groups. Targeted wildlife-friendly farming, supported by government funded agri-environment schemes, can help halt and reverse these declines, but to date the only successes have been for rare and localised species. The area of land receiving effective agri-environment measures may have helped slow the decline in nature, but it has been insufficient to halt and reverse this trend. [51]

The most recent *State of Nature* report published in 2023 [52] noted the unfortunate continuation of this general downward trend in the abundance and distribution of biodiversity, especially of that component not adapted to productivist environments:

> ...the UK's nature and environment continues, overall, to decline and degrade...the size of response and investment remains far from what is needed given the scale and pace of the crisis. (p. 3)

> The best available information suggests that nature-friendly farming needs to be implemented at a much wider scale to halt and reverse the decline of farmland nature. (p. 7)

> If we are to halt and reverse biodiversity decline we need not only to increase our efforts towards conservation and restoration, but also to tackle the drivers of biodiversity loss, especially in relation to our food system. That means making our food more sustainable and nature-friendly and adjusting our consumption to reduce demand for products that drive loss of nature. (p. 9)

These declines, again, can be attributed to structural causes arising from capitalist productivism. In response to these drastic declines in biodiversity, there is an urgent need to firstly conserve and enhance the remaining resource of semi-natural habitats through site buffering, linkage, and re-creation and to secondly address the decline in 'common' habitats and species in the 'wider countryside'. We will argue that this can be secured only through an integrated, *land sharing* perspective premised on the cessation of capitalist productivism (privileging the production of exchange value through farm capitalisation and intensification) and the adoption of the agroecological production of use values through 'nature-based' solutions [7,53].

4.2. Neoliberalism and the Land-Sparing Approach to Agri-Environmental Policy

The currently prevailing model of biodiversity conservation in the U.K., one that will continue and be reinforced with the adoption of neoliberally configured ELMS, is one in which nature is 'sequestered' on special sites/areas and accorded a role subordinate and opposed to the capitalist productivist impulse to maximise exchange value (growth) through capitalisation and intensification. In this de facto land sparing approach, biodiversity conservation is undertaken on a site-by-site, species-by-species basis and awarded a separate (usually paltry) budget for a series of discrete conservation activities that are juxtaposed to and must match the opportunity cost of the productivist capitalist enterprise [20]. Indeed, payment rates for ELMS are calculated not on the basis of the intrinsic value of the conservation resource but rather on the basis of 'profit foregone', that is, the amount of money that could be made by the capitalist enterprise were the biodiversity resource in question to be destroyed through 'improvement' (in other words, the 'opportunity cost' of foregoing productivist farming practice). The two aspects of this policy framework, a spatial/sectoral dichotomy between environmental and capitalist farming concerns and the expectation that biodiversity will be conserved only if the opportunity cost of agricultural 'improvement' is met, reflect both the productivist foundation of the state–capital nexus and, within this, the absolute property rights that farmers have been awarded to enable them to claim 'compensation' for not destroying nature [54].

The bankruptcy of this approach has been exposed over the last few decades by the continuing and inexorable decline of the U.K.'s biodiversity resource, exacerbated by persistent indirect subsidies to productivism in the form of the Basic Payment Scheme (only weakly linked to environmental outcomes) on the one hand and austerity-hit budgets of environmental agencies on the other. This continuing and alarming decline exposes not only the inadequacy of the land sparing approach to biodiversity conservation (nature cannot be conserved effectively on an isolated or fragmented basis or in relation to individual species alone—see [53,55]) but also the impossibility of effective biodiversity conservation when the structural causes of decline arising from capitalist productivism, whether 'political' or 'market' productivist in orientation, remain in place. Rather, change is required towards environmental (and social) sustainability in the character of that economic activity itself [20]. This might be defined as 'strong sustainability' [56]. This means that sustainability will not be secured through mitigating (in effect 'buying off') unsustainable agricultural practices, an approach embodied in the prevailing model of voluntary environmental incentive schemes (and to be perpetuated in the new ELMS), but will need to be secured by addressing (resolving) the structural causes of generic impacts, whether these derive from 'political' or 'market' productivism. At base, this means rendering food production itself agroecological so that there is no longer a dichotomy between food production and ecological sustainability.

A structural or generic issues analysis, a land sharing or whole countryside approach, and strong sustainability (a sustainable social metabolism) are mutually defining since each derives from or implies the other—a holistic and dialectical relationship between theoretical cause and political praxis. A land sharing and whole countryside approach has as its objective not only the conservation and enhancement of semi-natural habitats subsisting at the margins of productivist farming but, additionally, the transformation of its 'infield' practices into those of an agroecological orientation. The latter objective seeks to conserve and enhance not only characteristic biodiversity but also the resources of soil, water, and atmosphere and to provide nutritious food equitably to all citizens [53,55]. In short, a land sharing approach involves farming that satisfies the requirement for the *joint and sustainable production* of food, biodiversity, and soil/water/atmosphere. This simultaneous and equal concern for all the dimensions of sustainability is embodied in the principles of political agroecology and food sovereignty [4,7]. Sadly, agricultural and environmental policy, especially in England, the home of doctrinaire neoliberalism, is moving in precisely the opposite direction. The Agriculture Act of 2020 embodies the proposed abandonment of any pretension towards the joint management and multifunctional delivery of both food *and* nature; rather, the former is to be left to the tender mercies of the international 'law of comparative advantage', while state subvention and management is to be confined to the latter, shoring up the supposed 'market failure' of environmental provision through 'public goods for public services' [1][7].

5. Dietary and Food Security Impacts of the U.K. Food System
5.1. Dietary Impacts

If the climate change and adverse biodiversity impacts of the U.K. food system are highly concerning, so too are the socio-economic inequalities in diet and food access generated through its operation. Thus, the annual National Diet and Nutrition Survey produced by Public Health England and the Food Standards Agency demonstrates that adults on low incomes are more likely to have diets high in sugar and low in fibre, fruits, vegetables, and oily fish. Children from the least well-off twenty percent of families consume approximately thirty percent fewer fruits and vegetables, seventy-five percent less oily fish, and seventeen percent less fibre per day than children from the most affluent twenty percent [57]. These differences in diet have an important influence on health inequalities that correlate with socio-economic status. Populations resident in the most deprived decile of neighbourhoods are almost twice as likely to die from preventable causes by comparison to those in the wealthiest decile: they are 2.1 times more likely to die from preventable heart disease; 1.7 times more likely to die from preventable cancer; and three times more likely to have tooth decay at the age of five; children are nearly twice as likely to be overweight or obese at the age of eleven [58]. Since 2010, after which economic austerity policies began to be introduced, life expectancy has gone into reverse in the most deprived areas. Thus, women in the most deprived ten percent of neighbourhoods in England now die 3.6 months younger than they did in 2010, and their life expectancy is 7.7 years less than that of women in the wealthiest ten percent. The differential for men is 9.5 years [58,59]. For healthy life expectancy, there is an even greater disparity of nineteen years between the ten percent of poorest and richest [58,59].

The modern, industrial food diet of cheap 'junk' food (highly or ultra-processed food) possesses the singular but perverse quality of generating obesity and poor nutrition simultaneously. Thus, children in the poorest areas of England are both more overweight and significantly shorter at age ten–eleven than their peers in the richest areas [60]. The average five-year old in the U.K. is now shorter than his/her peers in nearly all other high-income western countries [61], an indication of the adverse health impacts of neoliberal austerity policies in the U.K. by comparison to more interventionist policies of most states elsewhere in the global North. Obesity can and does co-exist with outright hunger; the same households that eat poorly may find themselves unable to eat at all, a phenomenon that has been increasing with the recent 'cost-of-living' crisis in the U.K. caused by the Ukrainian conflict and commensurate increases in the cost of basic food items through inflationary pressure on artificial fertilisers (exacerbated by Brexit and exposing the vulnerability of the 'globalised' U.K. food system to external disruption [1]). Thus, data collected by the Department for Work and Pensions in 2019 found that, even before the COVID-19 pandemic, four percent of U.K. families experienced disrupted eating patterns or were obliged to cut back on food due to poverty [62]. Among recipients of Universal Credit (a general welfare payment), this proportion rose to twenty-six percent [62], a percentage that has increased still further with COVID-19 and the subsequent 'cost-of-living' crisis.

Sadly, poverty reinforces unhealthy food consumption, if and when such food is affordable and available. Unhealthy food is cheaper per calorie than healthy food and is more readily available in poorer neighbourhoods. There is a clear correlation between poverty and the density of fast-food outlets, for example [63]. Over three million people in the U.K. cannot reach any food stores that sell raw ingredients within fifteen minutes by public transport (which in the U.K. continues to deteriorate, with cancelled bus routes and less frequent services), and forty percent of lowest income households lack access to a car [63]. The lack of easy access to fresh ingredients or, in the case of nearly a million people, to a fridge in which to keep perishables, together with the increase in the cost of electricity and gas, compounded by the time and effort required to decide upon a menu and prepare a meal, act in tandem as strong deterrents to those with fewest resources and skills to cook 'from scratch' [31]. More serious still, access to affordable food of any kind is now beyond the means of a growing minority of the population—such 'absolute' food poverty is an increasing reality for those on low incomes or social security in the context of the 'perfect storm' of precarious incomes, a reducing welfare state, and inflationary pressure on food prices. As a result, the U.K. now has some two thousand food banks run by charities to supply free food to people in need. Nearly fifty percent of families with three or more children are now living below the poverty line, while the most recent cut in Universal Credit introduced in 2021 was predicted to drive a further half a million people into poverty and push the child poverty rate to one-third of all children [62,64].

The Food Foundation monitors moderate or severe hunger and malnutrition in the U.K. (food insecurity defined as insecure access to adequate amounts of nutritious food). Moderate or severe food insecurity are defined, more specifically, as the number of people in the previous month who had smaller meals or skipped meals; had been hungry but not eaten; or had not eaten for a whole day—each because of lack of access to or inability to afford food. In June 2023, the Food Foundation found that nine million adults in the U.K., or seventeen percent of households, experienced moderate or severe food insecurity, a massive rise from 7.3 percent in June 2021 [65]. Nearly one-quarter of households with children experienced food insecurity [65]. As suggested above, moderate or severe food insecurity in the U.K. is associated with obesity, since people who cannot afford or lack access to healthy and nutritious food eat unhealthily. Far from seeking to help low-income families to escape these life-debilitating disadvantages, however, the neoliberal policies of the incumbent U.K. government are simply exacerbating the conditions underlying dietary inequality and food poverty. In short, the wider social inequalities of neoliberal Britain are reflected in and exacerbated by inequalities of diet and food poverty.

5.2. Food Security Impacts

Turning to the issue of self-sufficiency and food security, we may define self-sufficiency as the ability to feed a nation from its own produce rather than from imports. Food security, for its part, has been defined as the ability to feed a nation *at a reasonable cost*, irrespective of the source of food or the manner of food production, even in the face of future shocks such as massive harvest failure or a general crisis of agricultural production caused by, for example, climate change [31]. Sadly, however, this definition, one operationalised by historical and contemporary U.K. governments alike, directs virtually no attention to the ecological costs entailed in supplying food to the nation 'at reasonable cost'—in other words, 'at reasonable cost' usually means 'at huge cost to the environment', 'cheapness' being predicated on ecological (and social) cost externalization, both domestically and overseas [1]. Stated otherwise, it is the 'hidden' subsidy afforded by ecological (and social) cost externalisation that enables food commodities, under the principle of capitalist 'comparative advantage', to be supplied 'at a reasonable cost' to consumers [4]. 'Food security', as currently configured, therefore generates longer-term food insecurity by undermining the ecological and biophysical basis of agricultural production [1]. Moreover, a capitalist food system cannot assure general access to food since this depends on the ability to pay—capitalist food producers have to balance 'cheapness' (affordability) with the imperative to generate profit. The state–capital nexus attempts to mitigate and balance these contradictions—the need to ensure the availability of food for consumers 'at reasonable cost' and the desire of capitalist producers to realise a profit—but, as a capitalist state, is ultimately constrained by capital's demand to realise surplus value as the 'bottom line' [17].

Food sovereignty and political agroecology propose, by contrast, to extricate food production, distribution, and consumption from capital's grip and place them in the hands of democratically controlled governance mechanisms to secure ecological sustainability and equality of access to wholesome food and to the means of food production [4,7]. National food self-sufficiency does not guarantee food security; but where food sovereignty and political agroecology are introduced within any polity, self-sufficiency is the only means of ensuring, so far as is feasible in relation to 'indigenous products', that food is produced, distributed, and consumed in conformity with the principles of ecological sustainability and social equity. While it is important to ensure that alternative supply is available in the event of harvest failure or other local crises, it is vital to repeat that 'food security', as currently defined and enacted by the U.K. government, is achieved only at cost of food insecurity over the longer-term as the ecological foundations of production are eroded and climate change accelerates [1].

Indeed, the U.K. state–capital nexus has, since the emergence of industrial capitalism in the 1840s, sought to deliver 'food security' on the basis of international capitalist supply, premised on the principle of 'comparative advantage' or 'least cost'—a principle that asks few questions concerning the ecological or social sustainability of food supply (only the two world wars and the aftermath of the latter have proven exceptions to the doctrinal dominance of comparative advantage) [4]. Consequently, food self-sufficiency in the U.K. fell progressively following the abolition of the protectionist Corn Laws in 1846 [1,4,18], reaching a nadir of thirty percent self-sufficiency on the eve of the Second World War. Wartime blockades and concerted efforts to increase domestic production saw self-sufficiency rise to seventy-five percent by the end of the Second World War (1945) [1]. The post-war period saw the introduction of 'political productivist' policies to boost self-sufficiency as discussed earlier in this paper. Thus, by the mid-1980s, when CAP subsidies and tariffs were at their zenith, U.K. self-sufficiency reached a peak of around eighty percent [66]. Subsequently, the fiscal costs and environmental impacts of 'political productivism' led the EU to reduce commodity support and tariffs; U.K. governments since Margaret Thatcher have been keen advocates again of freer trade, 'comparative advantage', and, consequently, 'globalisation' in agriculture, as with most sectors. U.K. domestic self-sufficiency has declined since then, a decline that has accelerated with Britain's departure from the EU and, on the basis of current trends and policies, is set to continue on this trajectory. Indeed, the present U.K.

government has such confidence in the operation of the capitalist market and the principle of 'comparative advantage' that it no longer sets a target for the amount of food that the U.K. should grow to feed itself [31]. It relies, rather, on two methods to assess national 'food security'.

First, it conducts occasional reviews to assess whether the U.K. possesses the means to restore necessary self-sufficiency in the event of food supplies from other countries being cut off completely. In 2009, DEFRA conducted a U.K. Food Security Assessment [67], concluding that the U.K. already grows much more of our own food than was the case before the Second World War—the U.K. would therefore be better placed than seventy years previously to restore self-sufficiency if needed (although self-sufficiency has continued to decline since that assessment was conducted). The 2009 assessment also indicated that the shortfall in self-sufficiency could be made good by a shift from livestock production to grains and vegetables. Presciently, from the perspective of this paper and the climate emergency-inspired need to transition to agroecology, the assessment suggested that the maximisation of calorie production would require a dramatic reduction in livestock production, with all crop production being used where possible for human consumption rather than for animal feed [67]. DEFRA estimated that, in such a scenario, the U.K. could produce more than enough calories per person per day (although this would be undertaken using synthetic fertilisers and fossil fuels of course—see below for discussion of agroecological transition and abjuring reliance on fossil fuels).

Second, DEFRA also conducts internal monitoring of food security, these reports being shorter but more wide-ranging than the 2009 Food Security Assessment. They assess the risk of various disruptions across the food system: how global harvests might change due to global heating and other stressors, the geographical diversity of U.K. food imports and the degree of exposure of the U.K. to harvest failure in any one region of the globe, etc. The 2020 Agriculture Act formalised DEFRA's food security reviews, creating a statutory duty to publish such a report at least every three years [31]. As noted earlier, however, the Act also cemented a commitment to increased reliance on cost-externalising 'comparative advantage', thus structurally reinforcing both the U.K.'s commitment and vulnerability to market-defined 'food security'—in effect, food insecurity [1].

In this way, industrial 'market' productivist agriculture, exacerbated by 'globalisation', continues to cause climate change, together with the other severe disbenefits enumerated above, in turn threatening longer-term food supply. Yet, rather than seeking to address this fundamental cause at home by increasing sustainable food self-sufficiency, the present U.K. government, in its quest for 'cheap' food through 'comparative advantage', is bound upon a course of increasing unsustainable 'globalisation' by its commitment to signing FTAs with countries with lower environmental standards than the U.K.'s. As the National Food Strategy [31] points out, it is fairly pointless trying to build a low-carbon, nature-friendly, and socially equitable food system in the U.K. if it is then undercut by imported food produced to lower standards. And yet, this is precisely what the present U.K. government is in the process of doing.

In the next section, we examine in more detail the policy discourses of the major politico-economic constituencies that are attempting, variously, to obfuscate, mitigate (symptom manage), or resolve the contradictions of the U.K. food system delineated above. This examination will serve to paint a picture of the current politico-economic context dominated by various shades of capitalism, which any transition to agroecology and degrowth in the U.K. must unavoidably confront.

6. Contested Policy Discourses Surrounding Agrarian Climate Change, Biodiversity Loss Mitigation, and Food Security Strategies

6.1. Net Zero: 'Societal Project' or 'Flanking Measure'?

Following mounting international pressure from the IPCC, and specific recommendations from the advisory U.K. Climate Change Committee (CCC) to address comprehensively the incontrovertible causes and impacts of anthropogenic climate change wrought by GHG emissions, the U.K. government under a Conservative Party administration announced a target of net zero for such emissions by 2050. The change to legislation came into force on 27 June 2019 and imposed a legally binding target on U.K. governments to achieve such a target by the stated date, thereby amending the Climate Change Act of 2008 (which had set a target of eighty percent reduction in GHG emissions compared to 1990 levels). Net zero means that the U.K.'s total GHG emissions will, by 2050, need to be equal to, or less than, the emissions the U.K. removes from the environment. It is proposed that this can be achieved by a combination of emission reduction and emission removal. GHG emissions can be removed by the natural environment (natural sinks/sequestration) or by using, the U.K. government asserts, technologies like carbon capture (usage) and storage (CC(U)S) [46], even though these are as yet unproven technologies in terms both of feasibility and safety.

Given that this legislation has been passed by a Conservative administration whose neoliberal or 'hegemonic' policies hold out little actual hope of meeting this emission reduction target (as is intimated above and further substantiated below), it is reasonable to surmise that the real purpose of the commitment is to construct a legitimating 'flanking measure' [15] to neutralise potential oppositional discourses, especially from influential environmental civil society groups, and to persuade doubters that something meaningful is being undertaken to avert climate crisis [68]. The adoption of a legally binding and very concrete target certainly sends a clear legitimating message of belief that current and proposed market dependent policies and techno-optimistic imaginaries will deliver on the emissions reduction objective if only 'rational' price signals can be liberated from the dead-hand of the state—and here 'liberation' from the 'sub-hegemonic' and 'state-interventionist' instincts of the EU is construed as a clear opportunity for neoliberals to demonstrate the putative efficiency of free markets in securing the 'optimal allocation of resources' in relation, inter alia, to climate change mitigation and biodiversity conservation.

This combination of immovable target and techno-market optimism appears designed, then, to deflect competing discourses entailing greater levels of market constraint and state interventionism such as are characteristic of the 'sub-hegemonic' narrative, for example. It is certainly true that there are many amongst the Conservative Party, the civil service, and neoclassically trained economists in environmental organisations who genuinely believe in the beneficence of the market (see, for example, [69]) and in its capacity, with the proviso of 'public money for public goods' and certain regulatory safeguards, to secure the required 'net-zero' emissions and improvement in the status of biodiversity. Currently, then, a combination of neoliberal and 'quasi-neoliberal' (that is market-optimist and market-oriented climate emergency discourses as defined earlier and discussed below) may be said to have entrenched hegemonic status in the U.K. Parliament, in state departments, in agri-food capital, amongst larger farmers, and amongst mainstream environmental pressure groups. To the extent that this 'combined' neoliberal discourse coheres and appears unassailable, it may indeed be considered to act as a 'societal project' or 'semantic fix' [68,70]. However, as post-Brexit optimism wains, and dissension again raises its disruptive head, it is perhaps more helpful to speak of the net-zero target as a 'flanking measure' [15], since this term suggests continuing underlying tension beneath this provisional hegemony. Booth [57] suggests the following:

> Attributing the "post-political" [that is, uncontested hegemony] *label to a field wherein even relatively aligned institutions are in tension is overly simplistic. The unfolding of these tensions in the coming decades must be attended to in a way that does not reify a monolithic state-capital nexus, but instead acknowledges the dynamism and lived nature of institutions and the class interests they represent, modulate, and materialise. These fissures are currently predominantly discursive in nature [due to provisional hegemony], but as the British countryside is reshaped in various forms by impending socio-ecological change, they will become socio-material gulfs.*

Booth [68] also notes that the U.K.'s net-zero by 2050 ambition has stimulated food system actors, additional to those pursuing a neoliberal course, to propose various 'imagined pathways to net-zero agriculture', pathways that may be variously seen as complementary to neoliberal discourse ('post-productivism' and 'alter-hegemony'), in tension with it ('sub-hegemony' or 'neo-mercantilism'), or diametrically opposed to it ('counter-hegemony' or 'radical' food sovereignty). We now examine the discursive content of these pathways, the class or interest group complexion of their proponents, and their policy proposals for the agri-food sector to secure net-zero by 2050 as well as ancillary objectives for biodiversity conservation and implications for food security.

6.2. Hegemonic and Quasi-Hegemonic Discourse

The first and **'hegemonic'** discourse is that of neoliberalism, conforming to the 'corporate driven, technological (optimist)' narrative identified by Borras et al. [71]. This is represented most fulsomely by many members of the Conservative Party, by state departments such as the Treasury and DEFRA, by many upstream and downstream fractions of agri-food capital, and by larger farms confident of their continued 'comparative advantage' without state subvention. However, we here combine this discourse with a 'climate emergency' narrative [71] characteristic of the U.K. CCC and certain statutory and non-statutory environmental bodies, for example. We do so because, although these bodies adopt rigorous natural scientific criteria to define the necessary parameters as the basis for policy action to mitigate climate change, together with biodiversity and natural resource decline, the recommended policy actions themselves are largely defined by neoclassical and neoliberal economic theory. Thus, while Booth [68] suggests that the U.K. CCC deploys a 'sub-hegemonic' discourse to differentiate it from the more doctrinaire neoliberalism of DEFRA, for example, we prefer to nominate the CCC as embodying a **'quasi-hegemonic'** narrative.

DEFRA may be taken as an exemplar of 'hegemonic' neoliberalism [11,12,68]. The overall pathway to net-zero that emerges from its departmental literature and statements is one reliant on 'faith in markets to solve complex socio-ecological problems' [68]. DEFRA accords great responsibility to individual farmers to attain net-zero through the prospect of innovation and the 'rationality' of the market. This basic competitive dynamo of supposedly beneficial change is to be supplemented by a measure of grant-based investment in both on-farm capacity and innovation as a means to bolster productivity improvements (the ratio of labour input to output) still further [72–74]. This stance is reiterated in an update of the Agricultural Transition Plan (January 2024) [75]. However, despite this optimism in the market, productivity improvements, and technological innovation on the part of individual farming entrepreneurs, DEFRA in actuality places central reliance on the new *publicly funded* Environmental Land Management Schemes (ELMS) to secure its net-zero vision, together with and through the mitigation of biodiversity and natural resource decline. While ELMS are legitimated according to the neoclassical doctrine of 'public goods' payments in the event of 'market failure' [13], and while disbursements are intended to be strictly delimited and competitive (with the possible exception of the Sustainable Farming Incentive (see below)), it is nonetheless the case that, in the absence of such public subvention and market intervention, there is virtually no prospect of achieving net-zero in the agri-food sector (see below).

The CCC, an exemplar of what we term 'quasi-hegemonic' discourse, is an independent, non-departmental body tasked with producing publicly available and independent expert advice on the government's efforts to meet its climate change targets [68]. Like reports from statutory environmental agencies such as Natural England, its published outputs in this area, notably *Land Use: Policies for a Net Zero UK* [41], are more technical and intended as objective analyses of potential pathways forward to ensure the government's environmental targets are secured. Further, the CCC's strategy of publishing various emissions reduction scenarios broadens its capacity to make proposals for more drastic change [41,68]. Given its formal independence, the CCC is able to recommend policy action further removed from norms of political conformity than a state department such as DEFRA. Consequently, CCC is able to deploy more of a 'climate emergency' discourse than would be feasible for DEFRA, for example, and this is indeed reflected in its unequivocal statements indicating the need for urgent and serious change in land use to meet net-zero: 'The UK's net-zero target will not be met without changes in how we use our land...Current policy measures will not deliver the required ambition...Throughout the UK there is an urgent need for a new approach: the legislative opportunities for real change are available and should progress immediately.' [41].

Despite this invocation of expeditious and meaningful policy action to curb climate change, however, 'the CCC...operates within the conjunctural orthodoxy and does little to challenge existing political "common sense" around the efficacy, rationality, and desirability of markets...' [68]. Like DEFRA, the CCC places huge faith in the market and technological innovation to render production and productivity improvements that will enable food to be produced with fewer GHG emissions and land to be released for designated carbon sequestration projects (afforestation, peatland creation, etc.) [41]. Consequently, we consider the CCC to remain in thrall to neoclassical and neoliberal doctrine despite its acknowledgement of the biophysical causes and impacts of climate change—hence our nomination of this discourse as 'quasi-hegemonic'. This thraldom in relation to neoclassical theory is exemplified by the following definitive statement of market optimism: 'Many of the environmental goods and services that land provides do not have a private market; their positive impacts are not priced and are under-supplied by the market. This has led to historic and ongoing degradation of land, soils, and water courses and loss of biodiversity.' [41]. Thus, according to neoclassical theory and the CCC, it is not capitalism and the capitalist market that generate environmental harm as a necessary part of their operation; it is in fact their very absence. The solution is not the abolition or restraining of capitalism, therefore, but rather the further privatisation and commodification of land and nature where these are 'excludable'—where exclusion (enclosure) is not feasible, as is supposedly the case with 'public goods' provision, then 'market failure' ensues. Such 'market failure' is then to be addressed through the provision of public subvention for 'public goods', in this case the new ELMS schemes. Like DEFRA, then, the CCC places central reliance on these schemes to deliver net-zero by 2050. Below, we examine the likely efficacy of ELMS in achieving this target and other biodiversity and natural resource benefits that are claimed by the Conservative government to flow from them.

These two discourses, 'hegemonic' and 'quasi-hegemonic' neoliberalism, comprise socio-technical imaginaries of net-zero futures through central reliance on an 'open future' of market and technological optimism born of a genuine ('quasi-hegemonic') or a rather more Machiavellian and class-centric ('hegemonic') belief in the 'truth' of neoclassical economic theory. Booth [68] suggests 'that this contemporary mode of discourse-oriented target governance does not in reality attempt to "make futures"... but only facilitate[s] the construction of future markets and market futures'. Again, there are certainly those, especially professionally trained neoclassical economists in the civil service, who appear genuinely to believe in the 'Promethean' qualities of capitalist rationality in the face of all challenges, climate change included ('quasi-hegemonic' discourse). It is nonetheless clear that the predominant and more 'Machiavellian', 'neoliberal' discourse of partisan class agents in parliament and agri-food capital is concerned primarily not to deliver the social change necessary drastically to reduce GHG emissions but rather, and as a 'flanking measure' [15], to create the illusion of government effort whilst simultaneously facilitating the 'permanence' [76] of current and increasingly 'disembedded' capitalist social–property relationships [12,68].

This lack of substantive effort to move meaningfully towards the net-zero by 2050 target is borne out by the woeful inadequacy of U.K. Nationally Determined Contributions (NDCs), which are intended to identify concrete rather than aspirational achievements and targets for GHG reduction by 2030. These NDCs currently fall far short, in terms of substantive policy actions, of the trajectory of GHG reductions that would be required to meet the 2050 aspiration, however. The NDC document (2020) identifies an ambitious target of sixty-eight percent reduction in GHG emissions by 2030 compared to 1990 levels, while a further target of seventy-eight percent reduction by 2035 compared to 1990 levels was announced in 2021 following recommendation from the CCC [77]. The NDC, however, provides virtually no substantive policy detail as to how this target might be secured. The section of the document entitled *Food Security and Policy* [78], for example, contains no detail other than to refer to the U.K.'s Agriculture Act 2020—this Act, however, simply embodies the empty techno-market optimism and heavy reliance on ELMS delineated above. Produced in conjunction with the NDC is the Adaptation Communication [79], designed to indicate how the U.K. will adapt and build resilience to current and future climate change. This document affords a little more policy detail than the NDC but does so simply by reiterating the heavy reliance on ELMS in agricultural policy to secure GHG emission reductions in the sector: 'A cornerstone feature of future agricultural policy, the new Environmental Land Management scheme, will provide a powerful vehicle for achieving the goals of the 25 Year Environment Plan and the commitment to net-zero emissions by 2050, while supporting our rural economy. Climate change adaptation and mitigation are core aims of the Environmental Land Management scheme.' [79].

Given this centrality of ELMS in the current U.K. government's neoliberal ('hegemonic' and 'quasi-hegemonic') discursive aspiration to secure in the agriculture sector the legally-binding reduction targets it has set, including for the mitigation of biodiversity and natural resource decline, we now turn to examine the likelihood of this eventuality. The U.K. Conservative government wishes to replace, through phased withdrawal, the inherited ('sub-hegemonic') support structures of the CAP with a neoliberal system that affords no direct support to farmers since 'production and trade-related' subvention is considered anathema to the 'free play' of market forces beloved of neoclassical theory [9,12]. Public subvention, as indicated, is thus to be confined to so-called 'public goods' payments since these are the supposed result of 'market failure' and therefore receive the imprimatur of neoclassical orthodoxy. In this way, it is proposed to restrict public subvention to ELMS, of which there will be three main components. Originally, these were proposed to comprise The Sustainable Farming Initiative (to support 'environmentally sustainable farming' across the landscape), Local Nature and Recovery (to support local environmental priorities and recovery), and Landscape Recovery (to support local environmental priorities and recovery, including 'rewilding') [80,81]. The first two schemes have now transmuted into the Sustainable Farming Incentive and Countryside Stewardship, while the third retains its original name [75]. All are supposed to contribute to meeting the net-zero target through actions and practices intended to reduce and sequester GHG emissions through and together with measures to mitigate biodiversity and natural resource decline.

These are to be 'supported' by a set of regulatory standards, the configuration of which remains as yet unclear [9]. ELMS are to be configured in a way that conforms to what Tilzcy [12] refers to as 'radical neoliberal' discourse, compatible with WTO 'green box' stipulations and as decoupled from agricultural production decisions as is feasible. As such, subvention will be voluntary and discretionary (competitive), meaning that, in contrast to current Pillar 1 direct payments, no 'automatic' entitlement to funding will be implied [81]. While the Sustainable Farming Incentive does appear to have features not wholly dissimilar to direct payments derived from CAP Pillar 1 (being a '"universal scheme", available to all farmers' [82]), thus representing something of a rowing back of radical neoliberalism under pressure from the likes of the NFU, the latter has nonetheless criticised the scheme as providing inadequate financial support and funding environmental actions at the expense of food production. In a response to the government's January 2024 changes to ELMS, the NFU indicated: 'With a minimum of 50% reduction in direct payments due in 2024, the tapering of payments to 2027 continues to be very concerning... It is imperative that the Sustainable Farming Incentive has sustainable food production at its core, with enough options that sit around productive farming. For this to happen it is absolutely vital that there's a better balance between policies that focus on enhancing food production as well as the environment' [83]. This statement clearly expresses the different emphases of 'hegemonic' neoliberal discourse—supporting the 'environment' through 'public good' payments whilst leaving food production (and food security/sustainability) to be determined by 'market forces'—and 'sub-hegemonic' market interventionist discourse—supporting the 'sustainable intensification' of productivist farming to secure national food security whilst subordinating environmental actions to this overriding objective. As such, both discourses a predicated on a 'land sparing' rather than a 'land sharing' approach.

Absent the financial 'cushion' afforded by current direct payments, and given the prevalently selective and competitive character of ELMS together with their focus on 'environment' rather than supporting 'food production', British farmers will find themselves competing against adverse pressures flowing from 'market productivism' embodied in the free trade agreements (FTAs) that the U.K. government is now committed to concluding with countries that often have significantly lower environmental and social standards than the U.K., and that will, therefore, exert greater downward pressure on prices, forcing farmers to further externalise ecological and social costs [9,20]. The FTA with Australia that came into force in 2023 is symptomatic of this trend, where adverse LUC (land clearance, soil degradation, and erosion) associated with 'cheap' sheep and cattle production represents considerable 'carbon leakage' [14,84]. Even more alarming are prospective FTAs with states such as Brazil, from which the U.K. already imports five percent of its beef, with huge implications for adverse LUC and, consequently, 'carbon leakage' [34].

Under the U.K. government's post-Brexit 'global Britain' scenario, therefore, enhanced competitive pressures will oblige farmers to accelerate 'market productivism' in an attempt to compensate for increased downward pressure on prices generated by cost externalising imports predicated on 'carbon leakage'. Indeed, the U.K. is already considering 'waivers' in relation to its climate change legislation to be written into FTAs. In the resulting competitive 'race to the bottom', the high opportunity costs of diverting land, investment, and management to GHG emission reductions and sequestration embodied in conservation farming or agroecology imply that the agri-environmental 'policy reach' of ELMS will be limited [9,14]. This will be the case particularly in respect of those farms described by DEFRA as 'very large' (the top twenty-five percent of farms) and those in the 'general cropping' (cereals and horticulture) and dairy sectors in which agricultural business activities are currently profitable and enterprises are not predominantly reliant on state subsidy in the form primarily if direct payments [85–87].

The new ELMS, unless endowed with very generous budgets, will, consequently, struggle to meet such opportunity costs on the approximately fifty percent of farmed area occupied by these farm categories. Throughout much of this area, therefore, in which farmers will be preoccupied with attaining further economies of scale in the face of the discontinuation of direct payments and enhanced exposure to overseas competition, the only means by which to secure compliance with environmental objectives and GHG emissions reduction targets will be through tighter regulation. For this very reason, however, such prospective regulations are currently the subject of intense contestation [9]. Under such downward price pressures, the motivation to rely increasingly on cost-externalising imports as 'cheaps' through Surplus Extraction 2, especially, will grow, and this will be particularly significant in the dairy, poultry, and pork production sectors heavily reliant on (LUC implicated) Latin American soya imports to secure profit margins. These same pressures will also induce further farm amalgamation and the perpetuation of production capitalisation, as machinery, including new robotics, and agrochemicals are substituted for human labour. While farmers will strive to reduce energy costs and consumption where feasible through 'ecological modernisation' and 'sustainable intensification', the underlying fossil fuel intensity of production due to high capitalisation and reliance on cost-externalising imports will thwart substantive progress towards GHG reduction targets in agriculture.

Alternatively, in those zones dominated by farms described by DEFRA [85–87] as 'mixed', 'lowland grazing livestock', and 'grazing livestock LFA' (less-favoured areas, generally indicating upland landscapes), predominantly located in the west and north of the U.K. and characterised by the majority of 'small' and 'medium' farms, businesses will struggle to survive the withdrawal of direct payments [31]. The commercial activities of farms in these categories (primarily sheep and beef cattle producers) are, on average, currently loss-making, and they remain solvent only thanks to state subsidy in the form of 'sub-hegemonic' direct payments and agri-environmental schemes [85–87]. Direct payments comprise the bulk of these payments, contributing up to one hundred percent of income in the case of many LFA farms, and their eventual withdrawal after 2028 spells the demise of many of these farms unless new ELMS subvention can make good the shortfall, an outcome that currently appears very unlikely [9] (despite the supposed 'universality' of the Sustainable Farm Incentive). This is so in part because ELMS payments are, in accordance with neoliberal and neoclassical dictates, calibrated to secure specific environmental outcomes that are deliberately decoupled from farm income considerations in order to minimise production and trade 'distortion' [81], predicated on the flawed notion that it is possible to dichotomise agricultural production from its negative or positive ecological impacts [9,12]. With future state disbursements via ELMS no longer designed to assist farm solvency, the pressures for smaller farms to sell up and for larger farms or other enterprises to absorb them will be compelling. Consequently, it seems unlikely that many of these farms will still be in existence to put into practice the Conservative government's much-vaunted ELMS. Farms in this pastoral sector have been characterised by the gradual process of extensification noted earlier [34], and their demise will lead, additionally, to the 'carbon leakage' of ruminant production to locations overseas, notably Australia and Brazil, where the GHG emission intensity of production, largely through ecologically devastating LUC, is much higher than in the U.K.

The result, overall, will be a dichotomous countryside, with agriculturally competitive (arable, general cropping, dairy, pig, and poultry) farms in the lowlands dominated by 'de-natured' market productivism and the (sheep, beef cattle) pastoral zones of the west and north reverting to 'de-socialised' 'wilderness' [56], converting to managed 're-wilding', or pursuing 'ranching-style' scale-economies. The former will intensify their dependence on the importation feed and inputs from overseas under increased competitive pressure facilitated by the FTAs currently scheduled for conclusion or negotiation with states, such as Brazil, happy to export agricultural commodities at huge ecological cost. The 'carbon leakage' of these enterprises is therefore likely to increase further. The solvency of these highly capitalised farm businesses will be thus ever more dependent on Surplus Extraction Mechanism 2, while their high GHG emissions intensity will rely, with the increasing loss of sequestration capacity in the U.K.'s intensively farmed landscapes, on the diminishing capacity degrading ecosystems overseas to continue to sequester carbon as the basis of Surplus Extraction Mechanism 3.

As for the pastoral zone of the west and north, potential gains in GHG sequestration through 'rewilding' programmes are likely to be seriously compromised by 'carbon leakage' overseas with the migration of beef and sheep production to high carbon emission locations such as Australia (sheep and beef) and Brazil (beef). 'Cheaper' beef and lamb for British consumers will be secured through the outsourcing of increased GHG emissions, with accumulation for the industry secured through Surplus Extraction Mechanism 2. In addition to failing to meet targets for GHG emission reductions in the agriculture sector, the current U.K. government's neoliberal land-use policies will, for the same reasons, also fail to meet its stated ambition for the restoration of farmland biodiversity and for the rehabilitation of soils degraded by decades of agrochemical applications. Moreover, the imperative of meeting food security needs consistent with these other objectives is wholly neglected. Consequently, questions concerning the co-production of food and biodiversity without recourse to fossil fuels whilst building sequestration capacity, together with the supply of and access to locally grown and nutritious food for all, are wholly antithetical to this radical neoliberal scenario espoused by the current U.K. government [1,9].

6.3. Alter-Hegemonic Discourse

Second, proponents of '**alter-hegemonic**' discourse, which perhaps may be seen in some respects to be part of the 'climate justice' narrative, advocate the putatively oppositional paradigm of 'post-productivism'. This asserts that the market power of 'corporate' food interests (and here their discourse typically constructs binary between overweening corporate power and the generalised interests of 'ordinary' citizens) can be countered by exploiting the turn by consumers away from industrial food provisioning in favour of quality, organic, local, and 're-territorialised' food production [4,14]. While 'alter-hegemonic' advocates do emphasise important facets of production that are key to ecologically and climate stabilising agriculture, their paradigm, with its reliance on 'reflexive consumerism', does little to confront market dependence, capitalist relationships of production, or the imperial mode of living. Thus, the turn to so-called 'economies of scope' and niche markets and therefore to dependency on middle-class consumption as the principal revenue stream for 'post-productivist' farmers is likely to afford only temporary respite from competitive pressures as more producers enter the field of quality production with the loss of direct subsidies [4,14]. Downward pressure on prices and capital concentration are likely outcomes of this process, while pivotal dependence on higher end income for consumption of quality produce, sustained by the imperial mode of living through the main drivers of the U.K. economy premised on Surplus Extraction 1, cast considerable doubt on both the longer-term viability and the environmental/social justice claims of this 'alternative' paradigm.

This means both that producers of 'quality food' remain highly market-dependent and subject to the pressures of capitalist competition and that, where there is a significant shift to 'post-productivism' while demand for cheaper wage foods remains undiminished, productivism, as the source of those wage foods, must be undergoing extra-territorial leakage [4,14]. Such leakage, as we have seen, is typically to the periphery either in the form of finished products (for example, beef and sheep) or in the form of feed and food ingredients for further productivist elaboration in the U.K. In other words, so long as it remains subject to capitalist relations of production, 'post-productivism' implies the existence of a 'spatio-temporal' fix that externalises the costs of productivism onto the periphery. This entails 'carbon leakage', the loss of biodiversity through damaging LUC, and the thwarting of food self-sufficiency in those peripheries subject to such ecological imperialism through Surplus Extraction 2. Such 'alter-hegemony' is thus a concomitant of continued reliance on globalised food supply [14]. As such, it fails to ask the key question posed by political agroecology and food sovereignty: How can the supply of food staples for general consumption rather than merely the supply of niche markets for higher income groups be undertaken primarily within the U.K. on an ecologically sustainable, climate stabilising, and socially equitable basis? By failing to pose this question, 'alter-hegemonic' 'post-productivism' remains parasitic upon an extractive frontier of market productivism in the periphery by token of its reluctance to problematise the wider imperial mode of living, of which it comprises a key legitimating element [4]. It is a legitimating element because, rather than challenging hegemonic neoliberalism, it *complements* it as a form of *environmental* neoliberalism through its reliance on market dependence and its advocacy of neoliberally configured ELMS.

6.4. Sub-Hegemonic Discourse

Third, advocates of **'sub-hegemonic'** or 'political productivist' discourse [19,20,22,39,68] comprise those farming constituencies which are likely to struggle to survive with the cessation of direct payments and/or currently commercially viable enterprises that stand to suffer attenuated profit margins with increased overseas competition arising from the conclusion of new FTAs. These constituencies are represented principally by the National Farmers Union (England and Wales) and NFUS (Scotland). Both bodies adopt an 'assured income' and 'neo-mercantilist' imaginary of the future of British farming and food, harking back to the heyday of post-war national developmentalism 11,15,19], and this permeates their discursive efforts to map a 'pathway' to net-zero by 2040 [68], a more ambitious target than that set by the U.K. government. The approach of both organisations emphasises the bolstering and 'greening' of national production by means of state-backed programmes, subsidies, and high standards that, in theory, prevent 'carbon leakage' to competitors overseas [68,88]. This discourse differs clearly from the neoliberal, 'hegemonic' vision of a transition to net-zero founded on the market rationality of individual farming entrepreneurs and sits more squarely within the 'climate justice' narrative of a technologically-driven but state-funded, 'green new deal' imaginary of the opposition Labour Party (and the Scottish National Party in Scotland).

For the NFU and NFUS (like DEFRA and the CCC), the solution to securing net-zero emissions in agriculture (and more land for biodiversity conservation) is the production of more food on less land by deploying improved yet 'sustainable' farming methods—in other words, 'sustainable intensification' or 'ecological modernisation'. This then permits more marginal land to be 'spared' for carbon sequestration through afforestation or biofuel crops or for biodiversity enhancement through 're-wilding' [68,88]. This discourse, then, as with DEFRA and the CCC, embodies a 'land sparing' rather than a 'land sharing' approach. But it is a *state-assisted*, intensified, expansionist, and technologically driven vision designed to feed the nation through capital-intensive though family-farm-based *domestic* production. The NFU and NFUS are confident, in their techno-optimism, that research will support the transition to low-carbon farming methods within an unchanged productivist configuration, with biotechnologies such as gene editing [89] being part of the innovation repertoire necessary to secure net-zero. Another facet of this, as with DEFRA's and the CCC's techno-optimism, is faith in the capacity of carbon capture and storage (CCS), with the NFU suggesting that, within its projected pathway to net-zero, the production of crops for bioenergy, carbon capture, and storage (BECCS) will be especially important, representing over half of predicted annual emissions reductions [88]. There is little hint here, however, of the considerable technical and environmental challenges and uncertainties surrounding the various forms of BECCS technologies [68]. Thus, rather than challenging productivism, it is techno-fix 'solutions' in the form, for example, of selective breeding programmes, crop genetic improvement, engineered feed additives, and slurry acidification (to reduce CH_4 and N_2O emissions) that will facilitate the transition to net-zero. Like DEFRA and the CCC, but this time with faith not in 'markets' but rather in state-supported investment, the NFU and the NFUS assert their confidence 'in the linear onward march of technological progress and the continued evolution of the agricultural metabolization of nature through science and innovation' [68].

Tilzey [4] has, elsewhere, referred to this 'sub-hegemonic' discourse as 'neo-productivism', a form of agrarian capitalism whose rationale is the need reliably to supply mass food consumption demand in the imperium at affordable prices and supported by government interventions to foster domestic production and efficient food supply systems but with the additional safeguard of neo-imperialist actions to secure the continuing flow of 'cheap' feed, 'flex-crop' ingredients, and energy from the global South. So, while there is a greater emphasis than with neoliberalism on the production of 'finished' food goods in the imperium and commensurate state support for a wider constituency of the domestic farming community, 'neo-productivism' is still reliant centrally on the importation of cheap feed and food ingredients from the periphery to undergird the national production of 'affordable' food items for the global Northern consumer. Despite the much-vaunted claims of 'ecological modernisation' described above, neo-productivism is likely to retain much of its GHG emissions intensity at home and will, with certainty, involve considerable 'carbon leakage' through its reliance, by means of Surplus Extraction 2, on the importation of feed/food ingredients and energy 'cheaps' from the periphery. As Tilzey [4] suggests, 'neo-productivism may partially address the first contradiction of capital as under-consumption crisis in the global North, but it cannot, in the new, ecologically constrained, conjuncture, do this without encountering the second contradiction, the impacts of which will be felt differentially in the South through the new wave of extractivism'.

All three discourses described above, namely neoliberal 'hegemonic' (and 'quasi-hegemonic' environmental neoliberalism), 'alter-hegemonic' 'post-productivism', and 'sub-hegemonic' neo-productivism, thus replicate, directly or indirectly, the imperial mode of living by continued reliance on 'cheaps' extracted from the periphery (Surplus

Extraction 1 and 2), consequent carbon leakage to these zones of super-exploitation, and differential dependence on peripheral sinks to absorb GHG emissions not sequestered in the U.K. (Surplus Extraction 3).

6.5. Counter-Hegemonic Discourse

Finally, we may identify what Borras et al. [71] define as 'structural transformation' narratives. We suggest that these comprise **'counter-hegemonic'** responses, such as 'radical' food sovereignty [4], espousing political agroecology as a means to secure multiple ecological and social objectives in *synergy* in the form of national food self-sufficiency, with equity, climate change stabilisation, biodiversity enhancement, and soil conservation being integral components of this strategy. Proponents of 'radical' food sovereignty or political agroecology recognise the need to address the structural foundations of capitalism in order to address the climate crisis as an integral element of the ecological precarity and social inequity wrought by capitalist food systems, whether these are articulated by 'hegemonic', 'sub-hegemonic', or indeed 'alter-hegemonic' discourses. Agroecological production can meet humanity's food needs while 'cooling the planet', they argue, so long as production is designed to meet socially determined and fundamental use value needs rather than the capitalist imperative of surplus value realisation through exchange value [4,39,89]. As noted earlier, the concept of 'market dependency' is pivotal here [4,9,15,39,90]. This concept not only considers the commodification of agricultural *inputs* to be essential to capitalist or market-dependent agriculture but also the compulsion to sell *outputs* into competitive markets in order to secure the economic reproduction of the producer. Market dependency focuses on *what* is produced by farmers, asserting that when producers rely on the sale of outputs into competitive markets, even when local and 're-territorialised' as per 'alter-hegemonic' advocacy, exchange value imperatives determine not only the methods of production but also the choice of food (or indeed non-food) commodities produced and who has access to them [39]. Such market imperatives impel a preoccupation with exchange value realisation rather than the satisfaction of social needs and ecological sustainability as determined through substantive and deliberative food democracy [9,17].

7. Outlining a Policy Framework for Agroecology, Food Sovereignty, and Degrowth in the U.K.

7.1. Transforming the Agri-Food System through Political Agroecology, 'Radical' Food Sovereignty, and Degrowth

'Radical' food sovereignty advocates identify an urgent need in the U.K. for a policy framework that strongly integrates, coordinates, and synergises farming, food, environment (including climate stabilisation), health, and social equity. This represents a key element of a programme of degrowth where this is defined as 'an equitable downscaling of production and consumption that will reduce societies' throughput of energy and raw materials... Degrowth signifies a society with a smaller metabolism, but more importantly, a society with a metabolism which has a different structure and serves new functions' [91] (p. 3). Here, this different structure is envisaged to be necessarily non-capitalist, abjuring market-dependency, imperial reliance on surplus extraction from overseas, and reversing primitive accumulation to re-connect people equitably with the fundamental means of production, most importantly land [10]. The new functions are democratic and equitable control of essential productive resources for the satisfaction of fundamental human needs in alignment with those of more-than-human nature. A sustainable social metabolism and ecological sustainability imply that social equity and human development, as measured by the Human Development Index (HDI) of the UNDP, need to be fulfilled at far lower levels of resource consumption than are currently characteristic of the U.K. and the global North in general. This need to fulfil HDI criteria (in other words, fundamental human needs) whilst keeping resource consumption and waste deposition to a minimum has been defined by the WWF [92]. In its *Living Planet Report*, the WWF indicates that the progress of states towards 'sustainable development' (or a sustainable social metabolism) can be assessed by using the UNDP's HDI as an indicator of human wellbeing and the ecological footprint of states as a measure of demand on the biosphere. The HDI is calculated from life expectancy, literacy and education, and per capita GDP. The UNDP considers an HDI value of more than 0.8 to be 'high human development'. Meanwhile, an ecological footprint lower than 1.8 global hectares per person, the average biocapacity available per person on the planet, could denote sustainability at the global level [92] (p. 19). Successful 'sustainable development' (sustainable social metabolism) requires that the world, on average, meets at a minimum these two criteria. Unfortunately, the global North achieves its generally high HDI only by imposing disproportionately large ecological footprint (10 global hectares in the case of the USA, only slightly less in the case of the EU and the U.K.) on the rest of the world, expressed in the imperial mode of living. A sustainable social metabolism in the U.K., as for the global North in general, would require, then, a drastic programme of degrowth through scaling back levels of resource and energy consumption and waste deposition perhaps by a factor of up to five [92].

Translating these desiderata for a sustainable social metabolism into a transformed agri-food system in the U.K. will mean, fundamentally, producing sufficient and nutritious food for all from domestic resources, importing, as a general rule, only 'non-indigenous' foods and founding such production on agroecological principles involving the cutting and, ideally, elimination of net GHG emissions, the sequestration of unavoidable GHG production by 'nature-based' means, and the conservation and enhancement of biodiversity, soils, and peatlands. The latter could be secured primarily through livestock production reduction and extensification, especially of sheep (currently numbering fifteen million in the U.K.), thereby releasing currently lost opportunities for woodland, sylvo-pastoral, and peatland conservation, expansion, and creation [93]. Concomitantly, production, distribution, and consumption of food require to be undertaken on a democratically defined basis that ensures the equitable and secure provision of healthy diets [9].[8] The basic parameters of this climate-friendly food system would comprise the elimination of grain-based meat production (for example, the cessation in the use of barley and wheat for animal feed), a proscription on the use of synthetic fertilisers and agrochemicals, the transition away from fossil fuel-based production, and the termination of imports of 'indigenous' produce and of livestock feed such as soya as part of a focus on food sovereignty through national self-sufficiency. This would, as we demonstrate below, require a significant shift from (especially grain-fed) meat diets towards vegetarianism.

7.2. Integrating ELMS and Agroecological Food Production

In this policy framework, ELMS payments would be integrated into support for agroecological production such that there would be co-production of food and agri-environmental benefits, including climate change mitigation. This 'land sharing' approach would be starkly different from the dichotomous 'land sparing' paradigm underpinning 'hegemonic' and 'sub-hegemonic' discourses. ELMS would therefore seamlessly align with support policy for agroecology since the two would be supporting entirely compatible rather than opposed agri-environmental and food policies. Under ELMS, within this agroecological, food sovereignty policy frame, farm management options would address three basic situations, from 'higher' to 'lower' tiers of ecological sensitivity: first, sensitive and irreplaceable sites, involving the maintenance and enhancement of semi-natural habitats; second, diversion/reversion involving the expansion and creation of semi-natural habitats; and third, agroecological production focused on the most fertile land (and least sensitive from a biodiversity perspective). Again, in stark contrast to the neoliberally configured ELMS of 'hegemonic' discourse, all farms delivering these benefits would, within a 'counter-hegemonic' policy frame, have an entitlement to an area payment, graduated according to tier, and subject to degressivity in the lowest tier for farms above a certain hectarage [9]. A strong regulatory baseline would prescribe statutory standards of land management and farming according to agroecological principles, including the proscription of synthetic fertilisers and agrochemicals.

In addition to these ELMS area payments, supporting the ecological and climate stabilisation dimensions of agroecology, stimulus to agroecological food production could be provided by the transformation and expansion of the current Basic Payment Scheme into a Basic Food Payment Scheme (or, more specifically, an Agroecological Area Payment Scheme)—all farms, including those under five hectares currently ineligible for BPS, would now qualify for this new payment, contingent upon an agroecological audit of the farm and accompanying recommendations for conversion to and optimal production of appropriate agroecological produce. Again, payment could be degressive for farms above a certain hectarage, although these proposals should be accompanied by a programme of land redistribution (see below). At least initially, agroecological production might receive an additional stimulus through guaranteed prices, with food then purchased by local/regional public authorities, thereby effectively severing capitalist market dependency and competition. As part of this new U.K. food policy framework, the social security system should include the provision of free, healthy, and nutritious food as a basic part of the welfare package—requiring recipients, where unemployed, to participate in socially and environmentally useful community work and training to facilitate productive participation in the national 'green transition'. Such free provision of agroecologically produced food would also apply to state-sector schools and to the National Health Service (NHS). Elimination of the food poverty and dietary inequalities detailed earlier in this paper should be part of the ambition of any responsible government, an ambition that should be part of a comprehensive national plan for a 'green new deal', including transition to agroecological production. Pending the provision of decent and rewarding livelihoods for all citizens as part of this transition, the alleviation of food poverty, insecurity, and dietary inequalities should be assisted by means of public food provision.

Given the increased labour and knowledge intensity of agroecological production and conservation management [93], there will be a need for a policy of voluntary but incentivised rural re-population, and a concomitant diminution in the size of landholdings both to encourage new entrants to farming and to reflect the more 'people-' and 'nature-centred' character of agroecology. This will necessitate a policy of land reform, proscribing ownership of land above certain size limits and redistributing the resulting surplus land to new entrants to farming. As noted, land proprietorship in the U.K. is currently extremely unequal, largely a legacy of unjust and undemocratic processes of primitive accumulation implemented by landlords and larger landholders between the sixteenth and nineteenth centuries [37,97]. For reasons of social justice but, more especially, for reasons of agroecological transition through re-peopling of the countryside, this inequality in land distribution demands redress.

7.3. Detailing a Sustainable Social Metabolism through Agroecological Production

How might agroecological production be configured to secure the real rather than aspirational elimination of GHG emissions, carbon sequestration, ecological sustainability, and self-sufficiency in 'indigenous' food production in the U.K.? Firstly, we need urgently to eliminate all grain-based livestock rearing and to confine livestock farming to pastureland where crops cannot be grown and that is free of synthetic fertilisers and agrochemicals. Secondly, neither conventional productivist nor 'rotational' organic production systems can generate the quantity of grain needed to supply U.K. consumption in a secure, climate stabilising, and ecologically sustainable way. This is true also of agroecological production where grain production is reliant on animal manures. This means essentially that arable and pasture must be rotated, implying, inter alia, that the potential for carbon sequestration on what would otherwise be permanent and extensive pasture is compromised, compounded by the adverse impacts of ploughing on soil biota (see [9,93,98] for further detail). Part of Poux and Schiavo's solution is to reduce U.K. consumption of cereals by some forty-five percent, permitting, they claim, some eleven percent of the U.K. to be devoted largely to carbon sequestration adequate to meet the U.K.'s GHG emission reduction commitments. The reduction of human grain consumption by this amount is likely to prove very challenging, however. A potential solution is to grow cereals in a way that does not rely on animal manures as does the modelling of Poux and Schiavo [93]. Grain can in fact be grown in an agroecologically based way that increases output whilst minimising or eliminating fossil fuel usage, enhancing biodiversity, *and* sequestering carbon on a greater scale than envisaged by Poux and Schiavo. This solution addresses the two main contradictions of 'rotational' organic production—the need for high soil fertility levels, requiring rotation with livestock systems to achieve these, and the use of modern grain varieties, particularly wheat, that require these high nutrient levels and have short stems, needing frequent rotation and tillage to control weeds. These modern wheat varieties, bred to respond to fossil fuel-based synthetic fertilisers, do not grow well in low-input agroecological systems [98,99].[9]

One of the main contradictions of 'rotational' organic systems, then, is the need for ploughing or tillage. However, 'heavy and frequent tillage negatively affects a soil's physical and biological properties and is probably the most important reason for decreases in soil structural quality...Tillage may also decrease soil organic matter, which may be further reduced by rising temperatures' [102] (p. 1440). Moreover, 'minimum tillage can improve soil structure and stability, resulting in better drainage and water-holding capacity, as well as enhancing microbial activity... These practices also reduce losses of soil organic matter and thus carbon losses, while improving soil structure and water retention and enabling more permanent soil cover. There is much potential for reduced tillage to mitigate GHG emissions...' [102] (p. 1440). Other research has demonstrated how conventional tillage decreases the abundance and biomass of earthworms, with severe knock-on implications for soil structure, drainage, the recycling of organic matter into the soil, and crop production [103,104].

Tillage is thus often needed to control weeds in organic farming. Many of these biotic problems could be resolved, however, through diversification strategies such as cultivar and species mixtures to reduce infection and spread of diseases and through plant traits which confer a high level of crop competitive ability against weeds [102,105]. In addition to disease, insect, and weed control and consequently reduced or eliminated pesticide inputs, nutrient conservation, soil fertility building through increased organic matter, and enhanced yield stability are some of the ecosystem services inherently conjoined to sustainable cereal production that can be secured by crop diversification. The introduction of crop variation over time and space stabilises these systems and includes growing heterogeneous varieties that can adapt to local and changing environments, extending from the landscape to the field scale (the latter using populations or mixtures of varieties within a field). 'In systems with more variable climate and reduced external inputs, crops will need to be able to cope with spatially and temporally more heterogeneous environmental conditions. Plant breeding will have to provide varieties that are adapted to these new needs in diversified agricultural systems, which will need innovative approaches. The requirements for such varieties are enormous, as they have to combine high yield with high levels of resistance and tolerance to pests and diseases, competitiveness with weeds, and an improved stand establishment with efficient use of nutrients, water, and light. *As new characteristics are needed, breeding will have to rely on the intensive use of genetic resources (landraces, exotic, and wild resources)*' (emphasis added) [99,102,106] (p. 1441).

Many of these required traits are based on a range of genes, that is, *polygenic* inheritance, rather than on single genes, that is, *monogenic* inheritance, and are thus greatly influenced by the environment, requiring phenotypic selection. In order to produce this required polygenic inheritance, one strategy has comprised the creation of diversity through the breeding of selected varieties of modern wheat for the above desired characteristics. This technique has been pioneered by Martin Wolfe and colleagues [99]. An alternative strategy, building on the research and recommendations of Wolfe, Ostergard, and colleagues, as above [99,106], is to use landraces or 'heritage' grains, drawing on the multitude of different wheat and other cereal varieties that, until the recent past, characterised the British and European landscapes, each adapted to local soil and climatic conditions. This has the large advantage of drawing upon *existing* (or very recently existing) polygenic inheritance, thus perpetuating or recreating agri-biodiversity, local adaptability, and resilience in the context of specific soil, biotic, and climatic conditions.

This strategy appears to be the optimum agroecological solution to the contradictions of both conventional and rotational organic cereal production. Landraces or 'heritage' grains have the great advantage of having taller stems to outcompete weeds, have higher nutritional value, and because of the need to avoid lodging (falling over due to heavy seed head in modern varieties) have lower nutrient demands than modern varieties [98,107].[10] The key here is to grow genetically diverse 'heritage' grains in the same fields, continuously, without animal manure or tillage, following a low-input approach known as continuous grain cropping (CGC) (also known as 'natural grain farming' or 'restorative continuous cropping') (see [98,107]). These cereals can be grown in this way as long as the crops are genetically diverse, have tall stems to help outcompete weeds, and all the stems are left in the field post harvest. The nitrogen removed with the grain each year is replaced by nitrogen fallout from the atmosphere, by the mineralisation of plant tissues above and below ground, and by the fixation of nitrogen by an under-sown layer of clover. Moreover, these varieties have much deeper root systems than modern varieties, enabling them to extract moisture from depth and develop, given zero tillage, complex associations with mycorrhizal fungi, greatly enhancing nutrient uptake in lower fertility soils [108]. These traits in turn confer much greater resilience in the face of weather extremes [98,107]. In short, CGC systems have the potential to greatly reduce or eliminate GHG emissions through zero application of synthetic fertiliser/agrochemicals, zero tillage, zero requirement for animal manure, and zero net oxidation of soil carbon while massively increasing sequestration through building up soil organic carbon with incorporation of cereal stems, clover, and weeds through zero tillage and through the development of carbon-rich mycorrhizal associations.[11]

CGC production yields about 2.5–3.0 t/ha even on fairly poor soils [98,107].[12] While it is necessary to note that these results, so far as it is possible to ascertain, have not as yet been published in a peer-reviewed journal despite their basis in long-term experimentation and that, therefore, further independent validation is required to confirm the long-term viability of the CGC system, they do suggest that current national demand could be met from approximately two million hectares of land (current field crop hectarage in the U.K., mostly cereals, is over six million, but a large percentage goes to feed animals and the crop land also needs to be rotated).[13] If diets were to become increasingly vegetarian/vegan, the area of CGC would need to expand further but would still be less than the current field crop hectarage. If, on a reasonable assumption, the U.K. could supply increased national demand, on the basis of increasingly vegetarian diets, from around five million hectares of land (this area of cropped land would include the agroecological production of the full range 'indigenous' crops, including those such as pulses and tubers, required for a varied and healthy human diet), this would still leave some twelve million hectares for alternative production (total farmed area in the U.K. is 17.6 million hectares [31]), including the production of extensively and agroecologically reared livestock/poultry, and greatly expanded provision for carbon sequestration. In this way, the remaining area of non-cultivated land could be devoted to extensive, grass-based livestock/dairy and other multifunctional uses involving carbon sequestration through agro-forestry and 're-wilding' [9]. The emphasis on the latter should be upon increasing the area of native woodland to considerably enhance carbon sequestration in order to meet the statutory net zero GHG ambition [32]. Under the proposal presented here, however, this contribution to net zero does not exclude the considerable contribution to GHG reduction and sequestration performed by areas devoted primarily to food production, both 'infield' through techniques such as CGC and also 'field edge', through the contribution of natural features such as hedgerows [9]. This is the essence of a land sharing approach, seeking to secure sustainability across the landscape, even if there are differences in emphasis between areas of high soil fertility on the one hand and areas of agriculturally marginal land more suited to 're-wilding' on the other.

Such a programme of 'counter-hegemony' through 'radical' food sovereignty, agroecology, and degrowth would, by securing net-zero GHG emissions, carbon sequestration, and food self-sufficiency, effectively eliminate, at least in the agriculture sector, the U.K.'s reliance upon the three surplus extractive mechanisms detailed earlier. It would, in short, eliminate the U.K.'s parasitic imperial mode of living. This would, then, enable the U.K. to secure effective self-sufficiency in indigenous food whilst simultaneously securing food equity, landscape-scale biodiversity, and biophysical resource conservation through agroecological land sharing and the realistic and expeditious rather than merely aspirational attainment of the net-zero GHG emission target. In short, this would enable the U.K. to secure a sustainable social metabolism, at least so far as the agri-food system is concerned.

7.4. Politico-Economic and Ideological Constraints Imposed by the 'Imperial Mode of Living'

Sadly, however, political awareness of, and political commitment to, such a programme of radical but necessary and *feasible* (not merely rhetorical) transformation is virtually non-existent in the U.K. largely, and ironically, due to the operation of the imperial mode of living itself. Thus, within the U.K. farming community, for example, such a programme of radical transformation would be likely to find support currently only amongst a very few interest groups such as the Land Workers' Alliance, affiliated to *La Via Campesina*, membership of which is numbered only in the hundreds. The unfortunate reality appears to be that the legitimacy and material basis of the imperial mode of living will likely require to be compromised *before* such an agroecological transition can occur. In other words, the necessary agroecological transition would seem to require the prior fracturing or at least severe attenuation of the imperial mode of living before it receives the necessary groundswell of political support. Were the contradictions of neoliberalism to continue to erode the livelihoods of the small and medium farm constituencies, compounded by the acceleration of the climate emergency, such current supporters of sub-hegemony and particularly of alter-hegemony could perhaps be persuaded of the merits of a political agroecological transition, with the proviso that such a transition were adequately funded. The greatest opposition seems likely to derive from the larger farm constituency and, given the pervasiveness of 'propertisation' [114], especially directed to any notion of land redistribution. Indeed, if developments in Germany, the Netherlands, and elsewhere are any guide, threats to absolute property rights, whether such rights are real or 'phantom' [114], appear to be leading to a shift in political affiliation to the right. The popularity of Brexit amongst the farming community (at least initially until its radical neoliberal intent became clearer) suggests that this rightward shift is also present in the U.K.

While the mounting impacts of austerity and now climate change are beginning, belatedly, to fracture neoliberal hegemony amongst the wider population, especially amongst a younger demographic, it seems likely that the bulk of popular opposition will be directed and co-opted into various forms of sub-hegemony, seeking to restore incomes, consumerism, and growth through a slightly more interventionist form of 'green' and 'redistributive' capitalism. While this may alleviate a degree of poverty amongst those who have suffered under austerity and make incipient moves to reduce fossil carbon dependency, it does very little to set the U.K. on a path of real sustainability, let alone of degrowth. Even here, the reluctance of the prospective new incumbents of government (The Labour Party) appear very unwilling to pursue any agenda that could be construed as anything more than mildly reformist. More likely, for the time being, then, is maintenance of something approximating the 'hegemonic' status quo in the field of agri-food, climate change, and ecological sustainability—through 'techno-fix' programmes of 'ecological modernisation', supplemented by 'creative carbon accounting'—than the likelihood of meeting the net-zero target recedes, whereby green-washed 'spatio-temporal' fixes are deployed, deferring the necessary actions and costs that the U.K. needs to take or absorb. For the farmed environment, this will likely involve a continued focus on 'infield' 'techno-fix' programmes, accompanied by 'field edge' or marginal land environmental initiatives supported by ELMS, perpetuating the dominant land sparing approach in which infield productivism is juxtaposed to 're-wilding'. This will serve to thwart the imperative for an integrated land sharing approach, one which synthesises the need to address simultaneously climate change, ecological sustainability, and food insecurity through a political agroecological and food sovereign transition.

8. Conclusions

By reference to empirical indices, this paper has laid out at length the essential unsustainability of the U.K. food system along the dimensions of climate change, biodiversity loss and decline, and food (in)security. It has also identified the causality underlying these indices of unsustainability, pointing to the prevalence of capitalism and especially neoliberalism in generating the multiple contradictions of the U.K. food system, including its adverse impacts on countries overseas through reliance on cost-externalising 'cheap' global supply. The paper has also shown how dominant 'hegemonic' and 'sub-hegemonic' politico-economic interests and their accompanying discourses have both generated these contradictions and shaped mitigatory responses to them, predominantly as symptom management. Since both are wedded to productivism and continued economic growth, it seems improbable, despite attempts—real or rhetorical—to decarbonise continued capital accumulation, that any real strides will be made in securing *integrated* solutions to the social and ecological contradictions of the U.K. food system. Rather, responses will be piecemeal, mitigatory (symptom management), and reliant on continued cost-externalisation overseas. Oppositional discourses are available, however. The first, a discourse we have nominated as 'alter-hegemony', strongly advocates a land sharing approach by integrating farming, biodiversity, carbon sequestration, and localised production-consumption relations at the territorial level—a form of bioregionalism [1]. As such, however, the problem is seen to be primarily one of scale rather than an issue of capitalist relations of production—inequalities and private social-property relationships are as entrenched at the local level as much as they are nationally and internationally. Failure to address social inequity 'locally' will lead to the persistence of food poverty and dietary inequalities, and ignoring the enduring presence of capitalist social property relationships will generate unavoidable pressures for continued growth and competition. It is salutary to recall that capitalism began 'locally' within the context of unequal class relationships (see [18]).

Given these shortcomings of 'alter-hegemony', we have argued that a second oppositional discourse, 'counter-hegemony', alone offers an integrated approach to simultaneously resolving the problems of food (in)security and social inequity on the one hand and ecological sustainability (subsuming climate change stabilisation) on the other. Such 'counter-hegemony', embodied in 'radical' food sovereignty, political agroecology, and degrowth, proposes the abrogation of capitalist social-property relationships. This entails the supersession of abstract capitalist market dependency (the rule of the market as an impersonal force) by means of concrete democratically (politically) determined systems of localised governance, overseeing equality of access to the means of food production and to the fruits of that production. This requires a rediscovery of political agency, solidarity, mutuality, and ways of nurturing our humanity by respecting non-human nature.

Funding: Research for sections of this paper was supported by Research England QR Strategic Priorities Fund.

Data Availability Statement: The raw data supporting the conclusions of this article will be made available by the authors on request. Data is publicly available due to privacy reason.

Conflicts of Interest: The author declares no conflicts of interest in this paper.

Notes

[1] By capitalist here we also mean market-dependent family farms, even though these may not employ off-farm labour. The peasantry (self-subsistent and semi-self-subsistent agrarian producers) had effectively disappeared from Britain by the mid-19th century (see [4] and [18] for more detail on the rise of agrarian capitalism in Britain).

Appropriationism and substitutionism refer to the undermining of discrete element of the agricultural production process, their transformation into industrial activities, and their re-incorporation into agriculture as inputs, for example, human labour by machinery, animal traction by the tractor, manure by synthetic fertilisers, etc. [23].

The 'imperial mode of living' refers to the normalisation of affluence, growth, and high levels of resource consumption characteristic of the global North (the imperium), predicated significantly upon the ideologically 'invisible' exploitation of the global South.

The authors (WWF and RSPB) assigned a risk score to each U.K. sourcing country based on their deforestation/conversion rates, labour rights, and rule of law indices. Scores varied from 0–12, with 11 or above being 'very high' risk and 9–10 being 'high' risk.

The uncompensated appropriation of land and resources by capital for wealth accumulation, involving the wholesale removal of the original inhabitants without absorption as labour into the subsequent agro-industrial or mining developments.

Key farmland and ground-nesting bird species, such as skylark, lapwing, and stone curlew, require no or very low vegetation when incubating eggs in order to see and avoid predators—autumn sown cereals are already too high in early spring to enable these species to incubate safely. Autumn-sown cereals are bred to respond to synthetic fertilisers and put on growth very quickly; traditional or 'landrace' cereals, even when germinating in the autumn, do not produce significant growth until the next spring, especially when grown in organic and no-till management systems—they may produce less per area than modern cereals grown with agrochemicals, but they can produce indefinitely and sustainably with no artificial inputs and generate no negative ecological externalities.

Despite the fine words expressed in DEFRA's Agricultural Transition Plan update of January 2024 and the improved payment offers and increased coverage/flexibility of the ELM schemes detailed therein, the essential principles of land sparing 'public goods' payments, embodying a dichotomy between productivist farming on the one hand and biodiversity conservation on the other, remain in place.

The characteristics of a diet consistent with public health, climate change stabilisation, and low environmental impact are already quite clear [94–96]. This is a diet that provides diversity, with a wide variety of foods consumed; achieves balance between energy intake and energy needs; is centred around minimally processed whole grains, tubers, and legumes, fruits, and vegetables; has moderate/small amounts of meat, dairy, unsalted seeds and nuts; has small quantities of fish from certified fisheries; has oils and fats with a beneficial omega 3:6 ratio such as rapeseed and olive oil; and is very limited consumption of foods high in fat, salt, and sugar and low in micronutrients.

A recent paper [100] appears at first sight to contradict this statement. Closer examination, however, shows this not to be the case. The study on which the paper is based only tests varying levels of agrochemical inputs on modern wheat varieties, with the 'lowest input' still at 110 kgNha^{-1}. This, however, is not 'low input' from an agroecological perspective, where the expectation is that *no* agrochemicals (or, more specifically, synthetic fertilisers) are employed. The modern varieties tested in this study would certainly not thrive under a zero-agrochemical regime. Moreover, the application of N at the lowest rates in the study would still prove toxic to most non-target plant species in the field or field edge (the great majority or wild plant species find even very low levels of N application toxic [101]), and this applies also to the soil biome—agroecology seeks to maximise the vigour of this soil biome by refraining from agrochemical use to support sustainable, resilient soils and hence sustainable and resilient cultivar production.

Although the work of John Letts as an academic archaeobotanist has been widely published in peer-reviewed journals, his long-standing experimental work with 'heritage' grains and CGC has, so far as it is possible to ascertain, not yet been similarly published (although it has been published in non-peer reviewed publications as per the 'Land' citation in the present paper). However, the agroecological foundations for his fieldwork and conclusions from it are supported by peer-reviewed research (see above), and his work has been funded through the EU Horizon 2020 Research and Innovation Programme under Grant Agreement No. 727848 and is summarised in the following link entitled 'Low input and organic heritage cereal production in South East England': http://cerere2020.eu/wp-content/uploads/2020/03/17_EN.pdf (accessed on 15 December 2023). Similar experimental fieldwork and findings have been undertaken and generated in the USA by Rogosa (funded by the USDA Sustainable Agriculture Research and Education Program), where einkorn, emmer, and other landrace wheats outperform modern wheats under organic conditions (that is, where synthetic fertilisers and pesticides are not applied) [107] (p.4).

It may be asked how CGC and agroecology are connected to related (or putatively related) production techniques such as conservation agriculture (CA) and circular agronomy. Concerning CA, this, according to the FAO's definition [109], is a farming system that can prevent losses of arable land while regenerating degraded lands. It promotes the maintenance of a permanent soil cover, minimum soil disturbance, and diversification of plant species. It enhances biodiversity and natural biological processes above and below the ground surface that contribute to increased water and nutrient-use efficiency and to improved and sustained crop production. CA principles are universally applicable to all agricultural landscapes and land uses with locally adapted practices. Soil interventions such as mechanical soil disturbance are reduced to an absolute minimum or avoided, and external inputs such as agrochemicals and plant nutrients of mineral or organic origin are applied optimally and in ways and quantities that do not interfere with or disrupt the biological processes. This definition is virtually identical to CGC and agroecology with the exception that these avoid agrochemicals altogether since they recognise the damage that agrochemicals cause to soil, soil biota, and non-target field and field edge plant species, thus compromising the underlying rationale of conservation agriculture itself.'Circular agronomy', for its part, aims to close nutrient cycles in the agri-food chain, aiming to improve the current carbon,

¹²   nitrogen, and phosphorus cycling in agro-ecosystems and related up- and downstream processes within the value chain of food production [110]. This, however, seems to be part of an 'ecological modernisation' agenda tied to capitalist productivism. Agroecology is based centrally on such circularity, of course, but without recourse to synthetic fertilisers or mineral supplements that generate major problems for the environment, soil health, and the longer-term sustainability of food production itself.

¹²   It may appear that these figures are contradicted by the long-term wheat yield trials at Rothamsted Experimental Station in the U.K. These trials show a yield of about 1 tonne/ha under continuous wheat cropping and 2 tonnes/ha when wheat is grown in rotation [111]. The results of CGC and Rothamsted are not directly comparable, however. This is because (a) the CGC method is no till, while the Rothamsted plots are ploughed annually; (b) CGC does rely (in part) on chopped straw and clover to retain fertility levels, so this is not directly comparable to the continuous cropping without fertiliser undertaken at Rothamsted. In other words, the 1 tonne/ha yield at Rothamsted is based on continuous cropping of wheat without any fertiliser application, which is not the same as CGC. A more meaningful comparison with CGC would be the continuous cropping with farmyard manure (FYM) application trial at Rothamsted, which demonstrates yields between 2 and 3 tonnes/ha for most of the experimental period (rising up to 6tonnes/ha after 1970 with change in wheat variety). But, as pointed out above, FYM relies on livestock which means diverting considerable areas of land to livestock production to retain the fertility of cropped areas.

¹³   In fact, 15 million tonnes of wheat are produced annually in the U.K., but only c. 5 million tonnes are milled to produce flour for human consumption—two-thirds of wheat produced is fed to animals. Animal feed grains are not suitable for flour milling, however. As argued above, all cereal production should be directed to human, not to livestock, consumption. However, this cereal should be produced on an agroecological basis without recourse to agrochemicals, synthetic fertilisers, or to livestock to provide the FYM for organic rotations. As argued above, this shift to non-rotational agroecological production is both necessary and feasible. In addition to the multiple disbenefits of conventional wheat production identified above, it also needs to be pointed out that modern varieties of wheat and conventionally milled wheat flour (through the Chorleywood method), together with the standard addition of sugar and other additives to bread so manufactured, has important negative health and nutritional impacts [112,113].

References

1. Lang, T. *Feeding Britain: Our Food Problems and How to Fix Them*; Pelican Books: London, UK, 2020.
2. Bhaskar, R. *Dialectic: The Pulse of Freedom*; Verso: London, UK, 1993.
3. Ollman, B. *Dance of the Dialectic: Steps in Marx's Method*; University of Illinois Press: Urbana, IL, USA, 2003.
4. Tilzey, M. *Political Ecology, Food Regimes, and Food Sovereignty: Crisis, Resistance, and Resilience*; Palgrave Macmillan: London, UK, 2018.
5. Kallis, G.; Paulson, S.; D'Alisa, G.; Demaria, F. *The Case for Degrowth*; Polity Press: Cambridge, UK, 2020.
6. Gerber, J.-F. Degrowth and Critical Agrarian Studies. *J. Peasant Stud.* **2020**, *47*, 235–264. [CrossRef]
7. De Molina, M.G.; Petersen, P.F.; Pena, F.G.; Caporal, F.R. *Political Agroecology: Advancing the Transition to Sustainable Food Systems*; CRC Press: New York, NY, USA, 2020.
8. Guerrero Lara, L.; van Oers, L.; Smessaert, J.; Spanier, J.; Raj, G.; Feola, G. Degrowth and Agri-Food Systems: A Research Agenda for the Critical Social Sciences. *Sustain. Sci.* **2023**, *18*, 1579–1594. [CrossRef]
9. Tilzey, M. The Political Ecology of Hedgerows and Their Relationship to Agroecology and Food Sovereignty in the UK. *Front. Sustain. Food Syst.* **2021**, *5*, 752293. [CrossRef]
10. Tilzey, M.; Sugden, F. *Peasants, Capitalism, and Imperialism in an Age of Politico-Ecological Crisis*; Routledge: London, UK, 2023.
11. Potter, C.; Tilzey, M. Agricultural Policy Discourses in the Post-Fordist Transition: Neoliberalism, Neomercantilism, and Multifunctionality. *Prog. Hum. Geogr.* **2005**, *29*, 581–601. [CrossRef]
12. Tilzey, M. Neoliberalism, the WTO, and New Modes of Agri-Environmental Governance in the EU, USA, Australia. *Int. J. Sociol. Agric. Food* **2006**, *14*, 1–28. [CrossRef]
13. Tilzey, M.; Potter, C. Neoliberalism, Neo-Mercantilism, and Multifunctionality: Contested Political Discourses in European Post-Fordist Rural Governance. In *International Perspectives on Rural Governance: New Power Relations in Rural Economies and Societies*; Cheshire, L., Higgins, V., Lawrence, G., Eds.; Routledge: Abingdon, UK, 2007; pp. 115–129.
14. Tilzey, M.; Potter, C. Productivism versus Post-Productivism? Modes of Agri-Environmental Governance in Post-Fordist Agricultural Transitions. In *Sustainable Rural Systems: Sustainable Agriculture and Rural Communities*; Robinson, G., Ed.; Routledge: London, UK, 2016; pp. 41–63.
15. Tilzey, M. Reintegrating Economy, Society, and Environment for Cooperative Futures: Polanyi, Marx, and Food Sovereignty. *J. Rural Stud.* **2017**, *53*, 317–334. [CrossRef]
16. Tilzey, M. 'Market Civilization' and Global Agri-Food: Understanding their Dynamics and (In)Coherence through Multiple Resistances. In *Resistance to the Neoliberal Agri-Food Regime: A Critical Analysis*; Bonanno, A., Wolf, S.A., Eds.; Routledge: London, UK, 2018; pp. 64–77.
17. Tilzey, M. Food Regimes, State, Capital, and Class: Friedmann and McMichael Revisited. *Sociol. Rural.* **2019**, *59*, 230–254. [CrossRef]
18. Tilzey, M.; Sugden, F.; Seddon, D. *Peasants, Capitalism, and the Work of Eric R. Wolf: Reviving Critical Agrarian Studies*; Routledge: London, UK, 2023.

19. Tilzey, M. Capitalism, imperialism, nationalism: Agrarian dynamics and resistance as radical food sovereignty. *Can. J. Dev. Stud.* **2020**, *41*, 381–398. [CrossRef]
20. Tilzey, M. Natural Areas, the Whole Countryside Approach, and Sustainable Agriculture. *Land Use Policy* **2000**, *17*, 279–294. [CrossRef]
21. Tilzey, M. Peasant Counter-Hegemony Towards Post-Capitalist Food Sovereignty: Facing Rural and Urban Precarity. In *Resourcing an Agroecological Urbanism: Political, Transformational, and Territorial Dimensions*; Tornaghi, C., Dehaene, M., Eds.; Routledge: London, UK, 2021; pp. 202–219.
22. Tilzey, M. From Neoliberalism to National Developmentalism? Contested Agrarian Imaginaries of a Post-Neoliberal Future for Food and Farming. *J. Agrar. Chang.* **2021**, *21*, 180–201. [CrossRef]
23. Goodman, D.; Sorj, B.; Wilkinson, J. *From Farming to Biotechnology: A Theory of Agro-Industrial Development*; Blackwell: Oxford, UK, 1987.
24. Gerber, J.-F. The Role of Indebtedness in the Evolution of Capitalism. *J. Peasant Stud.* **2014**, *41*, 729–747. [CrossRef]
25. Brand, U.; Wissen, M. *The Imperial Mode of Living: Everyday Life and the Ecological Crisis of Capitalism*; Verso: London, UK, 2021.
26. Tilzey, M. Authoritarian populism and neo-extractivism in Bolivia and Ecuador: The unresolved agrarian question and the prospects for food sovereignty as counter-hegemony. *J. Peasant Stud.* **2019**, *46*, 626–652. [CrossRef]
27. Bonanno, A.; Wolf, S.A. (Eds.) *Resistance to the Neoliberal Agri-Food Regime: A Critical Analysis*; Routledge: London, UK, 2018.
28. Otero, G. *The Neoliberal Diet: Healthy Profits, Unhealthy People*; University of Texas Press: Austin, TX, USA, 2018.
29. Sheingate, A. *The Rise of Agricultural Welfare State: Institutions and Interest Group Power in the United States, France, and Japan*; Princeton University Press: Princeton, NJ, USA, 2001.
30. Audsley, E.; Brander, M.; Chatterton, J.; Murphy-Bokern, D.; Webster, C.; Williams, A. *How Low Can We Go? An Assessment of Greenhouse Gas Emissions from the UK Food System and the Scope to Reduce Them by 2050*; FCRN-WWF-UK: Woking, UK, 2009. Available online: https://assets.wwf.org.uk/downloads/how_low_report_1.pdf (accessed on 22 May 2022).
31. National Food Strategy. The National Food Strategy: The Plan—July 2021. 2021. Available online: https://www.nationalfoodstrategy.org (accessed on 20 May 2022).
32. Harwatt, H.; Hayek, M.N. *Eating away at Climate Change with Negative Emissions: Repurposing UK Agricultural Land to Meet Climate Goals*; Harvard Law School: Cambridge, MA, USA, 2019. Available online: https://animal.law.harvard.edu/wp-content/uploads/Eating-Away-at-Climate-Change-with-Negative-Emissions%E2%80%93%E2%80%93Harwatt-Hayek.pdf (accessed on 3 September 2023).
33. Fan, J.; Rosenfeld, D.; Zhang, Y.; Giangrande, S.E.; Li, Z.; Machado, L.A.T.; Martin, S.T.; Yang, Y.; Wang, J.; Artaxo, P.; et al. Substantial Convection and Precipitation Enhancements by Ultrafine Aerosol Particles. *Science* **2018**, *359*, 411–418. [CrossRef]
34. De Ruiter, H.; Macdiarmid, J.; Matthews, R.; Kastner, T.; Lynd, L.; Smith, P. Total Global Agricultural Land Footprint Associated with UK Food Supply 1986-2011. *Glob. Environ. Chang.* **2017**, *43*, 72–81. [CrossRef]
35. World Wide Fund for Nature (WWF); Royal Society for the Protection of Birds (RSPB). Riskier Business: The UK's Overseas LandFootprint. 2020. Available online: https://www.wwf.org.uk/riskybusiness (accessed on 15 December 2023).
36. UK Roundtable on Sustainable Soya. Annual Progress Report, 2020. 2020. Available online: https://www.efeca.com/wp-content/uploads/2020/10/UK-RT-on-Sustainable-Soya-APR-19_20-final.pdf (accessed on 20 May 2022).
37. Hetherington, P. *Whose Land Is Our Land? The Use and Abuse of Britain's Forgotten Acres*; Policy Press: Bristol, UK, 2015.
38. Shrubsole, G. *Who Owns England? How We Lost Our Green and Pleasant Land and How to Take It Back*; William Collins: London, UK, 2019.
39. Wach, E. Market Dependency as Prohibitive of Agroecology and Food Sovereignty—A Case Study of the Agrarian Transition in the Scottish Highlands. *Sustainability* **2021**, *13*, 1927. [CrossRef]
40. DEFRA. *June Agricultural Census*; Department for Environment, Food and Rural Affairs: London, UK, 2017. Available online: https://assets.publishing.service.gov.uk/government/uploads/system/uploads/attachment_data/file/670004/structure-jun2017final-uk-21dec17.pdf (accessed on 19 May 2022).
41. Climate Change Committee. *Land Use: Policies for a Net Zero UK*; Climate Change Committee: London, UK, 2020. Available online: www.theccc.org.uk (accessed on 19 May 2022).
42. AHDB. *Brexit Prospects for UK Cereals and Oilseeds Trade*; Agriculture and Horticulture Development Board: Kenilworth, UK, 2019.
43. Poore, J.; Nemecek, T. Reducing Foods Environmental Impacts through Producers and Consumers. *Science* **2018**, *360*, 987–992. [CrossRef]
44. Higginbottom, A. 'Imperialist Rent' in Practice and Theory. *Globalizations* **2014**, *11*, 23–33. [CrossRef]
45. Smith, J. *Imperialism in the Twenty-First Century: Globalization, Super-Exploitation, and Capital's Final Crisis*; Monthly Review Press: New York, NY, USA, 2016.
46. Office of National Statistics. *Net Zero and the Different Official Measures of the UK's Greenhouse Gas Emissions*; Office of National Statistics: Newport, UK, 2019. Available online: https://www.ons.gov.uk/economy/environmentalaccounts/articles/netzeroandthedifferentofficialmeasuresoftheuksgreenhousegasemissions/2019-07-24 (accessed on 20 May 2022).
47. Meeus, J.H.A.; Wijermans, M.P.; Vroom, M.J. Agricultural Landscapes in Europe and their Transformation. *Landsc. Urban Plan.* **1990**, *18*, 289–352. [CrossRef]
48. Bignal, E.M.; McCracken, D.I. Nature Conservation and Pastoral Farming in the British Uplands. *Br. Wildl.* **1993**, *4*, 367–376.

49. Pain, D.J.; Pienkowski, M.W. *Farming and Birds in Europe: The Common Agricultural Policy and Its Implications for Bird Conservation*; Academic Press: London, UK, 1997.
50. Bignal, E.M.; McCracken, D.I. Low-intensity Farming Systems in the Conservation of the Countryside. *J. Appl. Ecol.* **1996**, *33*, 413–424. [CrossRef]
51. Hayhow, D.B.; Eaton, M.A.; Stanbury, A.J.; Burn, F.; Kirby, W.B.; Bailey, N.; Beckmann, B.; Bedford, J.; Boersch-Supan, P.H.; Coomber, F.; et al. The State of Nature; The State of Nature Partnership. 2019. Available online: https://nbn.org.uk/wp-content/uploads/2019/09/State-of-Nature-2019-UK-full-report.pdf (accessed on 20 May 2022).
52. Burns, F.; Mordue, S.; al Fulaij, N.; Boersch-Supan, P.H.; Boswell, J.; Boyd, R.J.; Bradfer-Lawrence, T.; de Ornellas, P.; de Palma, A.; de Zylva, P.; et al. State of Nature 2023; The State of Nature Partnership. 2023. Available online: https://stateofnature.org.uk (accessed on 15 January 2024).
53. Vandermeer, J.H. *The Ecology of Agroecosystems*; Jones and Bartlett Publishers: Boston, MA, USA, 2011.
54. Kelly, M. Habitat Protection, Ideology and the British Nature State: The Politics of the Wildlife and Countryside Act 1981. *Engl. Hist. Rev.* **2022**, *137*, 847–883. [CrossRef]
55. Perfecto, I.; Vandermeer, J.H.; Wright, A. *Nature's Matrix: Linking Agriculture, Conservation, and Food Sovereignty*; Earthscan: London, UK, 2009.
56. Tilzey, M. Conservation and Sustainability. In *Sustainability of Rural Systems: Geographical Perspectives*; Bowler, I., Bryant, R., Cocklin, C., Eds.; Springer: Dordrecht, The Netherlands, 2011; pp. 247–268.
57. Public Health England; Food Standards Agency. *National Diet and Nutrition Survey: Rolling Programme Years 9 to 11 (2016/2017 to 2018/2019)*. HMG, 2020. Available online: https://assets.publishing.service.gov.uk/government/uploads/system/uploads/attachment_data/file/943114/NDNS_UK_Y9-11_report.pdf (accessed on 15 January 2024).
58. Marmot, M.; Allen, J.; Boyce, T.; Goldblatt, P.; Morrison, J. *Health Equity in England: The Marmot Review 10 Years On*; Institute of Health Equity: London, UK, 2020. Available online: https://www.health.org.uk/publications/reports/the-marmot-review-10-years-on (accessed on 15 January 2024).
59. Office of National Statistics. *Health State Life Expectancies by National Deprivation Deciles England and Wales: 2017-2019*; Office of National Statistics: Newport, UK, 2021. Available online: https://www.ons.gov.uk/peoplepopulationandcommunity/healthandsocialcare/healthandlifeexpectancies/bulletins/healthstatelifeexpectanciesuk/2017to2019. (accessed on 15 January 2024).
60. Public Health England. *Height by Deprivation Decile in Children Aged 10 to 11*. HMG, 2020. Available online: https://www.gov.uk/government/publications/height-by-deprivation-decile-in-children-aged-10-to-11 (accessed on 16 January 2024).
61. National Risk Factor Collaboration (NCD-RisC). Height and body-mass index trajectories of school-aged children and adolescents from 1985-2019 in 200 countries and territories: A pooled analysis of 2181 population-based studies with 65 million participants. *Lancet* **2020**, *396*, 1511–1524. [CrossRef] [PubMed]
62. Department for Work and Pensions (DfWP). Family Resources Survey: Financial Year 2019 to 2020. HMG, 2021. Available online: https://www.gov.uk/government/statistics/family-resources-survey-financial-year-2019-to-2020/family-resources-survey-financial-year-2019-to-2020 (accessed on 16 January 2024).
63. Public Health England. Obesity and the Environment—Density of Fast Food Outlets at 31/12/2017. 2018. Available online: https://assets.publishing.service.gov.uk/media/5a749f6e40f0b616bcb17fe3/Fast_food_map_2016.pdf (accessed on 16 January 2024).
64. Child Poverty Action Group. Official Figures: The Number of Families Affected by Two-Child Limit Reaches 318,000. 2021. Available online: https://www.naht.org.uk/Portals/0/Images/End%20child%20poverty_slides%20for%20NAHT%20website.pdf?ver=2022-03-11-155106-560 (accessed on 16 January 2024).
65. Food Foundation. Food Insecurity Tracking. 2023. Available online: https://foodfoundation.org.uk/initiatives/food-insecurity-tracking#tabs/Round-13 (accessed on 14 January 2024).
66. DEFRA. *Sustainable Consumption Report: Follow-Up to the Green Food Project*; Department for Environment, Food and Rural Affairs: London, UK, 2013.
67. DEFRA. *UK Food Security Assessment*; Department for Environment, Food and Rural Affairs: London, UK, 2009.
68. Booth, R. Pathways, Targets, and Temporalities: Analysing English Agriculture's Net Zero Futures. *Environ. Plan. E Nat. Space* **2021**, *6*, 617–637. [CrossRef]
69. Tilzey, M. *Mapping UK Government Thinking on Globalization, Agriculture and Environment*; UK Food Group: London, UK, 2001. Available online: https://test.ukfg.org.uk/2000/mapping-uk-government-thinking-on-globalisation-agriculture-and-environment/ (accessed on 20 May 2022).
70. Jessop, B. *Putting Civil Society in Its Place: Governance, Metagovernance, and Subjectivity*; Bristol University Press: Bristol, UK, 2020.
71. Borras, S.M., Jr.; Scoones, I.; Baviskar, A.; Edelman, M.; Peluso, N.L.; Wolford, W. Climate Change and Agrarian Struggles: An Invitation to Contribute to a *JPS* Forum. *J. Peasant Stud.* **2022**, *49*, 1–28. [CrossRef]
72. DEFRA. *Farm Performance and Productivity: Analysis of Farm Business Survey*; Department for Environment, Food and Rural Affairs: London, UK, 2020. Available online: https://assets.publishing.service.gov.uk/government/uploads/system/uploads/attachment_data/file/955919/fbs-evidencepack-28jan21.pdf (accessed on 15 May 2022).
73. DEFRA. *Farming for the Future: Policy and Progress Update*; Department for Environment, Food and Rural Affairs: London, UK, 2020. Available online: https://assets.publishing.service.gov.uk/government/uploads/system/uploads/attachment_data/file/868041/future-farming-policy-update1.pdf (accessed on 15 May 2022).

74. DEFRA. *The Path to Sustainable Farming: An Agricultural Transition Plan 2021 to 2024*; Department for Environment, Food and Rural Affairs: London, UK, 2020. Available online: https://assets.publishing.service.gov.uk/government/uploads/system/uploads/attachment_data/file/954283/agricultural-transition-plan.pdf (accessed on 16 January 2024).
75. DEFRA. *Agricultural Transition Plan Update January 2024*; Department for Environment, Food and Rural Affairs: London, UK, 2024. Available online: https://www.gov.uk/government/publications/agricultural-transition-plan-2021-to-2024/agricultural-transition-plan-update-january-2024#annex-4-premium-payments (accessed on 25 January 2022).
76. Harvey, D. *Justice, Nature, and the Geography of Difference*; Blackwell: Oxford, UK, 1996.
77. The Guardian. UK to Toughen Emission Targets on Greenhouse Gas Emissions for Next 15 Years. *The Guardian*, 19 April 2021. Available online: https://www.theguardian.com/environment/2021/apr/19/uk-to-toughen-targets-on-greenhouse-gas-emissions-sources-say (accessed on 20 September 2023).
78. UK Government Nationally Determined Contribution. United Kingdom of Great Britain and Northern Ireland's Nationally Determined Contribution. 2020. Available online: https://www.gov.uk/government/publications/the-uks-nationally-determined-contribution-communication-to-the-unfccc (accessed on 15 May 2022).
79. UK Government Adaptation Communication. United Kingdom of Great Britain and Northern Ireland's Adaptation Communication to the United Nations Framework Convention on Climate Change. 2020. Available online: https://www.gov.uk/government/publications/the-uks-adaptation-communication-to-the-united-nations-framework-convention-on-climate-change-unfccc-2020 (accessed on 15 May 2022).
80. DEFRA. *Environmental Land Management Scheme: Overview*; Department for Environment, Food and Rural Affairs: London, UK, 2021. Available online: https://www.gov.uk/government/publications/environmental-land-management-schemes-overview/environmental-land-management-scheme-overview (accessed on 15 May 2022).
81. DEFRA. *Environmental Land Management Schemes: Payment Principles*; Department for Environment, Food and Rural Affairs: London, UK, 2021. Available online: https://www.gov.uk/government/publications/environmental-land-management-schemes-payment-principles/environmental-land-management-schemes-payment-principles (accessed on 15 May 2022).
82. House of Lords. *Environmental Management: Recent Changes to the Sustainable Farming and Countryside Stewardship Schemes*; House of Lords: London, UK, 2024. Available online: https://lordslibrary.parliament.uk/environmental-land-management-recent-changes-to-the-sustainable-farming-incentive-and-countryside-stewardship-schemes/ (accessed on 20 January 2024).
83. National Farmers Union. *NFU Responds to Latest Changes in Rollout of ELM Scheme*; National Farmers Union: Stoneleigh, UK, 2024. Available online: https://www.nfuonline.com/media-centre/releases/press-release-nfu-responds-to-latest-changes-in-roll-out-of-elm-scheme/ (accessed on 20 January 2024).
84. Kim, B.F.; Santo, R.E.; Scatterday, A.P.; Fry, J.P.; Synk, C.M.; Cebron, S.R.; Mekonnen, M.M.; Hoekstra, A.Y.; de Pee, S.; Bloem, M.W.; et al. Country-specific Dietary Shifts to Mitigate Climate and Water Crises. *Glob. Environ. Chang.* **2020**, *62*, 101926. [CrossRef]
85. DEFRA. *The Future of Farming and Environment Evidence Compendium*; Department for Environment Food & Rural Affairs: London, UK, 2019. Available online: https://assets.publishing.service.gov.uk/government/uploads/system/uploads/attachment_data/file/834432/evidence-compendium-26sep19.pdf (accessed on 15 September 2023).
86. DEFRA. *Agriculture in the UK Evidence Pack*; Department for Environment Food & Rural Affairs: London, UK, 2020. Available online: https://assets.publishing.service.gov.uk/government/uploads/system/uploads/attachment_data/file/955918/AUK-2019-evidencepack-28jan21.pdf (accessed on 20 May 2022).
87. DEFRA. *Agriculture in the UK Evidence Pack September 2022 Update*; Department for Environment, Food and Rural Affairs: London, UK, 2022. Available online: https://assets.publishing.service.gov.uk/media/6331b071e90e0711d5d595df/AUK_Evidence_Pack_2021_Sept22.pdf (accessed on 15 January 2024).
88. National Farmers Union. *Achieving Net Zero: Farming's 2040 Goal*; National Farmers Union: Stoneleigh, Warwickshire, UK, 2019. Available online: https://www.nfuonline.com/nfu-online/business/regulation/achieving-net-zero-farmings-2040-goal/ (accessed on 20 May 2022).
89. Vergara-Camus, L. *Land and Freedom: The MST, the Zapatistas and Peasant Alternatives to Neoliberalism*; Zed Press: London, UK, 2014.
90. Wood, E.M. The Question of Market Dependence. *J. Agrar. Chang.* **2002**, *2*, 50–87. [CrossRef]
91. D'Alisa, G.; Demaria, F.; Kallis, G. *Degrowth—A Vocabulary for a New Era*; Routledge: Abingdon, UK, 2014.
92. World Wide Fund for Nature. *Living Planet Report*; WWF: Gland, Switzerland, 2006.
93. Poux, X.; Schiavo, M. Modelling for an Agroecological UK in 2050—Findings from TYFARegio. 2021. Available online: https://www.iddri.org/en/publications-and-events/study/modelling-agroecological-uk-2050-findings-tyfa-regio (accessed on 15 May 2022).
94. Garnett, T. *What Is a Sustainable Diet? A Discussion Paper*; Food and Climate Research Network, Oxford Martin School, University of Oxford: Oxford, UK, 2014.
95. Gonzalez Fischer, C.; Garnett, T. *Plates, Pyramids, Planet: Developments in National Healthy and Sustainable Dietary Guidelines: A State of Play Assessment*; FAO: Rome, Italy; Food and Climate Research Network: Oxford, UK, 2016.
96. Mason, P.; Lang, T. *Sustainable Diets: How Ecological Nutrition Can Transform Consumption and the Food System*; Routledge: London, UK, 2017.
97. Hobsbawm, E.; Rude, G. *Captain Swing*; Verso: London, UK, 2014.
98. Letts, J. Continuous Grain Cropping. *Land* **2020**, *27*, 28–34. Available online: https://www.thelandmagazine.org.uk/articles/continuous-grain-cropping (accessed on 15 May 2022).

99. Wolfe, M.; Baresel, J.; Desclaux, D.; Goldringer, I.; Hoad, S.; Kovacs, G. Developments in Breeding Cereal for Organic Agriculture. *Euphytica* **2008**, *163*, 323–346. [CrossRef]
100. Voss-Fels, K.P.; Stahl, A.; Wittkop, B.; Lichthardt, C.; Nagler, S.; Rose, T.; Chen, T.-W.; Zetzsche, H.; Seddig, S.; Baig, M.M.; et al. Breeding Improves Wheat Productivity under Contrasting Agrochemical Input Levels. *Nat. Plants* **2019**, *5*, 706–714. [CrossRef]
101. Grime, J.P. *Plant Strategies and Vegetation Processes*; Wiley: Chichester, UK, 1979.
102. Østergård, H.; Finckh, M.R.; Fontaine, L.; Goldringer, I.; Hoad, S.P.; Kristensen, K.; van Bueren, E.T.L.; Mascher, F.; Munk, L.; Wolfe, M.S. Time for a Shift in Crop Production: Embracing Complexity through Diversity at All Levels. *J. Sci. Food Agric.* **2009**, *89*, 1439–1445. [CrossRef]
103. van Groeningen, J.W.; Lubbers, I.M.; Vos, H.M.J.; Brown, G.G.; De Deyn, G.B.; van Groenigen, K.J. Earthworms Increase Plant Production: A Meta-Analysis. *Sci. Rep.* **2014**, *4*, 6365. [CrossRef]
104. Briones, M.J.I.; Schmidt, O. Conventional Tillage Decreases the Abundance and Biomass of Earthworms and Alters their Community Structure in a Global Meta-Analysis. *Glob. Chang. Biol.* **2017**, *23*, 4396–4419. [CrossRef] [PubMed]
105. Mason, H.; Spaner, D. Competitive Ability of Wheat in Conventional and Organic Management Systems: A Review of the Literature. *Can. J. Plant Sci.* **2006**, *86*, 333–343. [CrossRef]
106. Murphy, K.; Lammer, D.; Lyon, S.; Brady, C.; Jones, S. Breeding for organic and low-inout farming systems: An evolutionary-participatory breeding method for inbred cereal grains. *Renew. Agric. Food Syst.* **2005**, *20*, 48–55. [CrossRef]
107. Rogosa, E. *Restoring Heritage Grains: The Culture, Biodiversity, and Resilience of Landrace Wheat*; Chelsea Green Publishing: Chelsea, VT, USA, 2016.
108. Tsiafouli, M.A.; Thébault, E.; Sgardelis, S.P.; de Ruiter, P.C.; van der Putten, W.H.; Birkhofer, K.; Hemerik, L.; de Vries, F.T.; Bardgett, R.D.; Brady, M.V.; et al. Intensive Agriculture Reduces Soil Biodiversity across Europe. *Glob. Chang. Biol.* **2015**, *21*, 973–985. [CrossRef] [PubMed]
109. FAO (Food and Agriculture Organization of the United Nations). *Conservation Agriculture*; FAO: Rome, Italy, 2022. Available online: https://www.fao.org/3/cb8350en/cb8350en.pdf (accessed on 2 April 2024).
110. Wageningen University. *Circular Agronomy*; Wageningen University: Wageningen, The Netherlands, 2018. Available online: https://www.wur.nl/en/project/circular-agronomy.htm (accessed on 2 April 2024).
111. MacDonald, A.J. (Ed.) *Guide to the Classical and Other Long-term Experiments, Datasets and Sample Archive*; Rothamsted Research: Harpenden, UK, 2018. Available online: https://www.rothamsted.ac.uk/sites/default/files/national-capability/long-term-experiments/Web_LTE%20Guidebook_2019%20Final2.pdf (accessed on 2 April 2024).
112. Shewry, P.; Pellney, T.K.; Lovegrove, A. Is Modern Wheat Bad for Health? *Nat. Plants* **2016**, *2*, 16097. [CrossRef] [PubMed]
113. Shewry, P.R.; Hassall, K.L.; Grausgruber, H.; Andersson, A.A.M.; Lampi, A.-M.; Piironen, V.; Rakszegi, M.; Ward, J.L.; Lovegrove, A. Do Modern Types of Wheat Have Lower Quality for Human Health? *Nutr. Bull.* **2020**, *45*, 362–373. [CrossRef]
114. Von Redecker, E. Ownership's Shadow: Neoauthoritarianism as Defence of Phantom Possession. *Crit. Times* **2020**, *3*, 33–67.

Disclaimer/Publisher's Note: The statements, opinions and data contained in all publications are solely those of the individual author(s) and contributor(s) and not of MDPI and/or the editor(s). MDPI and/or the editor(s) disclaim responsibility for any injury to people or property resulting from any ideas, methods, instructions or products referred to in the content.

Review

Exploring the Ecological Effects of Rural Land Use Changes: A Bibliometric Overview

Haojun Xie [1], Quan Sun [1,*] and Wei Song [2,3,*]

1. School of Agriculture, Ningxia University, Yinchuan 750021, China; 12021131171@stu.nxu.edu.cn
2. Key Laboratory of Land Surface Pattern and Simulation, Institute of Geographic Sciences and Natural Resources Research, Chinese Academy of Sciences, Beijing 100101, China
3. Hebei Collaborative Innovation Center for Urban-Rural Integration Development, Shijiazhuang 050061, China
* Correspondence: sqnxu@sina.com (Q.S.); songw@igsnrr.ac.cn (W.S.)

Abstract: Land use change is a significant contributor to global environmental change. The expansion of urban areas has increasingly impacted rural ecological environments, in particular the shift from agro-ecosystems to urban ecosystems, leading to alterations in land use patterns. Rural land use has led to economic, social, and environmental problems, including poor economic efficiency, emissions of pollutants, and increased environmental crises. The research of alterations in rural land use and their consequential environmental ramifications has garnered escalating attention, evolving into an indispensable subject of inquiry within pertinent academic disciplines. This study aims to obtain a comprehensive understanding of the ecological impacts of rural land use change. We examined 1237 literature sources through the Web of Science database and conducted a bibliometric analysis utilizing the Bibliometrix tool. Secondly, based on the results of bibliometric analysis, we conducted a review study on the impact of rural land use changes on the ecological environment, clarified the current research status in this field, and looked forward to future research directions. The study's findings indicate that there has been a steady rise in publication volume from 1982 to 2023 and a significant potential for growth. The top three journals by publication volume are *Sustainability*, *Land Use Policy*, and *Land*. (2) A total of 4768 scholars from 95 countries or regions have contributed publications in this domain, notably led by researchers and institutions predominantly based in China. Developed nations, exemplified by the U.S., exhibit a notable citation frequency and robust research prowess within this field. (3) Land use, urbanization, China, ecosystem services, biodiversity, and remote sensing emerge as keywords of elevated frequency within the field, indicative of the scholarly emphasis on these subjects. (4) Studies in this domain are directed towards evaluating the effects on intrinsic components of the environment, including but not limited to soil quality, atmospheric conditions, water resources, and biodiversity. The implementation of sustainable rural land use strategies is essential for the realization of rural development and environmental protection. In future research efforts, the use of remote sensing technology holds immense potential as a robust technical tool for investigating both land use change and rural ecology, offering viable strategies for addressing environmental challenges in specific, localized regions. The results of this study can assist in comprehending the current state and direction of research in this field.

Keywords: rural land use; land use change; urbanization; ecological environment; bibliometric

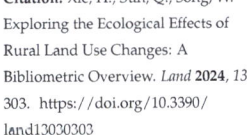

Citation: Xie, H.; Sun, Q.; Song, W. Exploring the Ecological Effects of Rural Land Use Changes: A Bibliometric Overview. *Land* **2024**, *13*, 303. https://doi.org/10.3390/land13030303

Academic Editors: Le Yu and Pengyu Hao

Received: 25 January 2024
Revised: 24 February 2024
Accepted: 25 February 2024
Published: 28 February 2024

Copyright: © 2024 by the authors. Licensee MDPI, Basel, Switzerland. This article is an open access article distributed under the terms and conditions of the Creative Commons Attribution (CC BY) license (https://creativecommons.org/licenses/by/4.0/).

1. Introduction

Alterations in land use practices are being recognized as a significant contributor to broad-scale environmental changes on a global level [1–3]. These shifts in land utilization carry extensive and enduring consequences for the viability of our planet's ecosystems [4–6]. The ability of land to support diverse functions is paramount to maintaining ecological balance [7]. Furthermore, it is critically important for achieving sustainable living conditions for communities across the globe [8]. Urbanization is reshaping the terrestrial landscape at a rapid rate with intricate and multifaceted processes that transform rural areas into urban centers and repurpose agricultural land for alternative industrial use [9,10]. The expansion of urban areas is occurring at an unprecedented velocity, encroaching upon previously agricultural terrains, woodlands, and a range of pristine natural habitats [11,12]. Land management decisions and the policies overseeing our natural resources have the potential to have a pronounced direct or indirect impact on ecological phenomena and processes [13]. The interplay between human endeavors and the natural environment clearly results in outcomes that are often permanent, imprinting lasting changes on our environmental systems [14].

The phenomenon of urbanization is seen around the world and is widely accepted as a change that cannot be turned back [15,16]. This movement of people and resources towards urban areas results in a clustering of human populations, a congregation of social networks and assets, and intensified industrial and commercial endeavors [17]. Nonetheless, this same concentration offers up vulnerabilities, making cities and their dense populations more susceptible to a gamut of risks and potential disasters [18]. The swift pace at which urban areas have expanded has brought about profound changes to their natural surroundings, particularly in regions once dominated by farming and agriculture [19]. What were previously landscapes sculpted by agricultural needs are now increasingly becoming part of the urban footprint, transforming into what we know as urban ecosystems [20]. This shift in how rural land is utilized in the wake of development carries with it a set of complex difficulties [21]. These difficulties have both tangible and intangible facets manifesting as lessened economic productivity, particularly in regions transitioning away from agriculture without adequate investment in new industries. Pollution has soared as well, with the prevalence of contaminants rising to levels that are not only harmful to the environment but also to human health [22]. The environmental situation has deteriorated as a result, with damage ranging from the loss of biodiversity to the depletion of natural resources. This degradation further complicates the prospects of sustainable development, as ecological considerations are often sidelined by the pressing need for urban expansion. Overall, the urbanization trend binds both opportunity and adversity. As population centers grow and draw in more human and financial capital, the challenges posed—economic, societal, and environmental—also magnify. These include, but are not limited to, reduced effectiveness in traditional economic sectors, an upsurge in pollution, and acceleration in the decline of environmental health [10,23–25], underscoring an urgent need for policy interventions and sustainable planning.

The study of global change by an increasing number of researchers from various fields has shifted towards analyzing the relationship between land use and ecosystems [26,27]. This research focuses on the ecological consequences that result from changes in land use and how they interact with the ecosystem. The analysis of alterations in rural land use and their ecological repercussions has increasingly gained attention, becoming a crucial topic of investigation in fields like geography, environmental science, landscape ecology, and ecological economics [28,29]. Previous research has thoroughly evaluated many aspects of modifications in rural land use and ecosystems. Laan et al. investigated the ecological consequences of changes in land use associated with agricultural production in the South African region, with a focus on specific amelioration measures [30]. Benoit et al. developed new techniques for assessing the effects of land use on the ecological environment, with the objective of identifying the degree of interdependence among various components [31]. However, many of these investigations focused on isolated aspects within the field without providing a comprehensive overview of the current state of development. Additionally, they did not provide predictions for future trends, thus limiting the progress of the field.

The expanding community of scientists from diverse fields is now concentrating on how the use of land affects the natural environments we live in and depend upon [26]. These experts are honing in on the intricate ways that our decisions about land development, agriculture, and conservation ripple through ecosystems. With each new building, farm, or protected area, there arise important questions about the balance of nature and the foreseeable effects of our footprint on Earth. This body of work has become pivotal, touching on several disciplines, including geography with its spatial awareness, environmental science with its dedication to holistic health, landscape ecology's understanding of spatial patterns, and ecological economics with its fusion of ecology and financial insights [28,29]. Extensive examinations have shed light on many facets of how altering rural landscapes impacts natural ecosystems. In the South African context, for instance, research conducted by Laan and colleagues peeled away the layers of complexity to reveal how shifts in land dedicated to farming activities are influencing the natural order and what measures could possibly soften these impacts [30]. Similarly, the efforts of Benoit and team shed light on innovative methodologies to quantify the footprint of human land use on the environment, aiming to untangle the web of relationships between diverse components of these ecosystems [31]. Despite the depth of these investigations, many have tunneled into specific niches without stepping back to consider the broader picture of progression within this vital area of study. Furthermore, there is an observable gap in the body of research regarding foresight, as few studies venture to predict where our current path may lead us in terms of environmental integrity. This shortfall highlights a potential stasis in the advancement of knowledge around land use and ecosystem dynamics, marking out a clear path for future research endeavors.

Bibliometric analysis is a method that relies on the adept use of literature data visualization instruments for the purpose of data mining, engaging in thorough network analysis, and effectively mapping to concisely articulate the breadth of knowledge that spans across various related subject areas [32]. In contrast to other methodologies, the strength of bibliometric analysis lies in its provision of a lens that is objective and systematic, essential for the in-depth examination of the prevailing hotspots and the emergent trends within the research field at large [33]. Standing on the cusp of contemporary academic study, the method has seen significant development and has been widely employed to scrutinize and identify the prevailing disciplinary tendencies among scholars and researchers alike. Among the tools commonly harnessed in the pursuit of bibliometric analysis, some stand out due to their high frequency of use and recognized effectiveness, including but not limited to CiteSpace, VOSviewer, and Bibliometrix [34,35]. These tools have gained acclaim and are frequently referenced within academic circles, evidencing their prominent roles in the pursuit of bibliometric study and analysis.

Bibliometrix, an open-source bibliometric analysis tool, was developed in 2017 by Massimo Aria et al. [36]. Unlike other bibliometric software, Bibliometrix can import and convert data from various databases, such as the Web of Science and Scopus. Additionally, it offers a wider range of literature analysis functions and visualization alternatives [36,37]. Currently, many researchers utilize bibliometric analysis in various fields, such as agriculture, ecology, and geography. In their study, Xu et al. conducted data extraction and statistical analysis by searching the Web of Science database for research papers related to land reclamation published between 2000 and 2020. Their research aims to identify future research directions in this field [38]. Li et al. conducted a bibliometric analysis to quantitatively examine the literature on rural settlements in the Web of Science core collection database from 1973 to 2021. Their study identified the current state of research and future trends in the field [39].

There is widespread global interest in rural land use and ecology, but there is still a lack of relevant review studies. Consequently, this study systematically sifts through potentially pertinent literature on the subject of rural land use alteration and ecology, utilizing the Web of Science (WOS) core database. The identified literature is meticulously organized and scrutinized employing the Bibliometrix R language package. This research aims to analyze its research dynamics and development trends more objectively, elucidating the advancements in existing studies and enhancing comprehension of future trends. To achieve the research objectives of this study, the following questions will be addressed:

(1) What are the trends in the production of relevant literature in this field?
(2) How are academic journals, authors, research countries, and institutions in this field of research developing?

(3) How have the research focal points and topics within this field progressed and advanced?
(4) What are the ecological impacts of changing land use in rural areas?
(5) What are the prospects for future directions of research in this field?

2. Data Sources and Methods

2.1. Data Sources

Bibliometrics is an objective technique for efficiently obtaining and analyzing publication data. The Web of Science is a vast and comprehensive academic database that contains a wealth of literature. It is a crucial tool for retrieving global scholarly information and is widely recognized as one of the world's most valuable resources in this field. The data used in this study was obtained from the Web of Science Core Collection database, which is a reliable source for academic research. Literature on the relationship between rural land use and the ecological environment was retrieved using the search term "topic". Through repeated experiments by the authors, the search formula is as follows: (TS = (rural land use) OR TS = (rural land use change)) AND (TS = (ecological environment)). The data were collected in November 2023, and the search criteria were limited to articles and reviews published between 1982 and 2023. Additional search results were obtained by screening entries to ensure search quality. The outcomes were exported to "complete records and cited references" and saved in text format. A total of 1237 articles meeting the specified criteria were retrieved from the search.

2.2. Methods

With reference to the general steps of bibliometric analysis, the steps of bibliometric analysis conducted in this paper include data collection and cleaning, data analysis and visualization, and data interpretation (Figure 1). Data collection and cleaning have been described in Section 2.1.

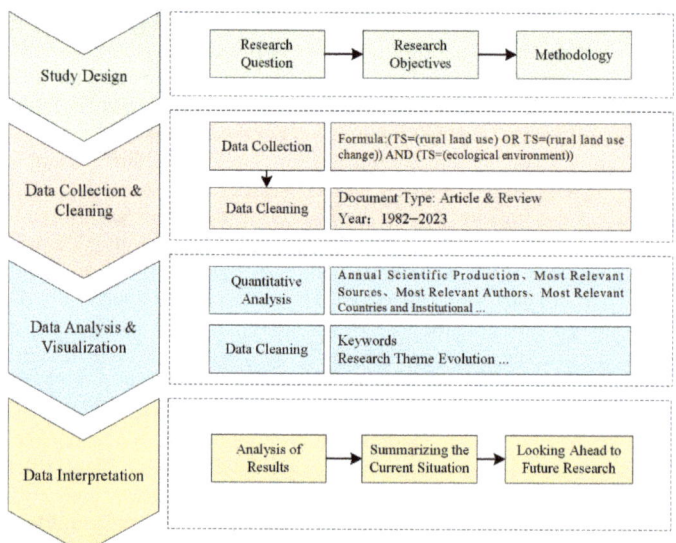

Figure 1. Research design and workflow.

Data Analysis and Visualization: To ascertain the primary research topics, the current research landscape, and the developmental trajectory of rural land use and the ecological environment, we employed Biblioshiny. Bibliometrix has introduced an online data analysis tool replete with numerous statistical methodologies and a diverse array of visualization charts. This tool is accessible for users, thereby fulfilling the fundamental needs of researchers [36,40]. For the field of rural land use and ecology, this paper will conduct a quantitative analysis of issuing journals and citations; a quantitative and collaborative network analysis of researchers, institutions, and countries; keyword analysis; and an examination of the evolution of research topics. In addition to utilizing the Biblioshiny software package, this study incorporates Origin software for graphing and charting to enhance data visualization.

Data Interpretation: While bibliometric software streamlines literature review research, it still necessitates comprehensive literature reading. To fully leverage the indispensable role of bibliometric analysis, the literature is meticulously reviewed and analyzed subsequent to the analysis. Consequently, it can offer a more objective, convenient, and accurate synthesis of the field's current state and its intricacies and predict future research directions.

3. Results of the Bibliometric Analysis
3.1. Quantitative Analysis of the Literature

The amount of published research in the fields of rural land use change and ecological environment can be indicative of the level of research focus and progress in the area. Figure 2 illustrates a fluctuating upward trend in the quantity of published papers in this field from 1982 to 2023.

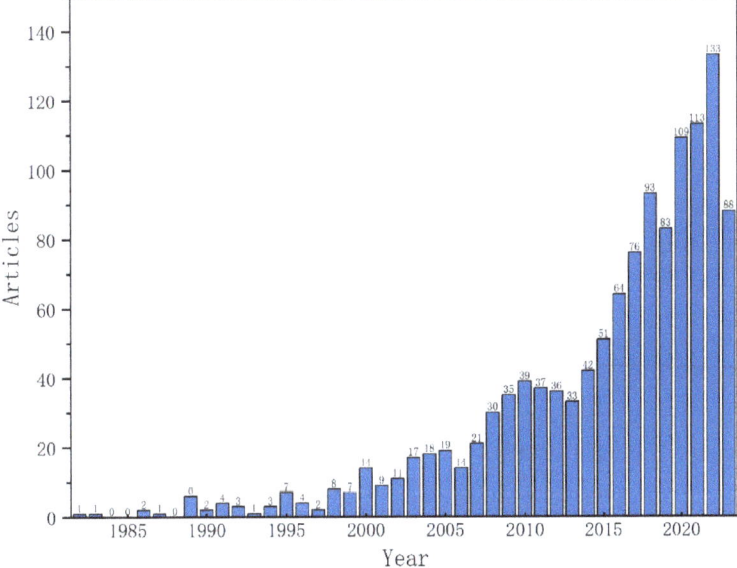

Figure 2. Annual publication trend in the field of rural land use change and ecological environment from 1982 to 2023.

This paper analyzes the trend of change in relevant studies by dividing them into three stages: 1982–2007, 2008–2014, and 2015–2023. During the 1982–2007 stage, there was a slow growth with a limited number of papers published, only amounting to 175, which represented 14.1% of the total number of publications. Conversely, the 2008–2014 stage witnessed significant growth, accompanied by a remarkable surge in the quantity of publications compared to the previous stage. Over the period, a total of 252 papers were produced, which represents 20.4% of the overall number of papers published. On average, 36 papers were published per year. The period from 2015 to present is the rapid growth stage, with a significant increase in the number of papers published reaching 810, representing 65.5% of the total number of papers published. The average number of articles per year in this phase reached 90, and relevant research in this area has entered a different phase.

3.2. Analysis of Journals

Based on the collected data, we tallied the number of articles published in each journal and identified the top ten journals based on the quantity of articles published between 1982 and 2023 (Table 1). According to the statistical information provided by the periodicals, the journals cover several subject areas, such as environmental sciences, environmental studies, ecology, and biodiversity conservation, among others. It is worth noting that the majority of journals publishing articles belong to the field of environmental sciences.

Table 1. Top ten journals with publications about rural land use change and ecological environment in the time period of 1982 to 2023.

Sources	Category	Number of Publications	H-Index
Sustainability	Environmental Sciences	59	12
Land Use Policy	Environmental Studies	47	23
Land	Environmental Studies	46	11
Ecological Indicators	Environmental Sciences	37	22
Science of The Total Environment	Environmental Sciences	32	20
Landscape and Urban Planning	Ecology	28	21
Journal of Environmental Management	Environmental Sciences	24	15
Landscape Ecology	Ecology	23	17
Urban Ecosystems	Biodiversity Conservation	23	10
International Journal of Environmental Research and Public Health	Environmental Sciences	21	7

The included journals each published 20 or more articles. *Sustainability* leads with 59 publications, followed by *Land Use Policy* with 47 papers, and *Land* with 46. The H-index, a quantitative measure developed by American physicist Jorge E. Hirsch, evaluates academic output. A higher H-index implies a more significant impact of the journal in the field. Among the journals listed, *Land Use Policy*, *Ecological Indicators*, *Landscape and Urban Planning*, and *Science of The Total Environment* exhibit a higher H-index, which infers their greater impact in the research field.

A trend analysis of the top five journals, based on the number of publications, indicates that *Science of The Total Environment* has focused the most on the field. It published its first article in 1991, followed by *Ecological Indicators* in 2004 and *Land Use Policy* in 2005 (Figure 3). *Land Use Policy* has been the most productive journal between 2009 and 2021 in terms of the number of publications. *Sustainability* and *Land*, on the other hand, began publishing their first papers in 2014 and 2018, respectively. However, the quantity of published articles has substantially increased over time. All five publications have consistently shown growth in the number of articles published from their inception until 2023. As of November 2023, *Sustainability*, *Land Use Policy*, and *Land* have released 59, 47, and 46 articles, respectively.

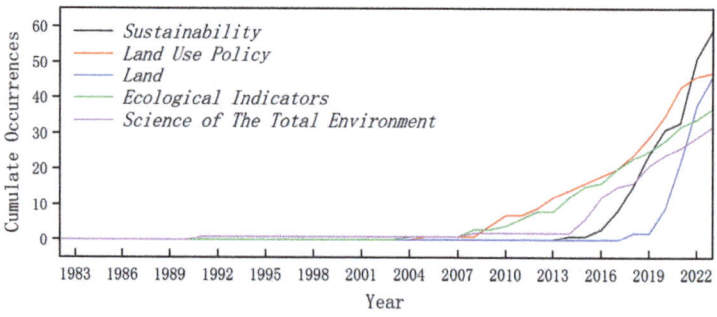

Figure 3. Trend in growth of journal sources in the field of rural land use change and ecological environment from 1982 to 2023.

3.3. Analysis of Key Researchers, Institutions, and Countries

3.3.1. Analysis of Key Researchers

The researchers in the field of rural land use change and ecological environment included 4768 participants. There are only nineteen authors who have published five or more papers. On the other hand, 91.2% of the total number of researchers, a total of 4350 individuals, have only published one paper, according to the study results. The results show that there are fewer researchers who have been working in this field for a long period of time.

In accordance with their contributions to the field in terms of publications (Figure 4), we counted the top ten scholars who had the greatest impact on the field. As of 2021, Liu Yansui, a researcher at the Institute of Geographic Sciences and Resources of the Chinese Academy of Sciences, holds the highest number of publications. Four of them are highly cited. As one of the leading scholars in human-land system science, Liu Yansui has spent a significant amount of time researching rural geography and land use, with his main focus on rural geography and land use as his main research focus. His work has significantly contributed to the academic impact of this field. Following him is Li Yurui of the Chinese Academy of Sciences Institute of Geographic Sciences and Resources. As the earliest author of this paper, Long Hualou, an expert in urban-rural development and land-use transformation at the Institute of Geographic Sciences and Resources of the Chinese Academy of Sciences, is a leading expert in these topics. In addition to researching transformational rural development, he is mainly involved in research on land use change and land-use transformation. Having a high level of academic influence in the field is part of his expertise.

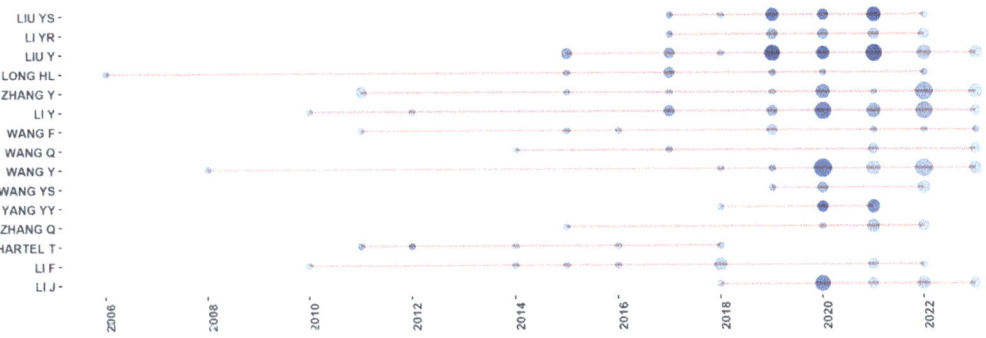

Figure 4. Authors' output of articles in the field of rural land use change and ecological environment over time. Note: The size of each circle indicates the number of publications generated, while the shade of color reflects the number of citations received.

Five clusters of authors were identified in the field of rural land use change and ecological environment (Figure 5). Liu Yansui's research has had a significant impact in this area, exploring topics such as China's transition from urban to rural development, new urbanization, as well as land use and reclamation engineering. Additionally, the research cluster led by E.C. Ellis at the University of Maryland in the United States is relatively robust and focuses primarily on global ecosystems, landscape ecology, and sustainable land management.

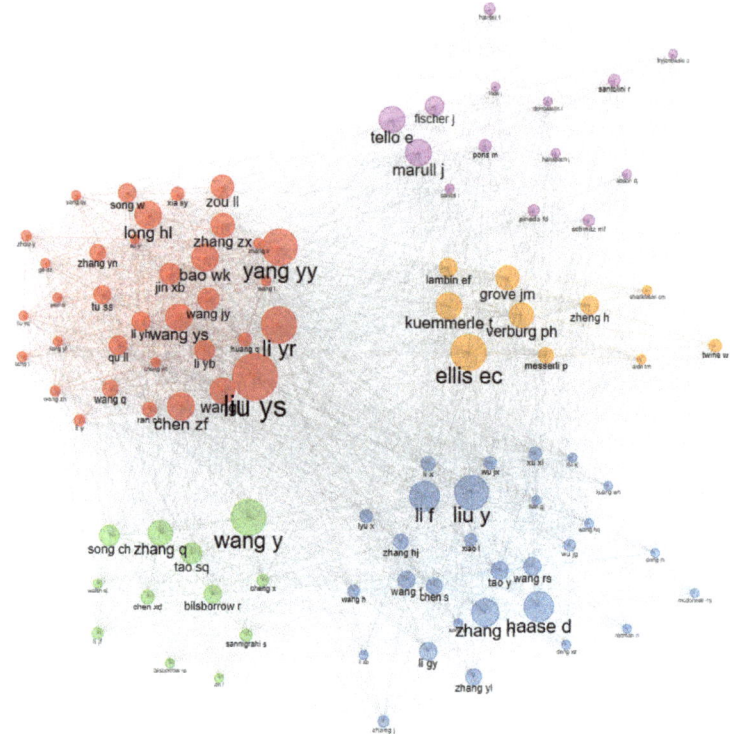

Figure 5. Collaborative network of authors in the fields of rural land use change and ecological environment. Note: The colored circles in the figure represent different cooperative network relationships, and the larger the circle, the more papers published by the scholar.

3.3.2. Analysis of the Main Countries and Institutions

The published papers from different countries enlighten the level of inquiry on rural land use change and its impact on the environment. The significance of these studies can be specific to certain regions. Based on the statistical results, 1691 institutions across 95 countries or regions are currently involved in research on rural land-use change and its ecological consequences. Table 2 presents the top ten publishing countries, with one Asian country (China) and three North American countries (the United States, Brazil, and Canada), along with five European countries (the United Kingdom, Germany, Italy, Spain, and France) and one Oceanian country (Australia).

Table 2. Dissertations published in major countries in the field of rural land use change and ecological environment from 1982 to 2023.

Country	Number of Publications	Total Citations	Average Article Citations
China	345	6393	18.53
USA	175	9783	55.90
United Kingdom	67	4204	62.75
Germany	56	2476	44.21
Brazil	50	796	15.92
Australia	43	2939	68.35
Italy	42	1392	33.14
Spain	37	1160	31.35
France	35	1427	40.77
Canada	30	998	33.27

China published the highest number of papers, with a total of 345, and came in second place after the United States in terms of citations, with 6393. However, the papers had a low average citation frequency of 18.53 times, suggesting that further improvement could increase interest. Australia boasts the highest average citation frequency for papers at 68.35 citations, followed by both the United Kingdom and the United States at 62.75 and 55.90 citations, respectively. Furthermore, the United States is ranked at the top of the list of many countries in terms of total citation frequency, with a staggering 9783 citations. This result suggests that developed countries possess robust research capabilities in rural land-use change and ecological environment research, resulting in highly influential research publications.

Our network analysis of collaborations among the highest yielding countries and regions indicates that research institutions in most countries cooperate with each other (Figure 6) and are grouped into four clusters: China and the United States (represented by red circles), France and Italy (represented by purple circles), Germany and New Zealand (represented by green circles), and the United Kingdom and Australia (represented by blue circles). In the context of changes in rural land use and the ecological environment, China and the United States have a closely intertwined relationship. Additionally, there are significant cooperative relationships between China and Japan, China and Brazil, and also between the United Kingdom and Australia. However, the United States has more extensive cooperative relationships with specific countries.

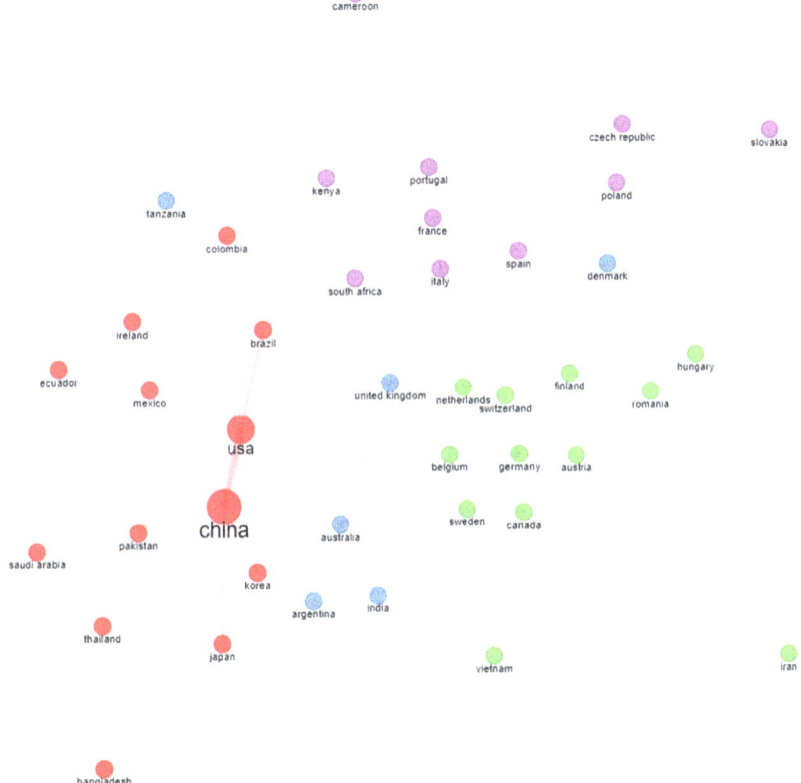

Figure 6. Network of national cooperation in the field of rural land use change and ecological environment from 1982 to 2023. Note: The colored circles in the figure represent different cooperative network relationships, and the larger the circle, the more papers published by the country.

We compiled the top ten academic institutions in the field of rural land use change and ecological environment with the highest number of publications (Table 3). Of these, six institutions are based in China, three in the United States, and one in Brazil. The top four institutions, based on publication numbers, are located in China. They are the Institute of Geographic Sciences and Natural Resources Research, the Chinese Academy of Sciences, Beijing Normal University, and two units of the Chinese Academy of Sciences. Their articles numbered 59, 41, 28, and 28, respectively.

Table 3. Number of published articles by major research institutions in the field of rural land use change and ecological environment from 1982 to 2023.

Affiliation	Number of Publications
Institute of Geographic Sciences and Natural Resources Research, CAS	59
University of Chinese Academy of Sciences	41
Beijing Normal University	28
Chinese Academy of Sciences	28
University of São Paulo	22
Arizona State University	19
Northwest A&F University	19
China University of Geosciences	18
University of Georgia	18
The University of North Carolina at Chapel Hill	16

In the context of cooperation between research institutions (Figure 7), the top four Chinese institutions maintain close relationships (indicated by red circles). The Institute of Geographic Sciences and Natural Resources Research, CAS, has the highest degree of centrality in collaboration, indicating the strongest impact on relationships within the collaborative network. These institutions exhibit significant publishing frequency and centrality and have served as the primary research centers in this field. Nanjing Agricultural University and Nanjing Normal University have established a strong domestic partnership. Additionally, closer international cooperation is underway between the China University of Geosciences and the University of North Carolina at Chapel Hill.

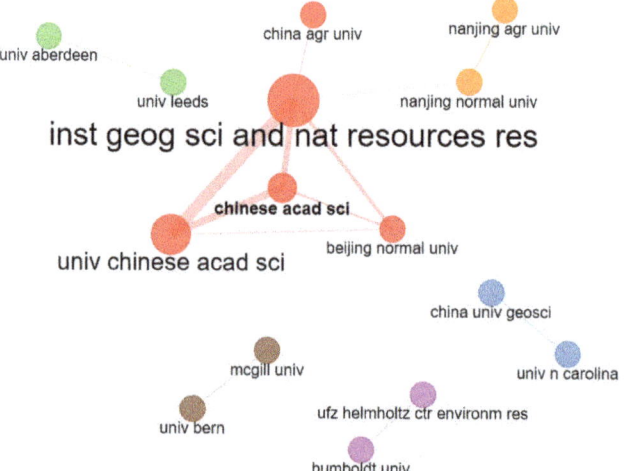

Figure 7. Collaborative network of relevant research institutions in the field of rural land use change and ecological environment from 1982 to 2023. Note: The colored circles in the figure represent different cooperative network relationships, and the larger the circle, the more papers published by the research institution.

3.4. Keyword Analysis

3.4.1. Analysis of High-Frequency Keywords

Keywords can provide an overview of the research topic and content and aid in identifying the knowledge structure of a field [41]. In this study, a word cloud was used to visualize the top 50 high-frequency keywords in this area (Figure 8). The top keywords and their frequency of occurrence in this field are as follows: land use (57), urbanization (49), China (41), ecosystem services (36), biodiversity (34), remote sensing (31), sustainability (28), land-use change (24), urban ecology (23), and climate change (22). Scholars in this field prioritize researching rural land use, urbanization, ecosystem services, biodiversity, sustainable development, and climate change.

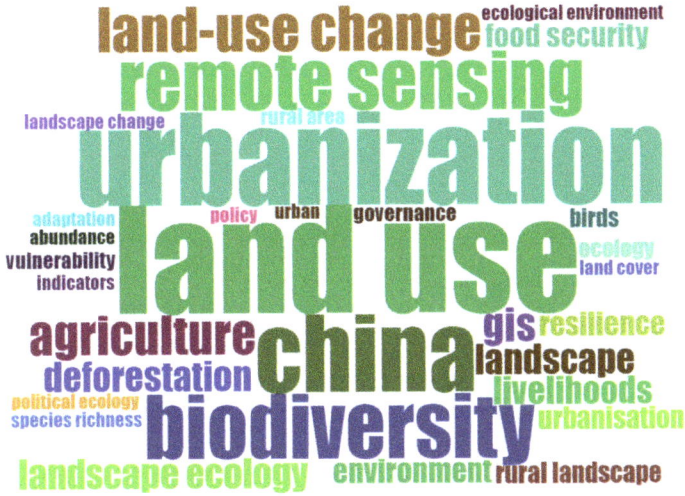

Figure 8. High-frequency keywords in the field of rural land use and ecological environment from 1982 to 2023.

3.4.2. Co-Occurrence Network Analysis of High-Frequency Keywords

A co-occurrence network analysis was conducted on commonly used keywords pertaining to rural environmental and land use change (Figure 9). This study's keyword co-occurrence network illustrates seven clusters that are centered around land use, urbanization, climate change, rural areas, biodiversity, urban ecology, and remote sensing as key nodes. These nodes are also more closely linked to each other.

3.5. Analysis of the Evolution of Research Hotspots

This paper presents the development of research themes in three phases, categorized by yearly publication frequency trends (Figure 10). The time period from 1982–2007 demonstrated slow growth, and themes emphasized agriculture, conservation, and land use. The time period from 2008–2014 exhibited steady growth, and research topics continued to expand, with a primary focus on conservation, biodiversity, and ecological restoration. Ecological protection, biodiversity, and ecological restoration are increasingly prominent research topics. The period from 2015 to the present has experienced rapid growth in this field, focusing on keywords such as land use, urbanization, climate change, and remote sensing. These keywords suggest that global changes, specifically urbanization and climate change, have impacted research in this area. Additionally, the utilization of remote sensing technology has made a substantial contribution to the advancement of this research area and has provided technical assistance for associated research endeavors.

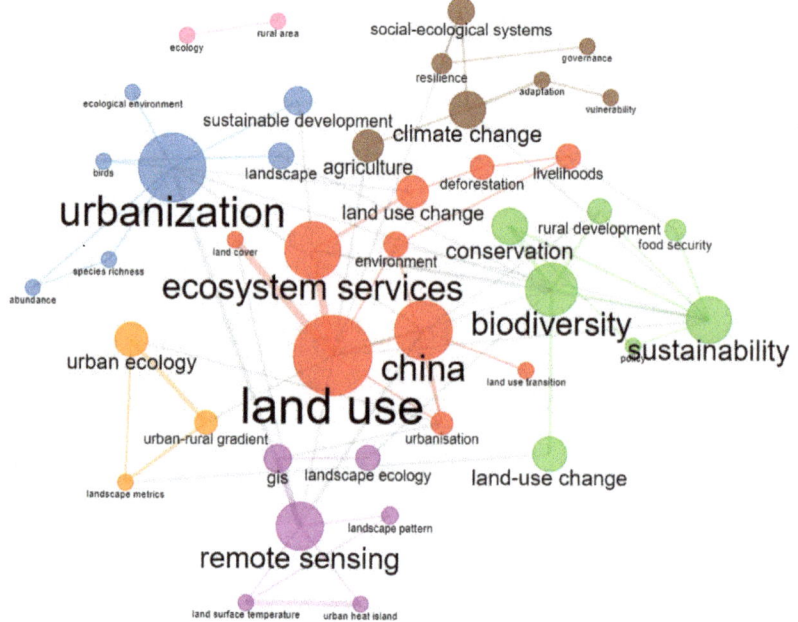

Figure 9. Co-occurrence network analysis of high-frequency keywords in the field of rural land use change and ecological environment from 1982 to 2023. Note: Different colors in the figure represent different clusters.

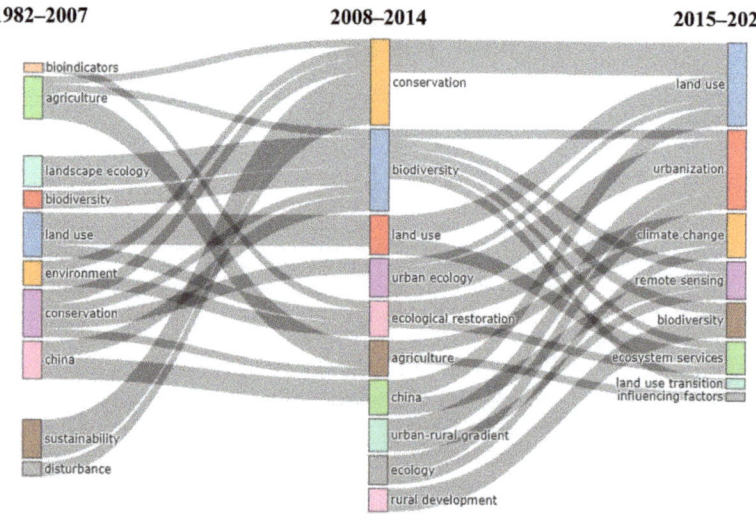

Figure 10. Evolution of research themes in the field of rural land use change and ecological environment from 1982 to 2023.

4. Review of the Ecological Effects of Rural Land Use Changes

Land use change is a core area of global environmental change research. It is of great significance to understand the impact of rural land use changes on the ecological environment. Judging from the results of the bibliometric analysis, land use change, urbanization, climate change, remote sensing technology, biodiversity, ecosystem services, etc. are all high-frequency keywords in this field and are also the subject headings for topic evolution. How rural land use changes affect the ecological environment deserves an in-depth study. In this section, we will conduct a review study in this field around subject headings and look forward to future research directions.

4.1. Urbanization and Rural Land Use Change

Urbanization is a significant aspect in bibliometric analyses and a prominent research theme in current studies. The concept of urbanization first appeared in the eighteenth century, after the industrial revolution in Europe. With economic development and social progress, urbanization has become a focus of attention and research for many scholars. Due to different starting points of research, different scholars have different interpretations of urbanization. This study considers urbanization from the perspective of territorial space.

4.1.1. Linkages between Urbanization and Rural Land Use Change

Urban development cannot be separated from land, and urbanization is always accompanied by the inevitable result of land expropriation and reduction of cultivated land [42]. Urbanization remains a substantial alteration in land use and land cover that is taking place globally [43]. Urbanization is an inevitable phenomenon and process of economic and social development [44], a transformation in the direction of "rural–urban", bringing about drastic changes in the structure of the rural economy and social relations, and having a broad and profound impact on rural society. The linkage between rural land use change and urbanization is close. In the strong impact of urbanization on rural society, rural land, as the most basic factor of production in rural areas, has borne the brunt of the impact, and the structure, pattern, and manner of its utilization have been affected in a comprehensive manner, which in turn has changed the process of natural, social, and economic development of the rural areas [45]. As land is converted to human use during urbanization, the process is frequently irreversible [46]. On the other hand, the changes in human activities have resulted in positive impacts on land use and the development of associated policies [47]. However, despite the many benefits of urbanization, the world's rapid urbanization faces growing resource scarcity and environmental degradation.

Urbanization refers to the process of changing from a rural society based on agriculture to an industrialized urban society [48]. This shift is typically achieved via the acquisition of arable land for urban development or by encouraging farmers to abandon farming in favor of more lucrative non-agricultural economic opportunities [49]. There are several categories of rural land use, which can be separated into three primary functions: productive land, living land, and ecology land [50]. The majority of research pertaining to rural land use change has concentrated on cropland. Urbanization usually leads to the depletion of cultivable land. During rapid economic development and urbanization construction, there has been a significant loss of arable land resources, particularly high-quality arable land surrounding towns and cities, which has become the primary aspect of arable land changes in recent years [15,51]. Developing nations are currently experiencing rapid urban growth compared to developed nations. China, as the biggest developing nation, is prominent in this area of research. The bibliometric results provide evidence to support this. China's urbanization rate increased from 17.9% in 1978 to 56.1% in 2015 [52]. However, studies indicate that in the first decade of the 21st century, China lost 4.37 million hectares of cultivated land [49], with over 80% of urbanization occurring on arable land issued by first-tier cities [53]. As a basic material for agricultural production, cultivated land plays an important role in natural evolution, human survival, the ecological cycle, resource economy, and other activities and is most widely and profoundly affected by human beings [54]. The loss of cultivated land is a product of direct or indirect human interference [55]. On the one hand, it is the loss of cultivated land use—the transformation of the type of cultivated land utilization [56]. A large amount of cultivated land has been utilized for non-agricultural and non-food purposes, and the resources of cultivated land have undergone great changes. On the other hand, there is the loss of cultivated land reserves—the risk of potential loss of cultivated land reserves. In rapidly urbanizing areas, high-quality cultivated land is rapidly lost with the outward expansion of urban space, and there is less and less land available for cultivated land reserve resources. Most of the potential cultivated land is usually of poor quality. In addition, problems such as land erosion, heavy metal pollution, and land salinization have led to an increasing decline in the quality of cultivated land, and the contradiction between people and land has become more prominent [57]. The decrease in arable land is primarily attributed to the growing demand for urban development. In the process of rapid urbanization, the demand for urban construction land remains high, while most of the high-quality cultivated land is located in the urban and rural periphery and near major transportation routes. Along with the increasing contradiction between economic growth and cultivated land protection, the risk of cultivated land being occupied has been increasing.

4.1.2. Protection of Cultivated Land in the Process of Urbanization

Urbanization has a strong correlation with economic development and is a process of agglomeration and integration of factors of production and economic activities. The process of urbanization is better promoted through the comprehensive benefits brought about by economic development. Cultivated land, as the basis of agriculture, provides most of the agricultural products for human beings. However, considering the occupation of cultivated land in the process of urbanization and the increasingly serious resource and environmental problems, it is necessary to take certain measures to strengthen the rational development and utilization of cultivated land. Only in this way can we better promote stable economic and social development.

First, land resources should be rationally allocated, and land-use efficiency should be improved. In the current process of urbanization, in order to realize the effective protection of cultivated land, it is necessary to rationally plan land-use zones, scientifically control the scale of urbanized land, and improve the efficiency of land use so as to realize the economical and intensive use of land. Second, it is essential to strengthen the monitoring of hot spots of arable land loss, especially the long-term time series monitoring method that can ensure spatial and temporal continuity [58], in order to strictly control the loss of high-quality arable land. Furthermore, the development of strategies and policies that effectively guide the pattern of urban expansion [59] is crucial for the protection of arable land. Policies aimed at protecting arable land by encouraging urban migration may paradoxically accelerate the occupation of arable land [60]. Therefore, it is important to rationally plan the development pattern of urbanization at the national level and make efforts to reform the land system.

4.2. Ecological Impacts of Rural Land Use Change

With the continuous development of the global economy, ecological and environmental issues have become increasingly important. Therefore, the ecological and environmental impacts of land use have been extensively studied, focusing on natural elements such as soil quality, atmospheric quality, water resources, biodiversity, and others [61,62]. With the application of related theories, techniques, and models in this field, the relevant studies have been gradually deepened, and the reliability has been greatly improved. At the beginning of this century, after the Global Land Project (GLP) was proposed, the focus of land use research gradually shifted to the impact of land use change on the ecological environment [63]. The evolution of research topics shows that, since 2015, research on the relationship between land use and ecosystem services has become more comprehensive and profound.

4.2.1. Impact of Rural Land Use Change on the Soil Environment

As the loose material layer on the earth's surface, soil serves as a crucial interface, facilitating the reciprocal exchange of energy between ecosystems and the external environment. Different land uses in this process exert diverse effects on the soil [64]. Reasonable land use can improve soil structure, increase soil resistance to external environmental changes, and maintain and improve soil quality [65,66]. The land has been transformed by urbanization and added to by human activities, and external intruders may enter the soil, thus causing a certain degree of pollution. For example, when forest lands are converted to agriculture, the soil often undergoes degradation. The removal of the forest canopy exposes soil to direct sunlight and rainfall, leading to increased soil erosion. Moreover, the loss of litterfall and root systems decreases organic matter input and soil cohesion, respectively, further exacerbating soil erosion and loss of fertility [65]. Irrational land use practices can cause problems such as soil ecological deterioration. The alterations in rural land use stemming from urbanization have markedly transformed the physicochemical environment of the soil, soil ecosystem services, and the communities of soil organisms [67].

With the population explosion and the expansion of the scope and intensity of human activities, the contradiction between the scarcity of the economic supply of land resources and the growth of their social demand is becoming increasingly acute [68]. Land use changes in areas of rapid economic development are dramatic, and important processes such as urbanization and industrialization threaten the stability and security of agricultural production. Changes in rural land use have caused many natural phenomena and changes in ecological processes, such as soil nutrients and soil heavy metal content. It has been shown that soil heavy metal accumulation is highly related to urbanization, industrialization, and land use changes, and that soil environmental quality in different areas within the same city may also have great spatial variability due to differences in land use patterns [65].

Soil microorganisms play a pivotal role in the soil environment, undertaking essential functions in ecological processes such as nutrient cycling, carbon sequestration, remediation of contaminated soil, and the decomposition of soil organic matter [69]. Significant variations have been observed in the composition, diversity, and network structure of soil bacteria in urban, peri-urban, and rural areas [70]. Urban peripheries, in particular, have emerged as hotspots of soil microbial diversity due to increased environmental heterogeneity resulting from disturbances, while rural soil bacterial communities exhibit intricate and stable structural networks [70]. The mechanisms influencing soil microbial communities necessitate further investigation to ensure the sustainability and health of soil ecosystems.

4.2.2. Impact of Rural Land Use Change on the Water Environment

Water resources constitute a fundamental component in sustaining the equilibrium of our ecological system. Alterations in land utilization, marked by an escalation in agricultural activities and construction, along with a decrease in forested and grassland areas, have multifaceted impacts on water resources. Such transformations can precipitate increased surface runoff in watershed areas, elevating both the likelihood and intensity of flooding. This escalation, in turn, augments the water volume within these regions. Modifications in the use of rural land further influence the discharge and dispersal of pollutants. Land designated predominantly for agriculture and construction, in conjunction with suboptimal farming practices, can lead to the substantial erosion of soil, water, nitrogen, and phosphorus and the introduction of further contaminants [71]. This degradation not only contaminates our water sources but also undermines aquatic habitats, reduces water storage capabilities, and elevates the expenses associated with water purification due to sedimentation. Additionally, the erosion of topsoil precipitates the depletion of nutrients, which, despite their importance to aquatic ecosystems, can engender excessive nutrient runoff [72]. Consequently, this may induce eutrophication, a phenomenon that depletes oxygen in water bodies, leading to the demise of fish and other aquatic organisms. Moreover, eroded soil can convey other detrimental pollutants, such as pesticides and heavy metals, from agricultural and industrial activities, further compromising water quality and posing risks to aquatic ecosystems and public health.

4.2.3. Impact of Rural Land Use Change on Biodiversity

The global issue of biological species diversity loss has garnered extensive research attention, emphasizing the paramount importance of biodiversity. Biodiversity-related research is currently a hotspot and focus of ecological research, which directly affects the stability and sustainability of ecosystems. Land use change affecting biodiversity has received great attention [73]. Land use change is the most direct manifestation of the impact of human activities on natural ecosystems on the Earth's surface, and it is the main process leading to species fragmentation and the loss of effective habitats, as well as the main driver threatening biodiversity [74]. The rural environment serves as a pivotal provider of habitats for diverse biological species [75]. The rapid evolution of rural environments has engendered notable transformations in rural land use, agricultural production, and tourism [76]. Consequently, these environmental shifts pose a potential threat to the survival of primitive biological species. Alterations in rural land use and heightened activities have profoundly modified traditional rural environments, resulting in a direct and detrimental impact on biodiversity. At the same time, environmental pollution, the increase in greenhouse gases, and the imbalance in carbon balance caused by land-use change have also indirectly affected biodiversity. Biodiversity is the basis for the harmonious development of mankind and nature, and with the continuous loss of biodiversity and changes in land use, the search for effective strategies and approaches to biodiversity must rely on in-depth research on the relationship between land-use changes and spatial and temporal variations in biodiversity [77] and requires the support of biodiversity conservation technologies [78].

4.2.4. Impact of Rural Land Use Change on Climate Change

The impact of rural land use change on climate change is a significant global phenomenon in the environmental field [79]. Climate change serves as the central theme of research in the current phase of investigation within this field. Changes in land use, such as forestry and cropland conversion, influence the emission of trace gases due to alterations in nutrient cycling and the distribution of organic matter [13]. Rural land use activities, encompassing land reclamation, cultivation, and wetland drainage, exert a substantial impact on carbon and nitrogen cycling. Moreover, in numerous developing countries, land reclamation is frequently associated with biomass burning, which releases substantial amounts of carbon dioxide into the atmosphere, leading to changes in greenhouse gas fluxes from agricultural lands globally [80]. Addressing climate change involves two primary objectives: mitigation and adaptation. Estimates indicate that the agriculture and land use sectors contribute to 30% of total greenhouse gas emissions globally [81]. There are widespread efforts globally to integrate the goals of mitigation and adaptation into regional and national policies.

Nature-based solutions (NbS) have become key in tackling the many challenges brought on by climate change [82]. These solutions bank on the natural abilities of ecosystems to both mitigate and adapt to environmental transformations, offering a sustainable and efficient way to manage rural land [83]. Preserving forests and extending the periods between harvests are standout examples of how NbS can boost carbon capture and storage, which is vital in the fight against climate change [84]. However, deploying NbS is not without its hurdles, such as disputes over land ownership, financial constraints, and the need for detailed planning and supportive policies. For the successful application of strategies like forest preservation and prolonged harvesting cycles, it is necessary to have cooperation between governments, NGOs, landowners, and local communities. Additionally, policies must encourage sustainable land management and the protection of ecosystems, along with financial support for landowners and communities to adopt these sustainable methods [85,86].

4.3. Global Innovations in Sustainable Land Use: Benefiting Rural Areas

Through the examination of scholarly research, we investigate novel approaches to encouraging sustainable practices in rural land use. It underscores four principal domains: green infrastructure integration, green space management, climate change adaptation, and agricultural policy. These strategies are pivotal for achieving equilibrium among rural advancement, environmental conservation, and societal well-being.

4.3.1. Green Infrastructure Integration

The integration of green infrastructure into the planning and development of rural landscapes represents a critical strategy for promoting sustainable land use and ameliorating its environmental repercussions [87]. This strategy adeptly marries the objectives of developmental growth with conservation principles, embedding sustainability within the architectural blueprint of transportation planning [88]. Such cohesion facilitates a symbiotic relationship between rural development aspirations and environmental preservation mandates, crucial for protecting natural habitats, reducing pollutants, and elevating the living standards of rural communities [89]. This holistic methodology is characterized by the sequential development of eco-friendly roads and pathways, the execution of integrative land use and transportation planning, and the advocacy for sustainable tourism and recreation pathways, cumulatively contributing to the edification of resilient rural infrastructure [90,91].

The prioritization of eco-friendly road and pathway designs underscores the need for thoroughfares that minimize ecological disruption [92]. These designs strive to mitigate habitat fragmentation, diminish runoff pollution, and integrate wildlife crossings to ensure animal safety. The deployment of wildlife overpasses and underpasses in Canada's Banff National Park serves as a prime example of initiatives aimed at reducing animal-vehicle collisions and conserving natural migratory paths [93,94]. Moreover, the concept of integrative land use and transportation planning highlights the importance of harmonizing development efforts with ecological sustainability goals and promoting the use of existing public and non-motorized transport infrastructures to safeguard open spaces and curb urban sprawl [95,96]. The urban growth boundary initiative in Portland, Oregon, encapsulates this strategy by constraining urban expansion and enhancing public transit efficacy, thereby supporting sustainable tourism and recreational endeavors [97,98]. Sustainable Tourism and Recreation Pathways advocate for the creation of transport infrastructures that bolster eco-tourism and leisure activities, contributing to both economic development and environmental preservation [99,100].

4.3.2. Green Space Management

Effective management of green spaces in rural areas necessitates a comprehensive approach that harmoniously balances agricultural productivity with the principles of ecological conservation. By adopting strategic measures for the implementation and stewardship of green spaces, rural communities can significantly augment biodiversity, enhance the quality of air and water, and contribute positively to the overall vitality of the ecosystem [101–103]. The three principal modalities through which these objectives can be realized include conservation agriculture, agroforestry systems, and wetland restoration and management.

Conservation agriculture, encompassing no-till farming, crop rotation, and cover crops, aims to minimize soil disturbance, thus preserving soil health, reducing erosion, and enhancing water retention [104]. The adoption of these practices in Brazil highlights their efficacy, demonstrating significant soil health improvements, erosion reduction, and increased crop yields, and underscores the vital role of sustainable practices in agricultural sustainability [105]. Agroforestry, integrating trees and shrubs into agricultural landscapes, fosters a beneficial synergy between agriculture and forestry, enriching biodiversity, soil structure, and carbon sequestration [106]. In Kenya, agroforestry initiatives by smallholder farmers, involving nitrogen-fixing trees, have notably improved soil fertility, agricultural productivity, and wildlife habitats, illustrating the environmental gains from effective green space management [107]. Furthermore, the restoration and judicious management of wetlands are crucial for enhancing water quality, supporting wildlife habitats, and providing natural flood management [108]. These instances highlight an integrated approach to green space management in rural settings, emphasizing the importance of conservation agriculture, agroforestry, and wetland restoration as pillars for sustainable land use and environmental conservation.

4.3.3. Climate Change Adaptation

Climate change adaptation in rural areas involves the implementation of strategies that promote sustainable land use and aim to reduce the negative impacts on the ecological environment [109]. Strategies are designed to enhance the resilience of agricultural systems, conserve biodiversity, manage water resources efficiently, and safeguard livelihoods against the challenges posed by a changing climate.

The advancement of climate-resilient crops plays a pivotal role in counteracting food production deficits caused by climate change by enhancing crop tolerance to extreme weather conditions, including drought and salinity [110]. This strategy improves the resilience of agricultural ecosystems by diversifying the portfolio of drought-resistant crops. For example, in sub-Saharan Africa, the adoption of drought-tolerant maize varieties developed through selective breeding has proven superior under arid conditions, securing food production and offering a solution to areas prone to unpredictable rainfall and drought [111]. Furthermore, the integration of agroforestry and sustainable farming practices boosts soil health, organic matter, and erosion while promoting natural vegetation recovery and carbon sequestration, essential for climate change mitigation [112]. Specifically, in the Sahel region, such as Niger, farmer-managed natural regeneration (FMNR) has gained traction. FMNR involves the regeneration and management of trees and shrubs from existing stumps, roots, and seeds, leading to enhancements in soil fertility, increased agricultural yields, and landscape restoration, marking it as an effective climate adaptation and sustainable land use strategy [113]. Additionally, managing rural water resources strengthens resilience to climate extremes like floods and droughts [114]. In Rajasthan, India, the resurgence of traditional rainwater harvesting techniques, through the construction of johads (small earthen check dams), has improved aquifer recharge and water availability for irrigation and domestic use, thereby securing water resources amidst the growing unpredictability of monsoon rains due to climate change [115].

4.3.4. Agricultural Policy

Agricultural policies serve an essential function in fostering sustainable land use within rural domains while concurrently mitigating adverse environmental impacts [116]. These policies are strategically structured to stimulate land use efficiency, conservation of natural resources, and maintenance of ecological equilibrium.

Offering subsidies and financial incentives encourages farmers to adopt sustainable practices by making them more affordable. These practices include enhancing soil health, reducing chemical use, and protecting biodiversity. An example is the European Union's Common Agricultural Policy, which supports eco-friendly farming to lower environmental damage and increase biodiversity [117]. Setting environmental protection standards forces farmers to follow certain practices, preventing deforestation and overuse of resources. Brazil's Forest Code, for instance, requires farmers in the Amazon to keep part of their land forested to reduce deforestation [118]. Land Use Planning and Zoning helps protect natural habitats and manage land use by setting restrictions in sensitive areas, like China's Ecological Red Lines Policy, which limits development in crucial ecological zones [119]. These policies balance economic growth with environmental preservation, ensuring sustainable land use in rural areas.

4.4. Future Research Directions

In the research on the impact of rural land use change on the ecological environment, various factors have led to research in this field. The object of the study is mostly focused on the larger macro-scale regional research, while the small-scale research is less focused, ignoring its importance in the process of urbanization. Choosing a typical small-scale area as the object of study is an effective way to analyze rural land use changes in depth, and it can also provide a basis for comprehensive large-scale research. The process of urbanization has led to a number of ecological and environmental problems in rural areas of developing countries. Strategies and measures to deal with these problems have not been studied in sufficient depth. For illustration, consider the application of scenario analysis in examining the ecological consequences of alterations in land use within rural regions. This methodology allows for the construction of a conceptual framework to envision potential future scenarios. It serves as an invaluable instrument for policymakers, urban planners, and researchers, facilitating the prediction and evaluation of ecological outcomes resulting from various land use and management strategies. The objective is to enhance our comprehension of potential environmental repercussions. In the preparation of a scenario analysis, forthcoming studies may investigate alternatives such as the Business-as-Usual scenario, wherein agricultural and urban expansion proceeds uninhibited, and the Governmental Intervention scenario, which concentrates on initiatives like sustainable land management, green infrastructure, reforestation programs, and a shift towards a low-carbon economy. Each scenario presents a distinct trajectory regarding the utilization of rural land and its ecological ramifications. The Business-as-Usual scenario underscores the grave dangers of inaction, highlighting the necessity for policy modifications. Conversely, the Governmental Intervention scenarios illustrate how targeted policies can facilitate sustainable land use and mitigate environmental degradation. Through the examination of these scenarios, we can discern proactive measures that can be adopted to preserve rural landscapes and ecosystem integrity over the long term, steering us towards more sustainable practices and investments.

In addition, with the wide application of GIS and remote sensing technologies in the fields of land use and ecosystems, strong technical support has been provided to the research in this field. Thus, remote sensing as a research topic in the current research phase can also show the importance of these technologies in conducting the research work. Future research should aim to improve the practical implementation of remote sensing and other technological advances while promoting the integration of different data sets, such as remote sensing, ecological, environmental, meteorological, and socio-economic data. In addition, it is imperative to improve our predictive and analytical capabilities and to move from environmental monitoring to comprehensive capabilities.

4.5. Limitations of the Research

This study has some limitations, which should be recognized. The results of the bibliometric analysis are highly dependent on the type of database chosen. Only the Web of Science Core Collection database was used as the data source for this study. The Web of Science database is one of the most influential databases, but it may not include all relevant publications in the field. The inclusion of other databases, such as the Scopus database and the China National Knowledge Infrastructure (CNKI) database, can provide a more comprehensive global perspective. In a follow-up study, we will explore the inclusion of other databases to expand the scope of our study based on the current discussion.

After decades of development of bibliometric analysis, both the theoretical system and the scope of application have been greatly developed and have received extensive attention from the academic community. However, bibliometric methods also have certain limitations. First of all, the measurement index of bibliometrics is only limited to the output of published papers, not the quality of papers. The methodology chosen in the process of bibliometric analysis may be influenced by the publication method of the journal, such as open access or subscription. Secondly, there is a lack of a standard for recognizing the number of published papers, citations, etc. as indicators of quality. This criterion is difficult to accurately quantify. These problems need to be solved by conducting in-depth research on relevant theories and methods.

5. Conclusions

Urbanization induces multiple changes in rural areas and agricultural industries, leading to the emergence of non-agricultural industries and urban sprawls that reshape the global land surface. Land use change has profound implications for the sustainability of the global environment. To comprehensively understand the macroscopic impact of rural land use change on the ecological environment, this paper uses bibliometric analysis to systematically organize 1237 papers related to this field in the core database of the Web of Science. It elucidates the trend in the number of papers and journals published in this field, analyzes the main researchers, countries, and institutions, and captures the development of high-frequency keywords and research topics. On the basis of the results of the bibliometric analysis, we conducted an overview study of the impact of rural land-use change on the ecological environment and discussed future research directions. The following are the main conclusions:

During the period 1982–2023, there is a general upward trend in the number of publications in the field, which can be divided into three phases: a slow growth phase from 1982–2007, a steady growth phase from 2008–2014, and a rapid growth phase from 2015–2023. *Sustainability*, *Land Use Policy*, and *Land* are the journals with a high number of articles; *Land Use Policy*, *Ecological Indicators*, *Landscape and Urban Planning*, and *Science of The Total Environment* are the journals with a high impact in the field. Moreover, 4768 researchers from 95 countries and regions have published papers in this field, and the main researchers and institutions are from China. Developed countries, such as the United States, have a high citation frequency and strong research strengths in this field. The high-frequency keywords in this field include land use, urbanization, China, ecosystem services, biodiversity, remote sensing, and so on, reflecting the key issues that scholars focus on. Research in this field focuses on the impact on natural elements such as soil quality, atmospheric quality, water resources, and biodiversity. In addition, we reviewed strategies to promote sustainable rural land use, focusing on four main areas: green infrastructure integration, green space management, climate change adaptation, and agricultural policy. These strategies are essential for realizing rural development and environmental protection. In future research, remote sensing technology can provide strong technical support for research on land use change in typical small-scale areas and strategies for rural ecological and environmental problems.

Author Contributions: Conceptualization, methodology, software, and writing—original draft preparation, H.X.; writing—review and editing, Q.S. and W.S. All authors have read and agreed to the published version of the manuscript.

Funding: This research was funded by the Third Comprehensive Scientific Investigation in Xinjiang (grant number 2022xjkk0905) and the research and development fund of the Ningxia Hui Autonomous Region (grant number 2023BCF01023).

Data Availability Statement: All relevant data sets in this study are described in the manuscript.

Conflicts of Interest: The authors declare no conflicts of interest.

References

1. Watson, R.T.; Noble, I.R.; Bolin, B.; Ravindranath, N.H.; Verardo, D.J.; Dokken, D.J. *Land Use, Land-Use Change, and Forestry. Special Report of the Intergovernmental Panel on Climate Change*; Cambridge University Press: Cambridge, UK, 2000.
2. Dale, V.H.; Brown, S.; Haeuber, R.A.; Hobbs, N.T.; Huntly, N.; Naiman, R.J.; Riebsame, W.E.; Turner, M.G.; Valone, T.J. Ecological Principles and Guidelines for Managing the Use of Land. *Ecol. Appl.* **2000**, *10*, 639–670. [CrossRef]
3. Prestele, R.; Alexander, P.; Rounsevell, M.D.A.; Arneth, A.; Calvin, K.; Doelman, J.; Eitelberg, D.A.; Engström, K.; Fujimori, S.; Hasegawa, T.; et al. Hotspots of uncertainty in land-use and land-cover change projections: A global-scale model comparison. *Glob. Change Biol.* **2016**, *22*, 3967–3983. [CrossRef] [PubMed]
4. Pereira, H.M.; Leadley, P.W.; Proença, V.; Alkemade, R.; Scharlemann, J.P.; Fernandez-Manjarrés, J.F.; Araújo, M.B.; Balvanera, P.; Biggs, R.; Cheung, W.W. Scenarios for global biodiversity in the 21st century. *Science* **2010**, *330*, 1496–1501. [CrossRef] [PubMed]
5. Lambin, E.F.; Meyfroidt, P. Global land use change, economic globalization, and the looming land scarcity. *Proc. Natl. Acad. Sci. USA* **2011**, *108*, 3465–3472. [CrossRef] [PubMed]
6. Miao, L.J.; Zhu, F.; Sun, Z.L.; Moore, J.C.; Cui, X.F. China's Land-Use Changes during the Past 300 Years: A Historical Perspective. *Int. J. Environ. Res. Public Health* **2016**, *13*, 847. [CrossRef] [PubMed]
7. Fang, Z.; Ding, T.H.; Chen, J.Y.; Xue, S.; Zhou, Q.; Wang, Y.D.; Wang, Y.X.; Huang, Z.D.; Yang, S.L. Impacts of land use/land cover changes on ecosystem services in ecologically fragile regions. *Sci. Total Environ.* **2022**, *831*, 154967. [CrossRef] [PubMed]
8. Chiarella, C.; Meyfroidt, P.; Abeygunawardane, D.; Conforti, P. Balancing the trade-offs between land productivity, labor productivity and labor intensity. *AMBIO* **2023**, *52*, 1618–1634. [CrossRef] [PubMed]

9. Follmann, A. Geographies of peri-urbanization in the global south. *Geogr. Compass* **2022**, *16*, e12650. [CrossRef]
10. Li, X.; Wang, Y.; Song, Y. Unraveling land system vulnerability to rapid urbanization: An indicator-based vulnerability assessment for Wuhan, China. *Environ. Res.* **2022**, *211*, 112981. [CrossRef]
11. Xuemei, B.; Peijun, S.; Yansui, L. Realizing China's urban dream. *Nature* **2014**, *509*, 158–160.
12. He, C.; Okada, N.; Zhang, Q.; Shi, P.; Li, J. Modelling dynamic urban expansion processes incorporating a potential model with cellular automata. *Landsc. Urban Plan.* **2008**, *86*, 79–91. [CrossRef]
13. Dale, V.; Archer, S.; Chang, M.; Ojima, D. Ecological Impacts and Mitigation Strategies for Rural Land Management. *Ecol. Appl.* **2005**, *15*, 1879–1892. [CrossRef]
14. Tu, S.; Long, H.; Li, T.; Ge, D. The mechanism and models of villages and towns construction and rural development in China. *Econ. Geogr.* **2015**, *35*, 141–147.
15. Liu, Y. Introduction to land use and rural sustainability in China. *Land Use Policy* **2018**, *74*, 1–4. [CrossRef]
16. Yusuf, S.; Nabeshima, K.; Ha, W. Income and Health in Cities: The Messages from Stylized Facts. *J. Urban Health-Bull. New York Acad. Med.* **2007**, *84*, I35–I41. [CrossRef] [PubMed]
17. Sun, H.Z.; Zhao, J.; Liu, X.; Qiu, M.; Shen, H.; Guillas, S.; Giorio, C.; Staniaszek, Z.; Yu, P.; Wan, M.W.L.; et al. Antagonism between ambient ozone increase and urbanization-oriented population migration on Chinese cardiopulmonary mortality. *Innovation* **2023**, *4*, 100517. [CrossRef] [PubMed]
18. Hong, H.; Xie, D.; Liao, H.; Tu, B.; Yang, J. Land Use Efficiency and Total Factor Productivity—Distribution Dynamic Evolution of Rural Living Space in Chongqing, China. *Sustainability* **2017**, *9*, 444. [CrossRef]
19. Li, J.Y.; Chen, H.X.; Zhang, C.; Pan, T. Variations in ecosystem service value in response to land use/land cover changes in Central Asia from 1995–2035. *PeerJ* **2019**, *7*, e7665. [CrossRef]
20. Garden, J.G.; McAlpine, C.A.; Possingham, H.P.; Jones, D.N. Habitat structure is more important than vegetation composition for local-level management of native terrestrial reptile and small mammal species living in urban remnants: A case study from Brisbane, Australia. *Austral. Ecol.* **2007**, *32*, 669–685. [CrossRef]
21. Shan, L.P.; Yu, A.T.W.; Wu, Y.Z. Strategies for risk management in urban-rural conflict: Two case studies of land acquisition in urbanising China. *Habitat Int.* **2017**, *59*, 90–100. [CrossRef]
22. Huang, Y.L. Technology innovation and sustainability: Challenges and research needs. *Clean Technol. Environ. Policy* **2021**, *23*, 1663–1664. [CrossRef] [PubMed]
23. de Groot, R.S.; Alkemade, R.; Braat, L.; Hein, L.; Willemen, L. Challenges in integrating the concept of ecosystem services and values in landscape planning, management and decision making. *Ecol. Complex.* **2010**, *7*, 260–272. [CrossRef]
24. Zhang, H.L.; Zhang, Z.B.; Dong, J.H.; Gao, F.W.; Zhang, W.B.; Gong, W.M. Spatial production or sustainable development? An empirical research on the urbanization of less-developed regions based on the case of Hexi Corridor in China. *PLoS ONE* **2020**, *15*, e0235351. [CrossRef] [PubMed]
25. Ji, Z.X.; Wei, H.J.; Xue, D.; Liu, M.X.; Cai, E.X.; Chen, W.Q.; Feng, X.W.; Li, J.W.; Lu, J.; Guo, Y.L. Trade-Off and Projecting Effects of Land Use Change on Ecosystem Services under Different Policies Scenarios: A Case Study in Central China. *Int. J. Environ. Res. Public Health* **2021**, *18*, 3552. [CrossRef] [PubMed]
26. Allan, E.; Manning, P.; Alt, F.; Binkenstein, J.; Blaser, S.; Bluethgen, N.; Böhm, S.; Grassein, F.; Hölzel, N.; Klaus, V.H.; et al. Land use intensification alters ecosystem multifunctionality via loss of biodiversity and changes to functional composition. *Ecol. Lett.* **2015**, *18*, 834–843. [CrossRef] [PubMed]
27. Zheng, Y.; Yu, G.; Zhong, P.L.; Wang, Y.X. Integrated assessment of coastal ecological security based on land use change and ecosystem services in the Jiaozhou Bay, Shandong Peninsula, China. *J. Appl. Ecol.* **2018**, *29*, 4097–4105. [CrossRef]
28. Gogoi, P.P.; Vinoj, V.; Swain, D.; Roberts, G.; Dash, J.; Tripathy, S. Land use and land cover change effect on surface temperature over Eastern India. *Sci. Rep.* **2019**, *9*, 8859. [CrossRef]
29. López, S.; Wright, C.; Costanza, P. Environmental change in the equatorial Andes: Linking climate, land use, and land cover transformations. *Remote Sens. Appl. Soc. Environ.* **2017**, *8*, 291–303. [CrossRef]
30. Swanepoel, C.M.; van der Laan, M.; Weepener, H.L.; du Preez, C.C.; Annandale, J.G. Review and meta-analysis of organic matter in cultivated soils in southern Africa. *Nutr. Cycl. Agroecosystems* **2016**, *104*, 107–123. [CrossRef]
31. Dumont, B.; Ryschawy, J.; Duru, M.; Benoit, M.; Chatellier, V.; Delaby, L.; Donnars, C.; Dupraz, P.; Lemauviel-Lavenant, S.; Méda, B.; et al. Review: Associations among goods, impacts and ecosystem services provided by livestock farming. *Animal* **2019**, *13*, 1773–1784. [CrossRef]
32. Zhao, X.C.; Li, F.S.; Yan, Y.Z.; Zhang, Q. Biodiversity in Urban Green Space: A Bibliometric Review on the Current Research Field and Its Prospects. *Int. J. Environ. Res. Public Health* **2022**, *19*, 12544. [CrossRef] [PubMed]
33. Goncalves, A.F.A.; dos Santos, J.A.; Franca, L.C.D.; Campoe, O.C.; Altoe, T.F.; Scolforo, J.R.S. Use of the process-based models in forest research: A bibliometric review. *Cerne* **2021**, *27*, e-102769. [CrossRef]
34. Wang, W.; Lu, C. Visualization analysis of big data research based on Citespace. *Soft Comput.* **2020**, *24*, 8173–8186. [CrossRef]
35. Garfield, E. From the science of science to Scientometrics visualizing the history of science with HistCite software. *J. Informetr.* **2009**, *3*, 173–179. [CrossRef]
36. Aria, M.; Cuccurullo, C. bibliometrix: An R-tool for comprehensive science mapping analysis. *J. Informetr.* **2017**, *11*, 959–975. [CrossRef]
37. Gagolewski, M. Bibliometric impact assessment with R and the CITAN package. *J. Informetr.* **2011**, *5*, 678–692. [CrossRef]

38. Xu, X.; Chen, Q.; Zhu, Z. Evolutionary Overview of Land Consolidation Based on Bibliometric Analysis in Web of Science from 2000 to 2020. *Int. J. Environ. Res. Public Health* **2022**, *19*, 3218. [CrossRef]
39. Li, J.; Song, W. Review of rural settlement research based on bibliometric analysis. *Front. Environ. Sci.* **2023**, *10*, 1089438. [CrossRef]
40. Liu, L.; Zou, G.; Zuo, Q.; Li, S.; Bao, Z.; Jin, T.; Liu, D.; Du, L. It is still too early to promote biodegradable mulch film on a large scale: A bibliometric analysis. *Environ. Technol. Innov.* **2022**, *27*, 102487. [CrossRef]
41. Su, H.N.; Lee, P.C. Mapping knowledge structure by keyword co-occurrence: A first look at journal papers in Technology Foresight. *Scientometrics* **2010**, *85*, 65–79. [CrossRef]
42. Zhang, P.; Xia, L.; Sun, Z.; Zhang, T. Analysis of spatial and temporal changes and driving forces of arable land in the Weibei dry plateau region in China. *Sci. Rep.* **2023**, *13*, 20618. [CrossRef]
43. Liu, Y.; Yang, Y.; Li, Y.; Li, J. Conversion from rural settlements and arable land under rapid urbanization in Beijing during 1985–2010. *J. Rural. Stud.* **2017**, *51*, 141–150. [CrossRef]
44. Gu, L.; Yan, J.B.; Li, Y.R.; Gong, Z.W. Spatial-temporal evolution and correlation analysis between habitat quality and landscape patterns based on land use change in Shaanxi Province, China. *Ecol. Evol.* **2023**, *13*, e10657. [CrossRef]
45. Zhao, S.X.; Yin, M.M. Change of urban and rural construction land and driving factors of arable land occupation. *PLoS ONE* **2023**, *18*, e0286248. [CrossRef]
46. Deng, X.; Lin, Y.; Seto, K.C. Land-Use Competition between Food Production and Urban Expansion in China. In *Rethinking Global Land Use in an Urban Era*; MIT Press: Cambridge, MA, USA, 2014. [CrossRef]
47. Liu, Y.; Li, Y. Revitalize the world's countryside. *Nature* **2017**, *548*, 275–277. [CrossRef] [PubMed]
48. Yao, R.; Wang, L.; Gui, X.; Zheng, Y.; Zhang, H.; Huang, X. Urbanization Effects on Vegetation and Surface Urban Heat Islands in China's Yangtze River Basin. *Remote Sens.* **2017**, *9*, 540. [CrossRef]
49. Hou, H.; Wang, R.; Murayama, Y. Scenario-based modelling for urban sustainability focusing on changes in cropland under rapid urbanization: A case study of Hangzhou from 1990 to 2035. *Sci. Total Environ.* **2019**, *661*, 422–431. [CrossRef] [PubMed]
50. Pan, F.; Shu, N.; Wan, Q.; Huang, Q. Land Use Function Transition and Associated Ecosystem Service Value Effects Based on Production–Living–Ecological Space: A Case Study in the Three Gorges Reservoir Area. *Land* **2023**, *12*, 391. [CrossRef]
51. Liu, Y.; Liu, Y.; Chen, Y.; Long, H. The process and driving forces of rural hollowing in China under rapid urbanization. *J. Geogr. Sci.* **2010**, *20*, 876–888. [CrossRef]
52. Wu, Y.; Shan, L.; Guo, Z.; Peng, Y. Cultivated land protection policies in China facing 2030: Dynamic balance system versus basic farmland zoning. *Habitat Int.* **2017**, *69*, 126–138. [CrossRef]
53. Qiu, B.; Li, H.; Tang, Z.; Chen, C.; Berry, J. How cropland losses shaped by unbalanced urbanization process? *Land Use Policy* **2020**, *96*, 104715. [CrossRef]
54. Liu, S.S.; Peng, Y.P.; Xia, Z.Q.; Hu, Y.M.; Wang, G.X.; Zhu, A.X.; Liu, Z.H. The GA-BPNN-Based Evaluation of Cultivated Land Quality in the PSR Framework Using Gaofen-1 Satellite Data. *Sensors* **2019**, *19*, 5127. [CrossRef] [PubMed]
55. Ma, Y.; Kalantari, Z.; Destouni, G. Infectious Disease Sensitivity to Climate and Other Driver-Pressure Changes: Research Effort and Gaps for Lyme Disease and Cryptosporidiosis. *GeoHealth* **2023**, *7*, e2022GH000760. [CrossRef] [PubMed]
56. Tang, Y.; Chen, M.H. The Impact Mechanism and Spillover Effect of Digital Rural Construction on the Efficiency of Green Transformation for Cultivated Land Use in China. *Int. J. Environ. Res. Public Health* **2022**, *19*, 16159. [CrossRef]
57. Zhang, Q.; Han, G.L.; Liu, M.; Li, X.Q.; Wang, L.Q.; Liang, B. Distribution and Contamination Assessment of Soil Heavy Metals in the Jiulongjiang River Catchment, Southeast China. *Int. J. Environ. Res. Public Health* **2019**, *16*, 4674. [CrossRef] [PubMed]
58. Qiu, B.; Li, H.; Chen, C.; Tang, Z.; Zhang, K.; Berry, J. Tracking spatial–temporal landscape changes of impervious surface areas, bare lands, and inundation areas in China during 2001–2017. *Land Degrad. Dev.* **2019**, *30*, 1802–1812. [CrossRef]
59. Song, W.; Liu, M. Farmland Conversion Decreases Regional and National Land Quality in China. *Land Degrad. Dev.* **2017**, *28*, 459–471. [CrossRef]
60. Deng, X.; Huang, J.; Rozelle, S.; Zhang, J.; Li, Z. Impact of urbanization on cultivated land changes in China. *Land Use Policy* **2015**, *45*, 1–7. [CrossRef]
61. Lee, E.; Sagong, J.; Lee, Y. Influence of land use change on the waterbird community of Sihwa Lake, Republic of Korea. *Avian Res.* **2020**, *11*, 36. [CrossRef]
62. Bueno, A.S.; Peres, C.A. The role of baseline suitability in assessing the impacts of land-use change on biodiversity. *Biol. Conserv.* **2020**, *243*, 108396. [CrossRef]
63. Pătru-Stupariu, I.; Hossu, C.A.; Grădinaru, S.R.; Nita, A.; Stupariu, M.-S.; Huzui-Stoiculescu, A.; Gavrilidis, A.-A. A Review of Changes in Mountain Land Use and Ecosystem Services: From Theory to Practice. *Land* **2020**, *9*, 336. [CrossRef]
64. Caulfield, M.E.; Fonte, S.J.; Groot, J.C.J.; Vanek, S.J.; Sherwood, S.; Oyarzun, P.; Borja, R.M.; Dumble, S.; Tittonell, P. Agroecosystem patterns and land management co-develop through environment, management, and land-use interactions. *Ecosphere* **2020**, *11*, e03113. [CrossRef]
65. Fu, B.; Ma, K.; Zhou, H.; Chen, L. The effect of land use structure on the distribution of soil nutrients in the hilly area of the Loess Plateau, China. *Chin. Sci. Bull.* **1999**, *44*, 732–736. [CrossRef]
66. Lal, R.; Mokma, D.; Lowery, B. Relation between soil quality and erosion. In *Soil Quality and Soil Erosion*; CRC Press: Boca Raton, FL, USA, 1999; pp. 237–258.

67. Liu, R.; Wang, M.; Chen, W. The influence of urbanization on organic carbon sequestration and cycling in soils of Beijing. *Landsc. Urban Plan.* **2018**, *169*, 241–249. [CrossRef]
68. Shan, Y.J.; Wei, S.K.; Yuan, W.L.; Miao, Y. Evaluation and prediction of land ecological security in Shenzhen based on DPSIR-TOPSIS-GM(1,1) model. *PLoS ONE* **2022**, *17*, e0265810. [CrossRef] [PubMed]
69. Hemkemeyer, M.; Schwalb, S.A.; Heinze, S.; Joergensen, R.G.; Wichern, F. Functions of elements in soil microorganisms. *Microbiol. Res.* **2021**, *252*, 126832. [CrossRef] [PubMed]
70. Li, M.; Chen, L.; Zhao, F.; Tang, J.; Bu, Q.; Wang, X.; Yang, L. Effects of Urban–Rural Environmental Gradient on Soil Microbial Community in Rapidly Urbanizing Area. *Ecosyst. Health Sustain.* **2023**, *9*, 0118. [CrossRef]
71. Gao, J.; Wen, Y. Impact of Land Use Change on Runoff of Taihu Basin. *Acta Geogr. Sin.* **2002**, *57*, 194–200.
72. Li, J.; Chen, L.; Guo, X.; Fu, B. Effects of land use structure on non-point source pollution. *Chin. Environ. Sci.* **2000**, *20*, 506.
73. Nunez, S.; Alkemade, R. Exploring interaction effects from mechanisms between climate and land-use changes and the projected consequences on biodiversity. *Biodivers. Conserv.* **2021**, *30*, 3685–3696. [CrossRef]
74. Zhao, G.; Liu, J.; Kuang, W.; Ouyang, Z. Disturbance impacts of land use change on biodiversity conservation priority areas across China during 1990–2010. *Acta Geogr. Sin.* **2014**, *69*, 1640–1650.
75. Horak, J.; Peltanova, A.; Podavkova, A.; Safarova, L.; Bogusch, P.; Romportl, D.; Zasadil, P. Biodiversity responses to land use in traditional fruit orchards of a rural agricultural landscape. *Agric. Ecosyst. Environ.* **2013**, *178*, 71–77. [CrossRef]
76. Li, X. Study on Protection Ways of Biodiversity in Rural Environment Construction. *IOP Conf. Ser. Earth Environ. Sci.* **2019**, *252*, 042020. [CrossRef]
77. Bonthoux, S.; Barnagaud, J.Y.; Goulard, M.; Balent, G. Contrasting spatial and temporal responses of bird communities to landscape changes. *Oecologia* **2013**, *172*, 563–574. [CrossRef] [PubMed]
78. Kati, V.; Kassara, C.; Vrontisi, Z.; Moustakas, A. The biodiversity-wind energy-land use nexus in a global biodiversity hotspot. *Sci. Total Environ.* **2021**, *768*, 144471. [CrossRef]
79. Yan, D.; Schneider, U.A.; Schmid, E.; Huang, H.Q.; Pan, L.; Dilly, O. Interactions between land use change, regional development, and climate change in the Poyang Lake district from 1985 to 2035. *Agric. Syst.* **2013**, *119*, 10–21. [CrossRef]
80. Lindesay, J.A.; Andreae, M.O.; Goldammer, J.G.; Harris, G.; Annegarn, H.J.; Garstang, M.; Scholes, R.J.; van Wilgen, B.W. International geosphere-biosphere programme/international global atmospheric chemistry SAFARI-92 field experiment: Background and overview. *J. Geophys. Res. Atmos.* **1996**, *101*, 23521–23530. [CrossRef]
81. Tzilivakis, J.; Warner, D.J.; Green, A.; Lewis, K.A. Spatial and temporal variability of greenhouse gas emissions from rural development land use operations. *Mitig. Adapt. Strateg. Glob. Change* **2017**, *22*, 447–467. [CrossRef]
82. Seddon, N.; Chausson, A.; Berry, P.; Girardin, C.A.J.; Smith, A.; Turner, B. Understanding the value and limits of nature-based solutions to climate change and other global challenges. *Philos. Trans. R. Soc. B-Biol. Sci.* **2020**, *375*, 20190120. [CrossRef]
83. Alikhanova, S.; Bull, J.W. Review of Nature-based Solutions in Dryland Ecosystems: The Aral Sea Case Study. *Environ. Manag.* **2023**, *72*, 457–472. [CrossRef]
84. Huang, C.; He, H.S.; Liang, Y.; Wu, Z.W. Effects of climate change, fire and harvest on carbon storage of boreal forests in the Great Xing'an Mountains, China. *J. Appl. Ecol.* **2018**, *29*, 2088–2100. [CrossRef]
85. Duveiller, G.; Hooker, J.; Cescatti, A. The mark of vegetation change on Earth's surface energy balance. *Nat. Commun.* **2018**, *9*, 679. [CrossRef] [PubMed]
86. Grilli, E.; Carvalho, S.C.P.; Chiti, T.; Coppola, E.; D'Ascoli, R.; La Mantia, T.; Marzaioli, R.; Mastrocicco, M.; Pulido, F.; Rutigliano, F.A.; et al. Critical range of soil organic carbon in southern Europe lands under desertification risk. *J. Environ. Manag.* **2021**, *287*, 112285. [CrossRef] [PubMed]
87. Teixeira, C.P.; Fernandes, C.O.; Ahern, J.; Honrado, J.P.; Farinha-Marques, P. Urban ecological novelty assessment: Implications for urban green infrastructure planning and management. *Sci. Total Environ.* **2021**, *773*, 145121. [CrossRef] [PubMed]
88. Anderson, V.; Gough, W.A. Enabling Nature-Based Solutions to Build Back Better-An Environmental Regulatory Impact Analysis of Green Infrastructure in Ontario, Canada. *Buildings* **2022**, *12*, 61. [CrossRef] [PubMed]
89. Zhao, S.; Yin, M. Research on Rural Population/Arable Land/Rural Settlements Association Model and Coordinated Development Path: A Case Analysis of the Yellow River Basin (Henan Section). *Int. J. Environ. Res. Public Health* **2023**, *20*, 3833. [CrossRef] [PubMed]
90. Wang, R.B.; Bai, Y.; Alatalo, J.M.; Guo, G.M.; Yang, Z.Q.; Yang, Z.B.; Yang, W. Impacts of urbanization at city cluster scale on ecosystem services along an urban-rural gradient: A case study of Central Yunnan City Cluster, China. *Environ. Sci. Pollut. Res.* **2022**, *29*, 88852–88865. [CrossRef] [PubMed]
91. Peake, L.; Robb, C. Saving the ground beneath our feet: Establishing priorities and criteria for governing soil use and protection. *R. Soc. Open Sci.* **2021**, *8*, 201994. [CrossRef]
92. Snyman, S.; Bricker, K.S. Living on the Edge: Benefit-Sharing from Protected Area Tourism. *J. Sustain. Tour.* **2019**, *27*, 705–719. [CrossRef]
93. Gunson, K.E.; Mountrakis, G.; Quackenbush, L.J. Spatial Wildlife-Vehicle Collision Models: A Review of Current Work and Its Application to Transportation Mitigation Projects. *J. Environ. Manag.* **2011**, *92*, 1074–1082. [CrossRef]
94. Sawaya, M.A.; Kalinowski, S.T.; Clevenger, A.P. Genetic Connectivity for Two Bear Species at Wildlife Crossing Structures in Banff National Park. *Proc. R. Soc. B Biol. Sci.* **2014**, *281*, 20131705. [CrossRef] [PubMed]

95. Zeppel, H. Managing Cultural Values in Sustainable Tourism: Conflicts in Protected Areas. *Tour. Hosp. Res.* **2010**, *10*, 93–115. [CrossRef]
96. Comín, F.A. Planning the Development of Urban and Rural Areas: An Integrative Approach. In *Sustainable Cities and Communities*; Springer: Berlin/Heidelberg, Germany, 2020; pp. 468–478.
97. Jun, M.-J. The Effects of Portland's Urban Growth Boundary on Urban Development Patterns and Commuting. *Urban Stud.* **2004**, *41*, 1333–1348. [CrossRef]
98. Giovannoni, G. Urban Containment Planning: Is It Effective? The Case of Portland, OR. *Sustainability* **2021**, *13*, 12925. [CrossRef]
99. Wani, G.A.; Nagaraj, V. Effect of Sustainable Infrastructure and Service Delivery on Sustainable Tourism: Application of Kruskal Wallis Test (Non-Parametric). *Int. J. Sustain. Transp. Technol.* **2022**, *5*, 38–50. [CrossRef]
100. Marzo-Navarro, M.; Pedraja-Iglesias, M.; Vinzón, L. Sustainability Indicators of Rural Tourism from the Perspective of the Residents. In *Tourism and Sustainable Development Goals*; Routledge: Abingdon-on-Thames, UK, 2020; pp. 148–164.
101. de Pádua Andrade, C.O.; Carvalho, R.d.C.R.; Godinho, R.F.; Magri, R.A.F. Elaboração e aplicação de uma rota de trekking em uma área do Parque Nacional da Serra da Canastra. *Rev. Bras. Ecotur. (RBEcotur)* **2016**, *9*, 285–317. [CrossRef]
102. Fifanou, V.G.; Ousmane, C.; Gauthier, B.; Brice, S. Traditional Agroforestry Systems and Biodiversity Conservation in Benin (West Africa). *Agrofor. Syst.* **2011**, *82*, 1–13. [CrossRef]
103. Hidayat, N.; Sianipar, J. The Potential of Agroforestry in Supporting Food Security for Peatland Community–A Case Study in the Kalampangan Village, Central Kalimantan. *J. Ecol. Eng.* **2021**, *22*, 123–130.
104. Uchino, H.; Iwama, K.; Jitsuyama, Y.; Yudate, T.; Nakamura, S. Yield Losses of Soybean and Maize by Competition with Interseeded Cover Crops and Weeds in Organic-Based Cropping Systems. *Field Crops Res.* **2009**, *113*, 342–351. [CrossRef]
105. Triplett, G.B., Jr.; Dick, W.A. No-Tillage Crop Production: A Revolution in Agriculture! *Agron. J.* **2008**, *100*, S-153. [CrossRef]
106. Torralba, M.; Fagerholm, N.; Burgess, P.J.; Moreno, G.; Plieninger, T. Do European Agroforestry Systems Enhance Biodiversity and Ecosystem Services? A Meta-Analysis. *Agric. Ecosyst. Environ.* **2016**, *230*, 150–161. [CrossRef]
107. Fuchs, L.E.; Orero, L.; Ngoima, S.; Kuyah, S.; Neufeldt, H. Asset-Based Adaptation Project Promotes Tree and Shrub Diversity and above-Ground Carbon Stocks in Smallholder Agroforestry Systems in Western Kenya. *Front. For. Glob. Change* **2022**, *4*, 773170. [CrossRef]
108. Du Toit, M.; Du Preez, C.; Cilliers, S. Plant Diversity and Conservation Value of Wetlands along a Rural–Urban Gradient. *Bothalia-Afr. Biodivers. Conserv.* **2021**, *51*, 1–18. [CrossRef]
109. Dumenu, W.K.; Obeng, E.A. Climate Change and Rural Communities in Ghana: Social Vulnerability, Impacts, Adaptations and Policy Implications. *Environ. Sci. Policy* **2016**, *55*, 208–217. [CrossRef]
110. Beacham, A.M.; Hand, P.; Barker, G.C.; Denby, K.J.; Teakle, G.R.; Walley, P.G.; Monaghan, J.M. Addressing the Threat of Climate Change to Agriculture Requires Improving Crop Resilience to Short-Term Abiotic Stress. *Outlook Agric.* **2018**, *47*, 270–276. [CrossRef]
111. Nkegbe, P.K.; Abu, B.M.; Issahaku, H. Food Security in the Savannah Accelerated Development Authority Zone of Ghana: An Ordered Probit with Household Hunger Scale Approach. *Agric. Food Secur.* **2017**, *6*, 35. [CrossRef]
112. Tang, H.; Liu, Y.; Li, X.; Muhammad, A.; Huang, G. Carbon Sequestration of Cropland and Paddy Soils in China: Potential, Driving Factors, and Mechanisms. *Greenh. Gases Sci. Technol.* **2019**, *9*, 872–885. [CrossRef]
113. Chazdon, R.L.; Uriarte, M. Natural Regeneration in the Context of Large-Scale Forest and Landscape Restoration in the Tropics. *Biotropica* **2016**, *48*, 709–715. [CrossRef]
114. Kohlitz, J.; Chong, J.; Willetts, J. Rural Drinking Water Safety under Climate Change: The Importance of Addressing Physical, Social, and Environmental Dimensions. *Resources* **2020**, *9*, 77. [CrossRef]
115. Glendenning, C.; Vervoort, R. Hydrological Impacts of Rainwater Harvesting (RWH) in a Case Study Catchment: The Arvari River, Rajasthan, India: Part 2. Catchment-Scale Impacts. *Agric. Water Manag.* **2011**, *98*, 715–730. [CrossRef]
116. Dosso, F.; Idrissou, L.; Moussa, I.M. Innovativity in Legislative, Political and Organizational Frameworks of Sustainable Land Management in Benin. *Asian J. Agric. Ext. Econ. Sociol.* **2021**, *39*, 603–615. [CrossRef]
117. Donald, P.F.; Green, R.; Heath, M. Agricultural Intensification and the Collapse of Europe's Farmland Bird Populations. *Proc. R. Soc. London. Ser. B Biol. Sci.* **2001**, *268*, 25–29. [CrossRef]
118. Camara, G.; Simoes, R.; Ruivo, H.M.; Andrade, P.R.; Soterroni, A.C.; Ramos, F.M.; Ramos, R.G.; Scarabello, M.; Almeida, C.; Sanches, I.; et al. Impact of Land Tenure on Deforestation Control and Forest Restoration in Brazilian Amazonia. *Environ. Res. Lett.* **2023**, *18*, 065005. [CrossRef]
119. Hu, T.; Peng, J.; Liu, Y.; Wu, J.; Li, W.; Zhou, B. Evidence of Green Space Sparing to Ecosystem Service Improvement in Urban Regions: A Case Study of China's Ecological Red Line Policy. *J. Clean. Prod.* **2020**, *251*, 119678. [CrossRef]

Disclaimer/Publisher's Note: The statements, opinions and data contained in all publications are solely those of the individual author(s) and contributor(s) and not of MDPI and/or the editor(s). MDPI and/or the editor(s) disclaim responsibility for any injury to people or property resulting from any ideas, methods, instructions or products referred to in the content.

MDPI AG
Grosspeteranlage 5
4052 Basel
Switzerland
Tel.: +41 61 683 77 34

Land Editorial Office
E-mail: land@mdpi.com
www.mdpi.com/journal/land

Disclaimer/Publisher's Note: The statements, opinions and data contained in all publications are solely those of the individual author(s) and contributor(s) and not of MDPI and/or the editor(s). MDPI and/or the editor(s) disclaim responsibility for any injury to people or property resulting from any ideas, methods, instructions or products referred to in the content.